Efficient Joint Analysis of Surface Waves and Introduction to Vibration Analysis: Beyond the Clichés

Giancarlo Dal Moro

Efficient Joint Analysis of Surface Waves and Introduction to Vibration Analysis: Beyond the Clichés

Springer

Giancarlo Dal Moro
Institute of Rock Structure and Mechanics
Academy of Sciences of the Czech Republic
Prague, Czech Republic

Eliosoft
Udine, Italy

ISBN 978-3-030-46302-1 ISBN 978-3-030-46303-8 (eBook)
https://doi.org/10.1007/978-3-030-46303-8

This Springer imprint is published by the registered company Springer Nature Switzerland AG
The registered company address is: Gewerbestrasse 11, 6330 Cham, Switzerland

Astronomy is to geology as steeplejack is to what?
Slapstick, Kurt Vonnegut

Preface

Geology, n. The science of the earth's crust—to which, doubtless, will be added that of its interior whenever a man shall come up garrulous out of a well.
The devil's dictionary—Ambrose Bierce

Technical books (textbooks, manuals, guides, and so on) are supposed to be boring and aseptic as if these *qualities* would represent the necessary seal of a serious study. Their audience is usually limited to a pretty restricted niche also because of very specialized terminologies that limit any desirable cross-fertilization (geologists can read only about geology, seismologists about seismology, structural engineers about…, and so on). Even within the, so to speak, geophysical community (*sensu lato*), seismologists and applied geophysicists never seat at the same table and, in case that happens, it is usually to allege their own superiority with respect to the other group.

But this is not without consequences. Applied geophysicists, for instance, routinely use *phase* velocities while ignoring about the relevant opportunities of the *group* velocities.

Topics and case studies presented in this volume were selected from a series of academic and applied works accomplished in the last years, always considering that surface waves represent a valuable tool for the solution of numerous geotechnical and engineering problems and can be applied well beyond the definition of the soil class in terms of response spectra in seismic hazard studies.

Often, both in the academic and professional sectors, works are performed by merely applying a standard and rigid "protocol" that does not necessarily matches the actual characteristics of the problem to solve. This often happens when the fundamentals are not sufficiently clear and, consequently, instead of shaping a solution (acquisition and analysis) that meets the actual needs and fully respect the physics, we apply a sclerotic approach that does not necessarily suit the specific conditions and goals.

In very general terms, the (active and passive) methodologies illustrated in the book address a very precise question: how can we efficiently collect and analyze the seismic data necessary to define unambiguously the subsurface model in terms of shear-wave velocities? How can we be sure (reasonably speaking) that the retrieved model is correct? How can we set up a joint inversion capable of solving all the possible data and model ambiguities? Does *more* mean necessarily *better*?

The perspective adopted while writing the book is quite simple: once the theory is clear, we can do much more than what we might think (and with a limited field effort). The borders between *theory* and *practice* should slowly fade away and everything should converge into a clear and unitary vision of the wide range of applications that can be tackled even with a limited and simple field equipment.

This is the reason why we decided to include vibration analysis in a book otherwise focused on surface wave acquisition and analysis. On one side, vibration data can be recorded by the same kind of equipment we need for recording surface waves or, in more general terms, seismic data. On the other side, vibration analysis has a prominent importance in all the seismic-hazard national and international building codes that also require the analysis of surface waves for the definition of the shear-wave velocity profile.

Some of the datasets presented and analyzed in the book can be downloaded from the following links: https://doi.org/10.6084/m9.figshare.12376955 or, alternatively, http://download.winmasw.com/data/Data_Dissemination_Efficient_Surface_Waves_Vibrations_Springer2020_Dal_Moro.rar.

Hope you enjoy the reading and find the way to apply the methodologies described in the book in your daily research or professional work.

Prague, Czech Republic/Udine, Italy

Giancarlo Dal Moro

Contents

Abbreviations

ESAC	Extended spatial autoCorrelation
GHM	Gaussian-filtered horizontal motion
HS	Holistic analysis of surface waves
HVSR	Horizontal-to-vertical spectral ratio
MAAM	Miniature array analysis of microtremors
MASW	Multichannel analysis of surface waves
ReMi	Refraction microtremors
RPM	Rayleigh-wave particle motion
RVSR	Radial-to-vertical spectral ratio
SPAC	Spatial autocorrelation
SSR	Standard spectral ratio
SSRn	Standard spectral ratio from microtremor (n = noise) data

Introduction: A Miscellanea

1

The limits of my language mean the limits of my world.

Ludwig Wittgenstein

Abstract

This first chapter could be considered as a sort of warm up section aimed at clarifying a series of general issues that are often misunderstood or not sufficiently clear. Through a series of theoretical considerations and practical examples are illustrated a series of problems associated to the ambiguities inherently present in

G. Dal Moro, *Efficient Joint Analysis of Surface Waves and Introduction to Vibration Analysis: Beyond the Clichés*,
https://doi.org/10.1007/978-3-030-46303-8_1

the data, the non-uniqueness of the solution, the number of channels necessary to accomplish standard (and improved) MASW (*Multichannel Analysis of Surface Waves*) surveys, the acquisition of multi-component data, the trace polarity and much more.

This first chapter could be considered as a sort of warm up section aimed at clarifying a series of issues that are often not sufficiently clear and that are necessary to comprehend the next chapters. It does not provide the fundamentals about surface wave propagation (presented in Dal Moro 2014). As the book title clearly states, we intend go beyond the ordinary clichés of the *Multichannel Analysis of Surface Waves* (MASW) and in order to do that, we must assume that the fundamentals facts are already known from good introductory courses on seismology.

1.1 Shear-Wave Velocities: A Brief Overview

Shear-wave velocity (V_S) versus compressional-wave velocity (V_P)

The very first point to clarify and underline is that the V_S values are important not only in seismic hazard studies (the determination of the Vs30 parameter and the associated ground shaking computation in case of earthquake—e.g. Seed and Idriss 1971; Schnabel et al. 1972; Bard and Bouchon 1980a, b; Hays 1980; Jongmans and Campillo 1993; Bowden and Tsai 2017).

Compared to the compressional (P) waves, shear waves have a distinctive property that, from the practical point of view, makes them very important for any near-surface exploration aimed at retrieving the shallow subsurface model: unlike the P waves, shear waves are little influenced by the water content.

This means that while the presence of water strongly influence the V_P values, it produces only minor variations of the shear-wave velocities (an interesting case study about the influence of the fluid content on shear-wave velocities is reported for instance in West and Menke 2001).

This fact has concrete and important consequences. Let us clarify this point through a field dataset recorded in an area characterized by soft unconsolidated sediments.

Figure 1.1 reports the seismic traces recorded while using an ordinary 8-kg (vertical impact) sledgehammer as source and 23 vertical geophones. The waveforms are clearly dominated by the Rayleigh waves (the so-called *Ground Roll*) but some low-amplitude early arrivals related to the P-wave refraction are also visible. These low-amplitude early arrivals can be emphasized by means of a simple AGC (*Automatic Gain Control*). Figure 1.2b reports the first 0.15 s (after the application of an AGC): the P-wave arrivals are clearer and can be used to retrieve a simple 1D V_P profile (Fig. 1.2a).

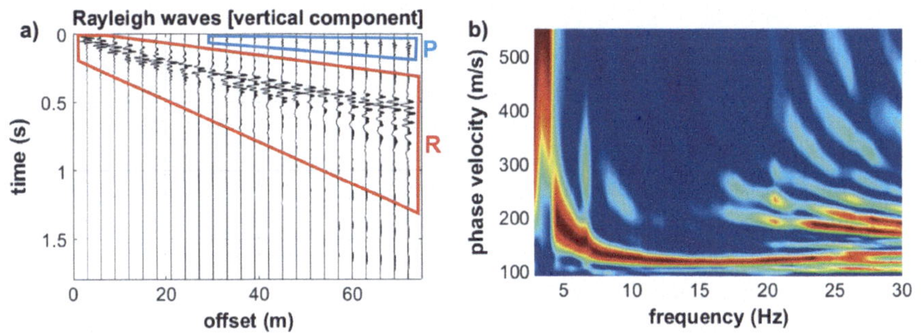

Fig. 1.1 An example of field dataset (vertical component) recorded on a small basin (about 25 m of soft sediments over the bedrock): **a** field traces (the polygon P highlights the P-wave refraction arrivals (see also Fig. 1.2) while the polygon R puts in evidence the Rayleigh waves); **b** phase velocity spectrum computed via phase shift (Park et al. 1998; Dal Moro et al. 2003)

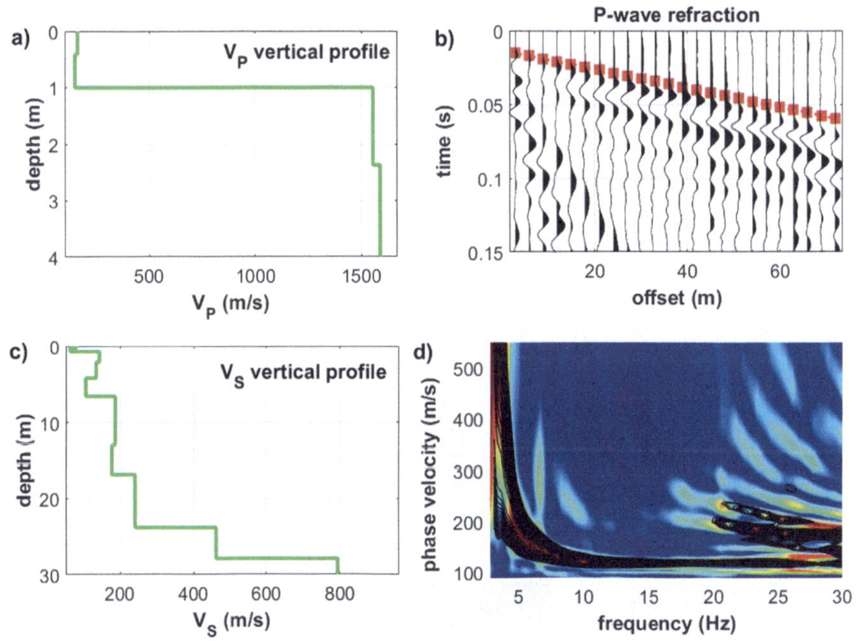

Fig. 1.2 Results of the joint modelling of the data presented in Fig. 1.1: **a** V_P profile; **b** P-wave refraction travel times (close up of the first 0.15 s); **c** V_S model; **d** observed and synthetic phase-velocity spectra (background colors refer to the field data, overlaying contour lines show the synthetic velocity spectrum (FVS [*Full Velocity Spectrum*] approach—see Chap. 2). As can be clearly seen, the V_S profile provides information also about the sediments below the water table while the P-wave refraction is unable to see anything below it

As a matter of fact, such a V_P model is useful to identify the depth of the water table responsible for the sudden V_P increase at a depth of about 1 m (V_P in water is about 1500–1600 m/s). Such a phenomenon actually prevents the P-wave refraction from providing any further information about the velocities of the sediments beneath such a (pretty shallow) "horizon".

On the other side, if we analyze the Rayleigh-wave dispersion (Figs. 1.1b and 1.2c and d), we can define the V_S values (clearly related to the characteristics of the sediments) well below the water table.

It must be underlined that, in near surface applications, water table represents a sort of physical limit/boundary for P-wave refraction studies since, in particular when we are working on unconsolidated sediments, layers below it cannot be easily characterized.

DownHole (DH) Seismics: Vertical Seismic Profile (VSP)

During a DH/VSP survey, in order to collect the data necessary for the identification of the SH-wave arrivals, it is necessary to produce SH waves by means of a shear source: the source is at the surface and the geophone is moving along the borehole (usually it is moved upward starting from the deepest point). Figures 1.3, 1.4 and 1.5 provide the basic information about the common DH set up used for near-surface applications (for a wider overview about all the possible methodologies that can be applied during a VSP survey see Hardage 1983; Galperin 1985; Nanda 2016).

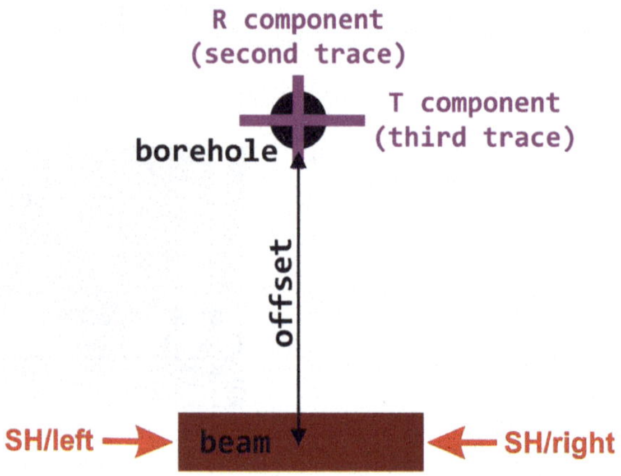

Fig. 1.3 Map view of a classical borehole set up for the generation and acquisition of seismic data useful for the identification of the SH-wave arrivals (see also Fig. 1.4). Although not strictly necessary, in order to better discriminate P-wave first arrivals, it can be useful to hit the beam both from the left and right sides. Data/traces are then subtracted so to cancel possible early (low amplitude) P-wave arrivals

Fig. 1.4 Field operation during a VSP/DH survey: **a** downhole survey for the generation and acquisition of SH waves. The wooden beam is secured/coupled to the ground thanks to the weight of the car (this way the beam does not slip away and most of the energy generated by the hammer impact is transformed into SH waves recorded by the geophone in the borehole). The survey took place in *San Severino Marche* (IT) after the 2016 central Italy seismic crisis; **b** data processing: several VSP software applications assume that the ray paths are linear but such approximation is not correct (see Fig. 1.5)

Table 1.1 Main pros and cons for the VSP (vertical seismic profiling) methodology

DH (borehole seismics) for the determination of the V_S profile	
Pros	Cons
• Potentially detailed reconstruction of the velocity model • If a Vertical-Impact (VF) source is also applied, it is possible to determine the V_P values and, consequently, the Poisson ratio; in case an oblique source is used (see Sect. 1.14), we can simultanously generate both P and SH waves (so as not to make necessary the double VF [for P waves] and HF [for SH waves] acquisitions)	• Expensive (digging a borehole is costly) and time consuming • Obtained information are very local and can be not fully representative of the *general* site conditions • Because of problems in the identification of the first arrivals when the geophone is near to the surface, shallow layers are usually poorly defined • Frequent and common problems in the borehole cementation can jeopardize the overall DH survey

Main pros and cons of the borehole methodologies are reported in Table 1.1.

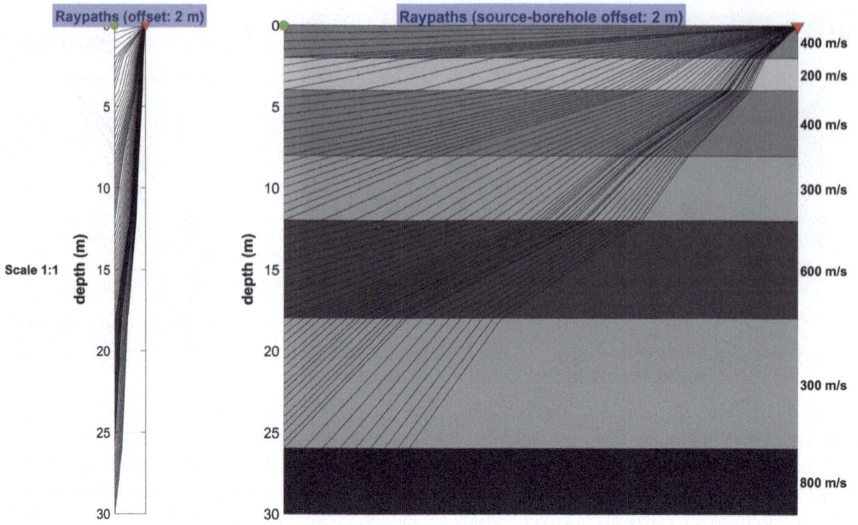

Fig. 1.5 The actual path of the body waves generated by the surface source follows a non-linear trajectory that depends on the wave refraction. The definition of the correct velocities would require the correct modelling of such a "non-linear" behavior (the actual ray path is not a straight line). This can be particularly important when abrupt velocity variations occur or when low-velocity or stiff layers are present. It must be also underlined that, if the distance between the source and the borehole (*offset*) is too small, complex wave phenomena can occur and prevent from the possibility to properly identify the transmitted waves. In the shown schemes, different *x* and *y* scales are used so to highlight the actual trajectories

1.2 Surface-Wave Analysis: Few Simple Introductory Aspects

It is well known that surface waves can be analyzed so to obtain information about the subsurface shear-wave velocities. This is possible for a very simple reason: the propagation velocities of Rayleigh, Love and Scholte waves are strictly related to the V_S profile (V_S and thickness of the layers).

The first and most important point to highlight is that surface-wave analysis can be performed through a vast number of techniques and the well-known MASW approach (with the analysis of the interpreted modal curves of the vertical component of Rayleigh waves) represents just one of the possible approaches (and, as we will see, not the best one). For a general overview on the classical multi-offset (and multi-component) MASW methodology (problems, solutions and case studies), see Dal Moro (2014).

As we know, the MASW acronym stands for *Multichannel Analysis of Surface Waves*. The goal is of course the analysis of surface-wave propagation for the determination of the vertical V_S profile.

The basic principle is that the lower the frequency (i.e. the longer the wavelength) the deeper the penetration (this is why, in near-surface applications, we need 4.5 Hz geophones).

As we briefly stated, surface-wave propagation mostly depends on V_S and thickness of the layers, being density and V_P relatively irrelevant (this means that densities and V_P cannot be obtained from surface-wave analysis).

The, so-to-speak, standard MASW approach is based on the analysis of the vertical-component of Rayleigh waves but, because of several issues, while considering the standard multi-offset approach it is strongly recommended to analyze Love waves as well. This way, it is possible to avoid several possible pitfalls and ambiguities that are inevitable in case we consider Rayleigh waves only (Safani et al. 2005; Dal Moro and Ferigo 2011; Dal Moro 2014) as well as gain important information about deep layers even by considering just the high frequencies (Dal Moro 2020).

From the practical point of view, the joint acquisition of Rayleigh and Love waves can be easily accomplished by means of just 12 horizontal geophones.

The procedure can be summarized in few lines (see also Figs. 1.6 and 1.7): we first orientate the horizontal geophones so to have their axis parallel to the array (Fig. 1.7b) and produce Rayleigh waves by means of a vertical-impact source. We then rotate the geophones by 90° and generate Love waves by means of a shear source (HF—Horizontal Force) (Fig. 1.7c—for further details see Appendix A and Dal Moro 2014, 2019a).

The acquisition of Love waves is thus extremely simple and the procedure is basically the same used for the data acquisition in SH-wave refraction studies (additional suggestions about the generation of shear waves are reported later on in Sect. 1.14).

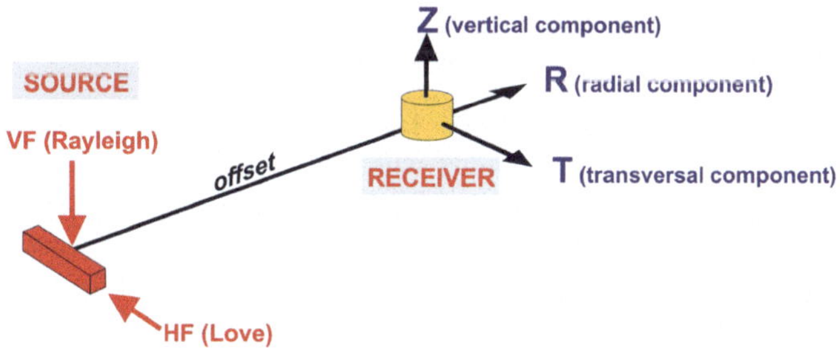

Fig. 1.6 Seismic components: the acquisition and processing of more than one component allows the joint inversion of several "objects" (observables) and, consequently, the determination of a well-constrained subsurface model that does not suffer from significant non-uniqueness of the solution. The whole book adopts this multi-component perspective

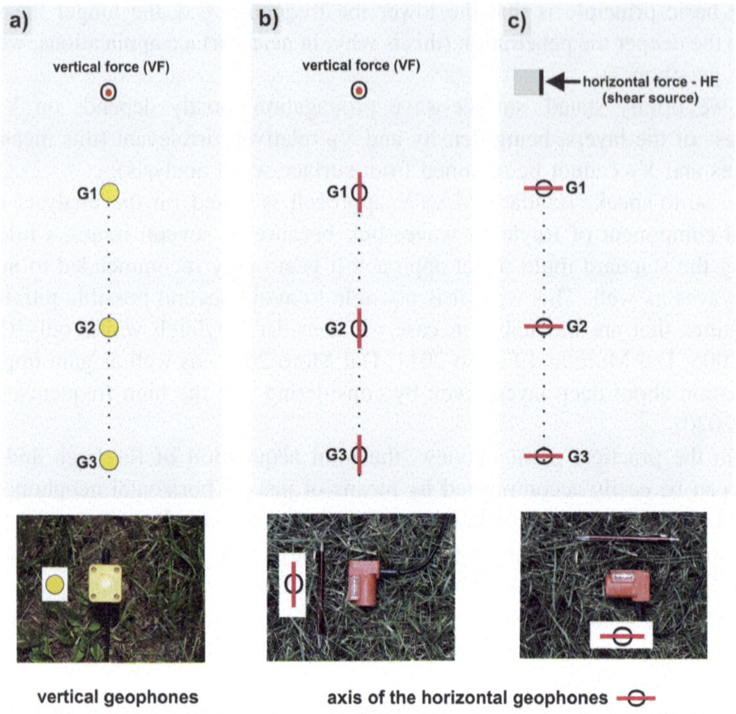

Fig. 1.7 Acquisition setting for the acquisition of multi-offset data: **a** standard acquisition of the vertical (Z) component (for P-wave refraction/reflection and Rayleigh-wave studies); **b** acquisition of the radial (R) component (mainly—but not solely—for the analysis of the radial component of Rayleigh waves); **c** acquisition of the transversal (T) component for shear-wave reflection/refraction studies and Love-wave analysis. In case we have a set of horizontal geophones we can record (and then analyze) both Love and Rayleigh (radial component) waves. See also Table 1.3

In case we need to investigate deeper layers (that cannot be sensed through the analysis of the surface waves recorded with ordinary, relatively-short, arrays), a 3-component (3C) geophone can be used for the determination of the *Horizontal-to-Vertical Spectral Ratio* (HVSR) to be analyzed jointly with surface-wave dispersion data (e.g. Arai and Tokimatsu 2005; Dal Moro 2010, 2015, 2019b).

The scheme in Table 1.2 summarizes the characteristics of the main active and passive techniques currently available for the determination of the V_S profiles.

Table 1.2 Main methodologies useful for the determination of the subsurface V_S profile from surface seismic data (overview, details and case studies in Dal Moro 2014, 2019b as well as in this very book)

	Technique	Pros	Cons	Notes
Active	Standard MASW (vertical-component of Rayleigh waves and modal dispersion analysis)	Pretty popular	Velocity spectra can be highly ambiguous (and, consequently, the picking erroneous) Solution is non-unique (see Sect. 1.3)	Dispersion can be analyzed considering the dispersion curves or according to the FVS approach (more detailed with respect to the standard approach based on the modal dispersion curves—see Chap. 2)
	Multi-component MASW	Requires the acquisition of at least 2 components (see Figs. 1.6 and 1.7)	It solves the ambiguities of the velocity spectra It strongly reduces the non-uniqueness of the solution	
	HS (*HoliSurface—Holistic analysis of Surface waves*)	Very simple acquisition setting (just one 3-component geophone)	Currently still not very popular	
	SH-wave refraction	Reconstruction of 2D sections	Complex field operations Limited investigated depth The presence of shallow stiff layers prevents from the correct identification of the deeper velocities (i.e. the identification of velocity inversions is problematic)	–
Passive	ReMi	–	Ambiguities in the determination of the *effective* dispersion curve (due to the linearity of the array which, in passive seismics, does not allow to handle the directivity of the signals)	Dispersion must be modelled according to the *effective* dispersion curve and **not** to the fundamental mode (details in Chaps. 3 and 4)
	ESAC (Extended Spatial AutoCorrelation)/ SPAC (SPatial AutoCorrelation)	Obtained *effective* dispersion curve does not suffer from the ambiguities typical of the ReMi approach	Complex field operations (necessary large *bidimensional* arrays—SPAC requires circular geometries)	
	MAAM (*Miniature Array Analysis of Microtremors*)	It requires just 4 or 6 (high-quality) vertical geophones and limited room (for near surface applications about 2–5 m—see Chap. 4)	Very sensitive to the quality of the equipment and to the accuracy of the acquisition procedures (see Dal Moro 2019b and Chap. 4 of this book)	
	HVSR (*Horizontal-to-Vertical Spectral Ratio*)	Simple acquisition procedures	Highly non-unique	Recommended to be used only together with dispersion data (see other methods)

Table 1.3 Nomenclature adopted for the different components (i.e. acquisition settings). See Figs. 1.6 and 1.7 as well as Dal Moro (2014)

Acronym	Geophone	Source	Component
ZVF	Vertical (Z) (Fig. 1.7a)	Vertical force (VF): sledgehammer, weight drop, vibroseis	Vertical component of Rayleigh waves
ZEX	Vertical (Z) (Fig. 1.7a)	Explosive source (EX)	Vertical component of Rayleigh waves
RVF	Horizontal with axis parallel/radial (R) to the array (Fig. 1.7b)	Vertical force (VF): sledgehammer, weight drop, vibroseis	Radial component of Rayleigh waves
REX	Horizontal with axis parallel/radial (R) to the array (Fig. 1.7b)	Explosive source (EX)	Radial component of Rayleigh waves
THF	Horizontal with axis perpendicular (transversal, T) to the array (Fig. 1.7c)	Shear source (horizontal force—HF)	Love waves

1.3 Non-uniqueness of the Solution: Problems and Solutions

The analysis of any kind of geophysical surface data inevitably suffers from the problem of the non-uniqueness of the solution (e.g. Scales et al. 2001).

A synthetic and conceptual representation of this well-known problem is schematized in Fig. 1.8. In this representation, the method/dataset/observable A (for instance the velocity spectrum of the vertical component of Rayleigh waves) can be explained by seven models (A–G) while the method/dataset/observable B (for instance the Love-wave velocity spectrum) by the E–M models. Only some of them (the models G, E and F) are in common and that means that, by considering these two methods/datasets/observables, we have now better constrained the solution, by excluding the models A–D (possible if we would use only the first method/dataset/observable) and the models H–M (that could be used to justify the second method/dataset/observable). This can continue to include more and more observables thus reducing the ambiguities that would otherwise taint and jeopardize any inversion procedure based on just one single observable.

In this conceptual scheme, the joint analysis of the three considered observables allows to identify the model F as the only one capable of explaining all the three observables.

The standard approaches (standard MASW, ReMi, ESAC/SPAC, MAAM etc.) are based on data recorded by a set of vertical geophones but the inevitable consequence is that, this way, we deal with just one observable (the dispersion of the vertical component of Rayleigh waves).

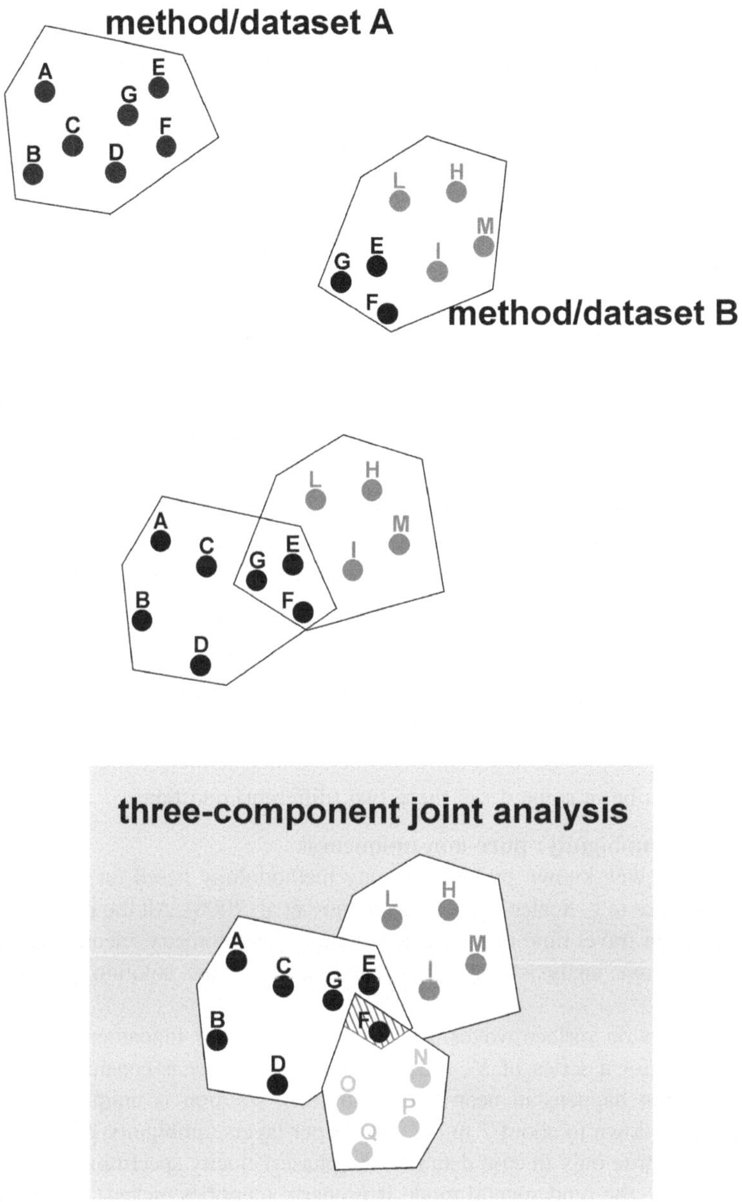

Fig. 1.8 Conceptual scheme that highlights the importance of the joint analysis as solution for the problems related to data ambiguity and non-uniqueness of the solution: only by using more "objects" (observables) it is possible to constrain a robust inversion process and obtain a reliable subsurface model

In order to obtain more "objects" to jointly invert, we can adopt several strategies (see for instance the procedure briefly illustrated in the previous section for the acquisition of both Rayleigh and Love waves and the techniques presented later on in this very book).

On the other side, we could decide to deal with Rayleigh waves only but record both the vertical (Z) and radial (R) components. This way, we would be able to jointly analyze three observables: the Z and R phase-velocity spectra and the RPM (Rayleigh-wave Particle Motion) frequency-offset surface (see Chap. 4 and Dal Moro et al. 2017a, b).

On the other side, it can be underlined that through the HS (*Holistic analysis of Surface waves*) approach (a methodology based on the active data recorded by a single 3-component geophone and processed according to the group velocities and the *Radial-to-Vertical Spectral Ratio* (RVSR) and/or the *Rayleigh-wave Particle Motion* (RPM) curves), it is possible to obtain the same result (Chap. 4 presents the details of this multi-component approach).

First and Second Order Ambiguities

As a matter of fact, the expression *non-uniqueness* is just a generic warning that, in a more analytical way, could or should be described in a deeper way. We can in fact identify (at least) two sources of ambiguities:

#1 inversion ambiguity: pure non-uniqueness
#2 data ambiguities: *impossible* velocity spectra

Let us describe in some detail these two (different) questions.

#1 Inversion ambiguity: pure non-uniqueness

This is a very well-known problem for any methodology based on data collected from the surface (e.g. Scales et al. 2001; Ivanov et al. 2006). All the methodologies (e.g., refraction-travel time analysis, gravimetry, magnetometry, electrical methods and surface-wave analysis) suffer from the fact that the solution is inherently non-unique.

Let us focus on surface waves. Figure 1.9 presents the fundamental mode dispersion curve for a series of V_S models (dispersion curve is considered down to 5 Hz—as often happens in near-surface studies): solution is unique (i.e. unambiguous) only down to about 7 m but, for deeper layers, ambiguity gets larger and larger. This is true only in case data (i.e. the phase-velocity spectrum) are correctly interpreted (i.e. the fundamental mode is properly identifies/picked) but in case of incorrect or inaccurate data interpretation, things get worse and worse (see next section).

In case we would include also lower frequencies (below the 5 Hz considered for the data reported in Fig. 1.9), we would slightly improve the solution but the problem would not be solved.

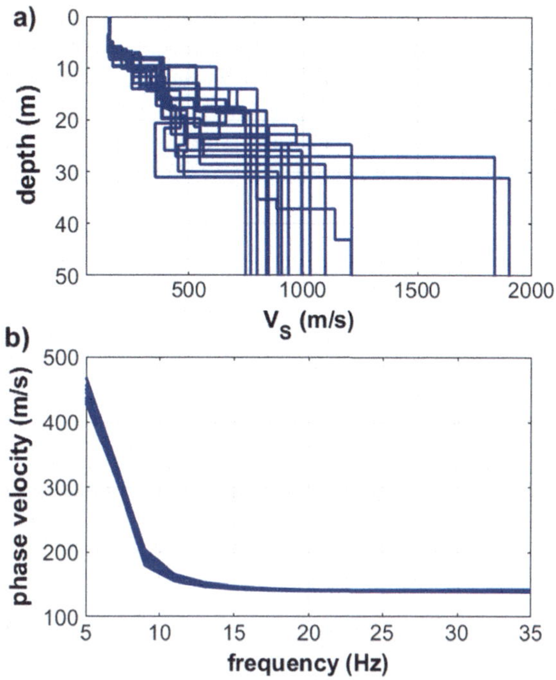

Fig. 1.9 Non-uniqueness of single-component dispersion analysis analyzed according to the standard approach based on the modal dispersion curves of a single component: **a** a series of V_S profiles (Poisson ratios are different for each model and layer); **b** their Rayleigh-wave fundamental-mode dispersion curves. The same dispersion curve can be associated to different subsurface models. After Dal Moro (2011)

Non-uniqueness is such that (we should pay great attention to this point) *the same dispersion curve can refer to different subsurface models*. The consequence is simple and straightforward: since the same curve can refer to *n* models, there is no way to understand which one is the correct one. Unfortunately the literature is plenty of exotic inversion strategies that somehow claim to solve such a point which is actually unsolvable since *the same dispersion curve can be attributed to different subsurface models*.

How can we solve this issue? Combining two things:

(1) analyzing surface-wave dispersion in a more efficient manner, i.e. not considering the ordinary (problematic and ambiguous) analysis based on the modal dispersion curves;
(2) jointly inverting several observables (e.g. analyzing both Rayleigh hand Love waves—but we can do much more and the whole book focuses on different possible strategies aimed at improving the analyses and, therefore, the reliability of the solution).

#2 Data ambiguities: *impossible* velocity spectra

The fundamental "object" (*observable*) considered while analyzing surface wave propagation is the velocity spectrum, i.e. a sort of frequency-velocity correlation matrix (depending on the methodology, we can deal with phase or group velocities —in Chap. 4 we will come back to this point).

In order to explain the intrinsic ambiguity of the velocity spectra, we can consider a standard phase-velocity spectrum as commonly analyzed for an "ordinary" MASW survey. Figure 1.10 presents a synthetic dataset about the vertical (Z) component of Rayleigh waves (caption provides all the details). The phase-velocity spectrum (Fig. 1.10c) is apparently simple (smooth and continuous without any weird feature) and thus seemingly suitable for the classical *picking* and *inversion* procedure.

The problem is that, once we plot the modal dispersion curves of the first two modes (Fig. 1.10d), we realize that such continuous signal does not belong to a single mode but is the combination of the fundamental mode (for frequencies higher than 40 Hz) and the first higher mode (for lower frequencies) (some authors call the point where two modes meet the *kissing* point—e.g. Gao et al. 2016).

It should be clear that, as a matter of fact, it is impossible to realize that the observed signal (Fig. 1.10c) is actually composed of two distinct modes and, by following the ordinary approach, you would pick the dispersive signal in Fig. 1.10c

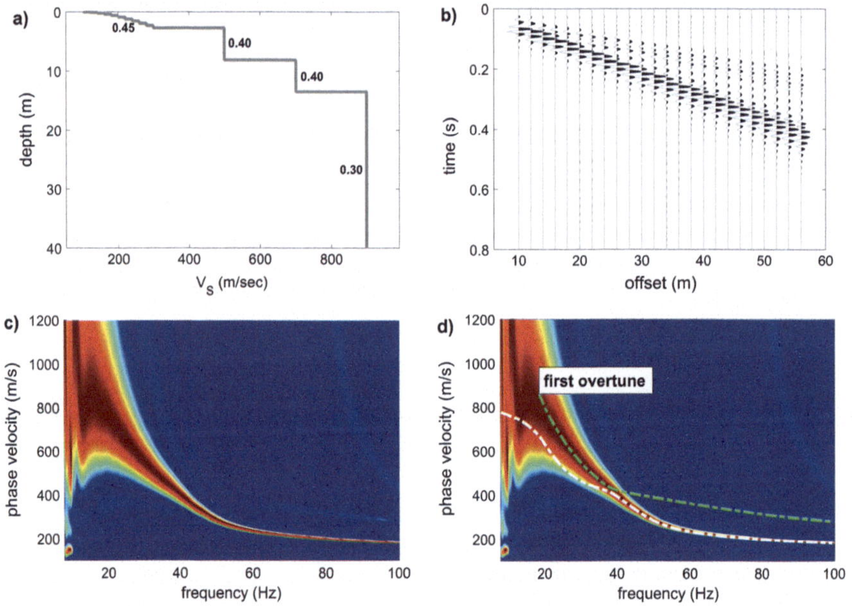

Fig. 1.10 Example of synthetic dataset (classical ZVF component) to put in evidence the intrinsic ambiguity of standard (single component) surface-wave data: **a** considered V_S model (numbers report the Poisson ratios); **b** synthetic traces computed according to Carcione (1992); **c** phase velocity spectrum computed from the synthetic traces; **d** phase velocity spectrum with the theoretical modal dispersion curves of the first two modes. This kind of data are *impossible* to solve since the phase-velocity spectrum cannot be correctly interpreted in terms of modal dispersion curves (see text for comments)

as the fundamental mode, thus eventually obtaining an erroneous V_S profile (with overestimated values).

It must be underlined that the inversion of the erroneously-picked dispersion curve would surely provide a seemingly-good misfit (any dispersion curve—right or wrong—can be associated to a subsurface model).

The practical consequence is that everything would seem simple and fine but, as a matter of fact, the obtained solution would be meaningless (with overestimated V_S values).

Things can be even more complex (see e.g. the case study #11 in Dal Moro 2014) and this means that, in some cases, the phase-velocity spectrum is ambiguous *per se* and, consequently, analyses accomplished according to the standard approach (picking of interpreted modal dispersion curves and inversion) necessarily fail.

The problem can be solved just through the joint inversion of more observables and the dispersion analysis accomplished by means of more sophisticate approaches that go beyond the standard approach based on the analyses of interpreted modal dispersion curves (that, as we just saw, sometimes cannot be applied).

The whole book is about that: trying to solve the issues related to surface-wave analysis.

Figure 1.11 presents an example of joint (simultaneous) acquisition of the two (Z and R) components of Rayleigh waves: two seismic cables are set parallel to each other (see also Fig. 1.12). One is equipped with vertical (Z) geophones, the other with radial geophones (R component). This way we record two datasets that can be processed so to obtain three observables: the two phase-velocity spectra and the RPM frequency-offset surface that describe the actual particle motion (prograde or retrograde) for each offset and frequency (in this introductory pages we just want to provide a scenario about the several possibilities actually available for surface-wave acquisition and analysis while in the next chapters we will see in detail how these methodologies work).

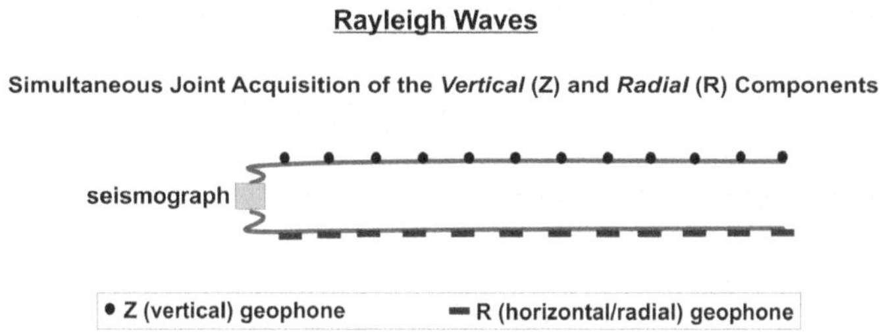

Fig. 1.11 Example of joint acquisition of multi-offset data for the vertical (Z) and radial (R) components (see also Fig. 1.7) useful for the joint analysis of the Z and R components also jointly with the frequency-offset RPM (Rayleigh-wave Particle Motion) surface (see also Fig. 1.12 and the case study reported in the Appendix G). More details in Dal Moro et al. (2017b)

Fig. 1.12 An urban multi-component (ZVF + RVF) surface-wave survey: a land-streamer equipped for the acquisition of both the vertical (yellow vertical geophones) and radial (green horizontal geophones) components (see Figs. 1.7 and 1.11). Courtesy of Lorenz Keller (*roXplore.ch*). Appendices B and G illustrate two case studies of this kind. Figure 1.13 presents an example of the joint analysis that can be accomplished while considering the data collected and can be considered a sort of preview of the topics and methodologies considered in detail in the next chapters

Figure 1.13 presents an example of this kind of joint analysis and refer to one of the several shots recorded for the reconstruction of a 2D section in an urban area (see also Figs. 1.12 and 1.14).

Needless to say that the Z and R data do not have to be recorded simultaneously and it is also possible to record first one component and then change the geophones so to record the second component.

1.4 MASW Tests?

We know very well the concept of *test* in geotechnics: a series of standardized operations that must be meticulously followed so to obtain a value that can be compared with other values obtained (for different sites or samples) by following exactly the same procedure. Geotechnical tests can be carried out by lab technicians

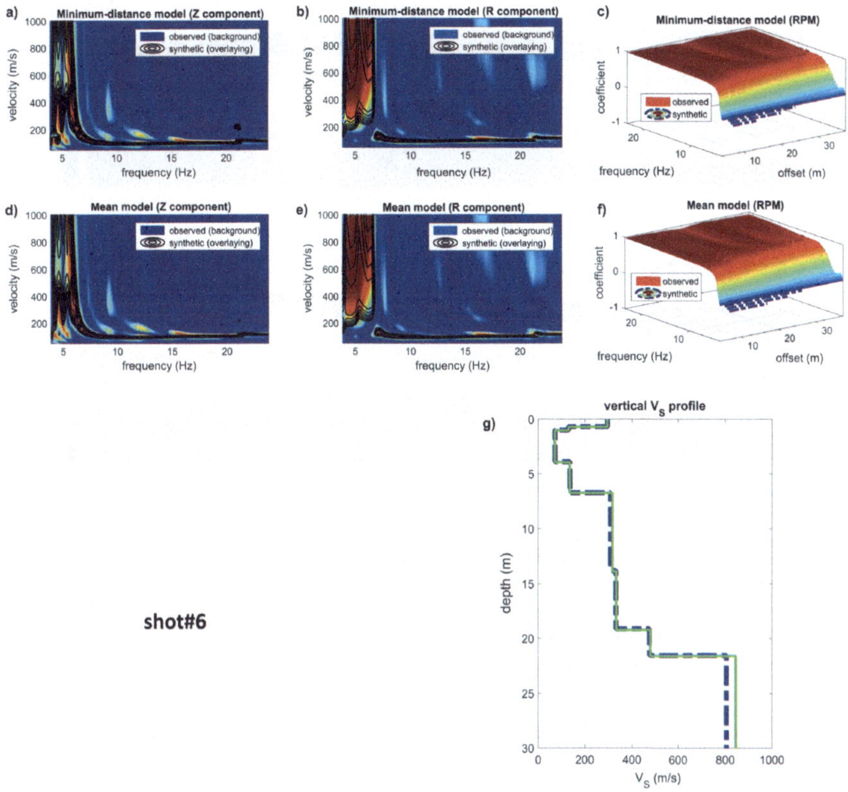

Fig. 1.13 Joint inversion of the Z and R phase-velocity spectra together with the RPM frequency-offset surface. Upper plots are about the so-called *minimum-distance model* (Dal Moro et al. 2019a): **a** vertical-component phase velocity spectra; **b** radial-component phase velocity spectra; **c** RPM frequency-offset surfaces. Lower plots refer to the *mean model*: **d** vertical-component phase velocity spectra; **e** radial-component phase velocity spectra; **f** RPM frequency-offset surfaces. The two V_S profiles are reported in the **g** plot. For the velocity spectra, the colors in the background represent the field data while the overlaying black contour lines pertain to the synthetic data of the identified models. For the RPM data, the synthetic surface is reported by dashed contour lines with the same color scale as the field data (since the agreement between the field and synthetic data is extremely good, the two surfaces are visually hardly separable). Details in Dal Moro et al. (2017b)

by following accurate protocols so to obtain values that can be compared to other values obtained by following exactly the same protocol.

In geophysics this is not possible for at least two reasons:

(1) during a field survey, nothing can be standardized;
(2) data analysis/processing always and necessarily requires the work of somebody with specific (theoretical and practical) skills in that specific methodology.

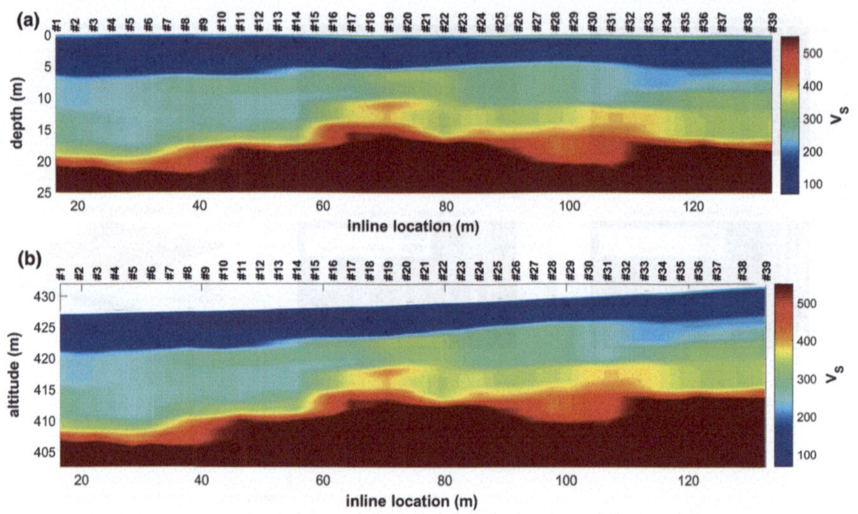

Fig. 1.14 2D V_S section obtained from the joint analysis of the Z + R components together with the RPM frequency-offset surface: **a** V_S values as a function of the inline position and depth from the surface; **b** V_S values as a function of the inline position and altitude (above sea level). Labels reported at the top of the two sections indicate the *shot number* (Fig. 1.13 reports the data about the shot#6). From Dal Moro et al. (2017b)

In geophysics, the concept of *test* simply does not (and cannot) exist. Just as we cannot speak of "reflection seismic tests" (it would be absurd and nobody ever used this kind of expressions), so we cannot speak of "MASW tests" (or HVSR/ESAC tests etc.).

Any geophysical survey (data acquisition and analysis) is a professional activity where a long series of "arbitrary" decisions (that will eventually influence the final outcome) are taken.

The accuracy of the final result necessarily depends on the decisions taken both during the data acquisition (components to record and related acquisition parameters) and data processing (how we analyze the chosen observables).

Since physical phenomena are necessarily exact (nature does not follow different laws according to a momentary whim), if the subsurface model obtained from a seismic survey is inaccurate or incorrect, the cause must be necessarily attributed to the fellows who recorded and analyzed the data since we cannot surely state that seismic waves propagated in a "wrong way".

The only way to properly collect and analyze geophysical data is to get a sufficiently wide and profound theoretical background on all the aspects involved in data acquisition and analysis. Acquisition guidelines and protocols are necessary only when we do not know what we are doing and ignore the physical principles of the considered phenomena. As they say in the States: *rules are for fools*.

We should rather learn the fundamentals and take all the decisions based on our knowledge of the phenomena we intend to record and analyze. And avoid to think that a *seismic survey* is a mere *geotechnical test*.

1.5 Some Legends and Some Opportunities: Number of Channels, *f-k* Versus *f-v* and Non-equally Spaced MASW

You blind guides! You strain out a gnat but swallow a camel.

From the *Gospel* according to Matthew, 23–24

This passage from the Gospel of Matthew applies to several issues regarding surface wave analysis and can be put in simple terms as it follows: what is the point in putting an exaggerated attention to extremely minor (irrelevant) questions when, at the same time, we are neglecting what really matters?

The first very classic issue is about the number of channels necessary to perform a standard MASW (multi-offset) survey. Let us address this point by means of a real-world example presented in Fig. 1.15. On the left column are shown the original traces (70 channels, array length of about 150 m) and the computed phase velocity spectrum (i.e. the *f-v* [*frequency-velocity*] matrix computed according to the phase-shift method—Park et al. 1998; Dal Moro et al. 2003). Moving to the right are reported the decimated data: in the central column the 35-trace data and on the right the 18-trace data (the total length of the array remains fundamentally the same: what change is the geophone distance).

Is there any significant difference in the phase velocity spectra obtained (with the phase shift method) while using 70, 35 or 18 traces? Absolutely nil. This is due to the fact that phase velocity is fundamentally a slope and a slope can be properly defined also considering a limited number of points (traces).

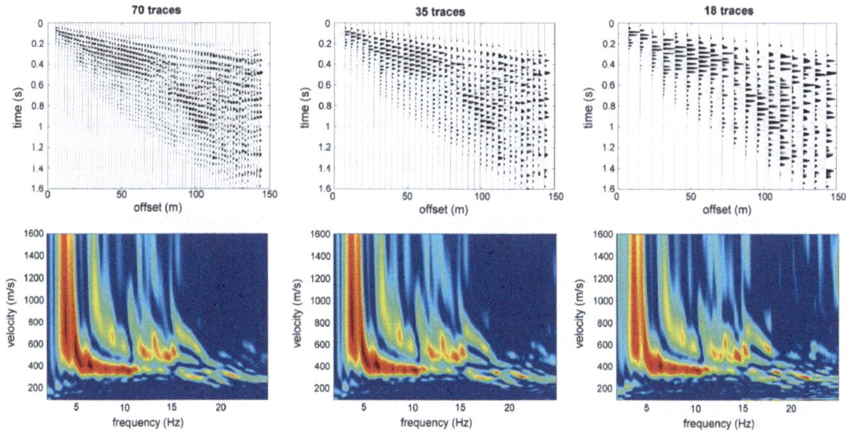

Fig. 1.15 Number of channels to be used for multi-channel surface wave analysis. On the left, the original 70-trace dataset and (below) the corresponding phase velocity spectrum (i.e. the *f-v* matrix); in the middle the data after removing half of the original traces (now we have 35 traces); on the right, the data after a further trace decimation (now we keep just 18 traces). Is there any significant difference in the velocity spectra obtained while using 70, 35 or 18 traces? From Dal Moro (2014)

What does that mean from a practical point of view?

Simply that what really matters is the overall length of the array (which affects the largest identifiable wavelength and, consequently, the maximum investigated depth since the accuracy of the dispersion curve in the low-frequency range is proportional to the length of the array), but not the amount of channels. For ordinary arrays with a length between 60 and 100 m, 9–12 channels are usually more than sufficient (see also Dal Moro et al. 2003 and the further examples presented in this volume, e.g. in Chap. 2).

And about the "vertical resolution"? Does it change when we modify the geophone distance? (one of the several urban legends claims that the vertical resolution depends on the geophone distance).

The answer (in form of rhetorical question) is very simple: if the phase velocity spectra obtained from data with difference geophone spacing are the same (e.g. Figs. 1.15 and further examples in Chap. 2) how can the "vertical resolution" be different?

The problem is the confused and chaotic mix of concepts that actually belong to different methodologies: while geophone spacing surely influences the resolution of refraction/reflection surveys, the same does not apply to surface wave analysis.

In surface wave analysis, the vertical resolution is an "intrinsic" fact ruled by the physical laws that describe the surface wave propagation and dispersion. While analyzing surface waves, the vertical resolution decreases with depth according to (roughly speaking) a sort of geometric progression: surface layers are well defined but getting deeper and deeper the thickness of the identifiable strata significantly increase. We should imagine that the minimum thickness that can be actually discriminated follows a progression like (roughly speaking) 0.4, 0.8, 1.5, 3, 5, 7 m, etc. (we intentionally avoided a perfectly-regular series cause the actual thickness also depends on how much the V_S values change).

So, in surface wave analysis, the geophone distance is not relevant but, a further question arises: is it really necessary to collect equally-spaced data? In some cases, logistical problems can prevent from deploying the geophones at fixed distances.

If we adopt the phase-shift approach, we can work with non-equally spaced data (this is not possible if we work with the *f-k* transform). In Fig. 1.16 are reported the data (seismic traces and respective phase-velocity spectra) for an equally-spaced dataset (upper plots) and for a non-equally spaced dataset obtained from the first one. The computed velocity spectra are fundamentally equivalent (the small differences are mainly due to the fact that the contribution of the different modes is a function of the offset and, in the second dataset, higher modes are more clearly visible).

There are therefore at least four reasons to prefer the velocity spectra computed via phase-shift instead of the *f-k* (*frequency-wavenumber*) spectra that some authors are sometimes still using:

(1) *f-k* spectra do not provide direct information about the velocities (in order to obtain them it is necessary to consider that $v = f/k$);
(2) because of this, in the *f-k* domain, the separation between different modes is extremely tricky;
(3) *f-k* spectra are significantly influenced by *spatial aliasing* (see Chap. 2);

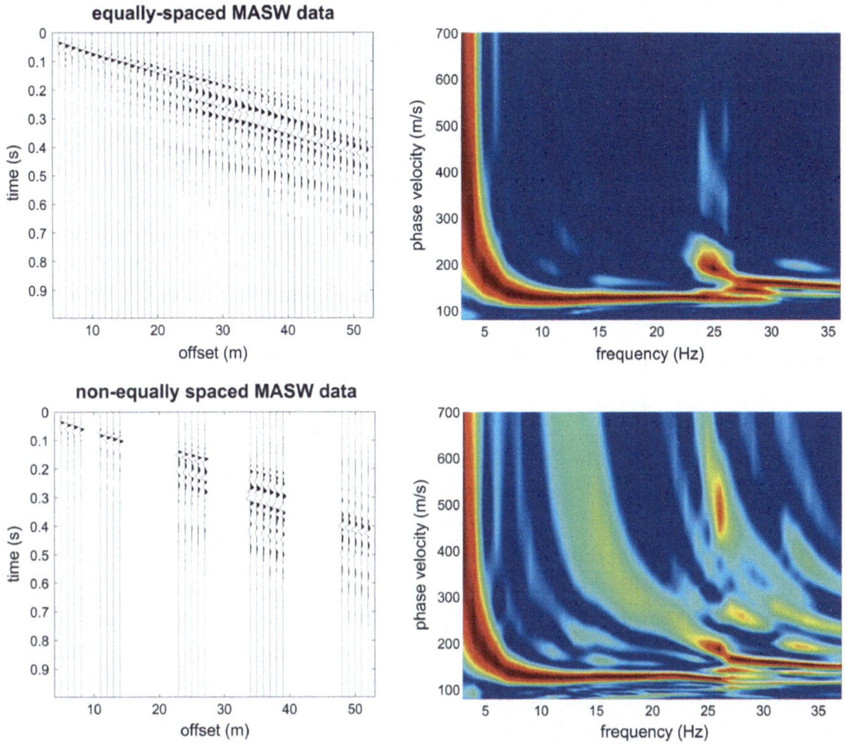

Fig. 1.16 Upper panel: seismic traces (ZVF component) and phase-velocity spectrum for a traditional (equally-spaced) dataset; lower panel: non-equally-spaced data obtained by removing some traces from the original (equally spaced) data reported in the upper panel. See text for comments

(4) *f-k* processing is based on a double Fourier transform and requires equally-spaced data while phase shift can be applied to non-equally spaced data (Fig. 1.16).

Joint analysis* versus *data integration

It must be emphasized that the expression *joint analysis* is sometimes used to indicate a completely-different procedure that, strictly speaking, should be called *data integration*.

Joint analysis refers to a procedure where a series of observables are considered together (within the same algorithm) in order to find a model that, in quantitative terms, represents the best compromise for all the considered data.

On the other side, *data integration* is a qualitative evaluation of the results obtained by ordinary (single-component) methodologies. An example can better clarify the difference. In case we analyze Rayleigh-wave dispersion (for instance in the 5–40 Hz frequency range) and obtain a certain V_S profile and then use the obtained V_S values to obtain some hints about the meaning of a HVSR peak observed at 1 Hz, we are doing a merely qualitative *data integration*.

On the other side, if we put the dispersive properties and the HVSR curve within the same (properly designed) inversion algorithm and obtain a solution that simultaneously matches both the observables, we are doing a (quantitative) *joint analysis*.

Some general notes:

1. when dealing with the HVSR curve (see Chap. 3), avoid using frequencies higher than about 15 Hz since they refer to extremely shallow features that are both irrelevant and (potentially) very local and not representative of the general conditions in the area;
2. if we (jointly) invert a picked modal dispersion curves (standard MASW approach) we are not inverting *data* but *data interpretations* and data interpretation can be erroneous (see previous pages) (geologists should consider very carefully the difference between *data* and *data interpretation*).

Standard MASW and ESAC/ReMi/SPAC provide information about the same thing (the vertical component of Rayleigh waves) and therefore cannot be used to perform a true joint inversion. Although passive data can provide clearer information about the lowest frequencies, those observables are fundamentally telling the same thing (Rayleigh-wave dispersion along the vertical component) while to perform a real joint analysis we need to deal with two independent observables (e.g. Rayleigh and Love waves, dispersion from ESAC and HVSR curve and so on).

1.6 Seismographs?

Among the several misunderstandings caused by a limited vocabulary (remember the opening quotations by Wittgenstein?), there are those related to the so-called *seismograph*.

As a matter of fact, the word *seismograph* does not indicate anything clear and specific and, in case we intend to perform certain types of acquisitions (and analyses), it is not possible to be vague and generic.

The most correct (and comprehensive) expression that should be used is *acquisition system* since, during a field survey, we are not using a seismograph but an *acquisition system*.

The *acquisition system* is in fact a set of devices that are used to record the data and, in short and very general terms, consists of:

- geophones;
- seismic cable(s);
- A/D (Analog-to-Digital) conversion (and storage) unit (which is what people usual call *seismograph*).

Actually, there is a fourth element that might seem somehow "intangible" but is absolutely decisive (it is perhaps the most important element): the acquisition software that manages the operations of the A/D unit and that represents the interface between the seismograph and the field operator (a poorly-designed acquisition software can reflect in a field nightmare).

In case the acquisition software is designed without adequately considering the kind of acquisitions we intend to perform, the entire *acquisition system* becomes scarcely useful.

Having the best geophones, a perfectly-shielded cable and the most sophisticated electronic components is completely useless when the acquisition software has been designed by somebody who is not sufficiently familiar with the needs of the state-of-the-art techniques.

On the other side, an excellent seismograph becomes useless when questionable geophones are used.

Furthermore: the same seismograph (i.e. A/D unit) and the same geophones can produce different data in case we are using different seismic cables.

Let us suppose (Fig. 1.17) we have two horizontal geophones (G1 and G2) with opposite polarities and two cables (A and B) with opposite polarities as well. Figure 1.17 shows the seismic traces obtained when three different geophone-cable combinations are adopted. It is clear that, depending on the polarities of the single elements (geophones and cables) and on the orientation of the geophone, the seismic traces have the correct or wrong polarity. Let us now consider the active data recorded by two geophones: one vertical and one horizontal (set radially with respect to the source) deployed at the same distance from the source.

With the two (Z and R) traces it is possible to compute the group velocity spectra of the Z and R components of Rayleigh waves. But there is another "object" that can be obtained from these two traces: the RPM (Rayleigh-wave Particle Motion) curve (a curve that is useful for further constraining the subsurface model—details will be discussed in Chap. 4).

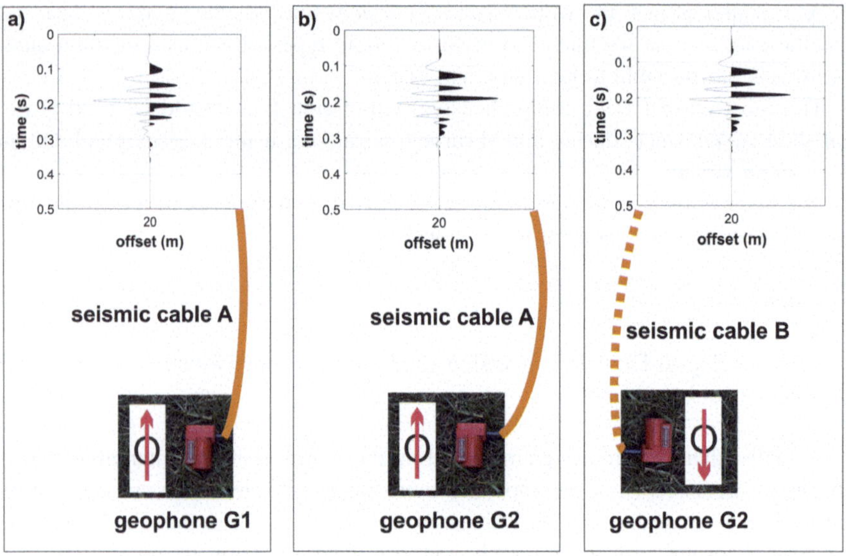

Fig. 1.17 Geophones, cables and signal polarity. In this case we have two types of cables (A and B) and two types of geophones (G1 and G2): different cables and different geophones (differently oriented) produce traces with different polarities (but only one is the correct one). How do you know how to properly orientate our geophones? See also Fig. 1.18 (further details in the text). After Dal Moro (2019b)

Depending on the polarity of the cable and on how we rotate the radial geophone, we obtain two different (specular) curves (see Fig. 1.18). Of course, only one curve is correct but in order to know which one is the right one, we cannot test just the geophone(s) but the whole *acquisition system* since the trace polarities depend on the combination of *all* the elements: the geophones (and their orientation), the seismic cable(s) and the seismograph itself.

Looking at the single elements without considering the *acquisition system* as a whole, is an error since the final outcome depends on the combination of all the elements.

Multi-offset and multi-component data shown in Fig. 1.19 are an example of poorly-designed acquisition system with several problems both about the trace polarities and the origin time (the trigger procedure was poorly managed by the A/D unit and the traces of the RVF component are clearly "cut"—data acquisition started too late and some data were lost).

Just few words about the *processing system* (i.e. the elements involved in the data analysis). This is clearly represented by the software used to analyze the data and by the characteristics of the computer (workstation). But there is a final

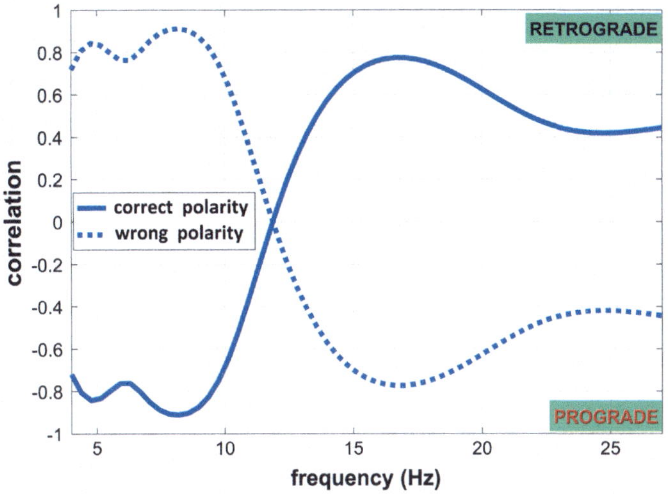

Fig. 1.18 Consequence of incorrect data polarity in the computation of the RPM curve. An *acquisition system* designed without considering the polarity issues can lead to misleading data. In this case, two RPM curves are shown: one (continuous line) correctly represents the particle motion induced by the Rayleigh wave propagation, while the other (dotted line) is incorrect (and would thus lead to meaningless analyses). See text for further comments

(critical) element: the theoretical background of the operator. A perfect acquisition system and a great software are useless if the fellow in charge of the data analysis does not master each single issue involved in the data acquisition and analysis.

1.7 Single-Component Geophones, Seismic Cables and Connectors

In the daily life, there are often some bad linguistic habits about the way vertical and horizontal geophones are called. They are in fact often called "P geophones" (or "P-wave geophones") and "S geophones" (or "S-wave geophones").

But vertical geophones are and must be called vertical geophones.

In near-surface applications they can be used for the acquisition of at least three types of waves/events:

(1) the vertical component of refracted P waves;
(2) the vertical component of reflected P waves;
(3) the Rayleigh-wave vertical component (both in active and passive techniques).

What can we do with a set of horizontal geophones? Depending on their orientation, we can record:

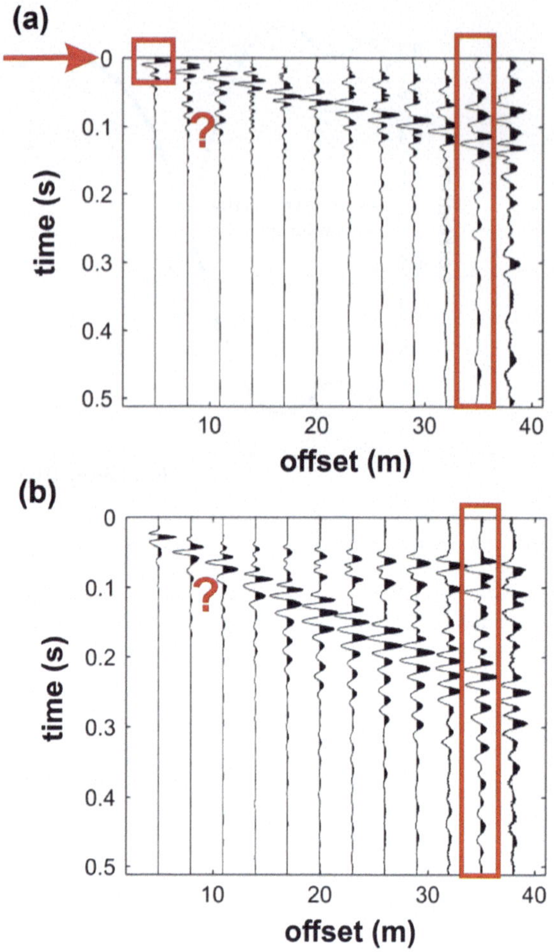

Fig. 1.19 A problematic joint dataset: **a** RVF data; **b** THF traces. A series of serious problems due to a poorly-designed *acquisition system* are apparent. RVF data clearly show a severe problem with the "origin time". Probably due to a delayed trigger signal, a data loss occurred (the first trace is clearly cut—compare with the signals of the THF traces shown below). For both the components the eleventh trace (highlighted by the red rectangle) clearly shows a wrong polarity (compare with the nearby traces). Maybe even the second or third traces might be inconsistent with the rest of the traces but, due to the proximity to the source and to the limited amount of traces, it is not possible to say anything clear about it

(1) Love waves (THF component);
(2) the radial component of Rayleigh waves (active seismics);
(3) the radial component of refracted P waves (shortly we will see something about it);
(4) SH-wave refraction;
(5) SH-wave reflection.

The response curve of a geophone provides information about two main facts: its eigen frequency and its sensitivity. We cannot enter into subtle technical details but it is important to underline that the knowledge of the response curve is a key point in case we intend to set up a good *acquisition system*. Together with an appropriate A/D unit it will be in fact possible to obtain data in physical units such as mm/s or cm/s (or whatever).

In case we do not know the response curve of the geophones or in case the seismograph modifies the output voltage from the geophones, it can be impossible to retrieve the information about the actual particle motion in terms of physical units.

Nowadays, there are really several types of geophones with different response curves (usually expressed in mV/cm/s or V/m/s, or whatever). Often geophones are classified according to their sensitivity. Geophones with a response curve of about 0.8 V/cm/s (or higher) are often called high-sensitivity geophones, while when the sensitivity is around 0.3 V/cm/s are referred to as low-sensitivity geophones. Of course the most appropriate geophones should be chosen so to match the characteristics of the seismograph and considering the specific kind of surveys we are interested in (there is no universal rule and high-sensitivity sensors are not necessarily better than low-sensitivity sensors).

For instance, while recording high-amplitude vibrations, high sensitivity geophones can be a problem if the seismograph has a limited dynamic range (in that case, data clipping/saturation can occur [see Appendix A]).

Although nowadays some cableless systems are available, for educational purposes we prefer to stick to the classical *acquisition system* consisting of geophones, cables and seismograph.

How are geophones connected to the seismic cable? Of course through a connector (or clip) that transmits the electrical signal generated by the geophone to the seismic cable, which will send it to the A/D unit (i.e. the unit that converts this electrical signal into numbers and stores them as seismic data/files—such a unit is what is commonly called *seismograph*).

There are two main types of clips: *split spring* (Fig. 1.20) and *Müller* (Fig. 1.21). The first one is definitely more common as it is quicker and easier to handles (*Müller* type has two clips to secure to the cable).

Fig. 1.20 *Split spring* clip

Fig. 1.21 *Mueller* (actually *Müller*) clip

It goes without saying that the seismic cable must be congruent with the clips of our geophones, since it is not possible to connect a split spring geophone to a cable built to work with Müller clips.

And how is the seismic cable connected to the seismograph? There are certainly several solutions. Among these, a very common one is the 27-pin Cannon connector (Fig. 1.22) that can be used to connect the cable to the seismograph, to connect two cables etc.

Fig. 1.22 Cannon 27-pin connectors (often called simply NK-27): female and male versions

Fig. 1.23 Two 3C geophones: the one on the left has a 27-pin Cannon connector and can be connected directly to the seismograph or to the end of a seismic cable (see Fig. 1.22), while the one on the right has three split spring clips (to be connected to three channels of the seismic cable)

Figure 1.23 shows the same 3-component geophone in two different versions: one equipped with a Cannon connector (on the left) and the other with three split spring clips. Of course the first could be connected directly to the seismograph while the second one needs a seismic cable. The best solution does not exist since, as usual, everything depends on the purpose of our surveys. What should be really understood is that everything must be coherent and consistent (in the light of our goals).

1.8 Geophones for MASW and Geophones for Refraction Surveys? Technical Notes and Commercial Legends

Sometimes, as a client asking for a seismic acquisition system, you are offered the purchase of two sets of geophones: 4.5 Hz geophones for MASW and 10 or 14 (or more) Hz geophones for refraction surveys.

If we know the fundamentals of seismics, it is easy to understand that this is pretty meaningless and would just represent a waste of money since, for common near-surface applications, one set of 4.5 Hz geophones is enough. Why? For at least two obvious reasons:

1. With the 4.5 Hz geophones we can *see* everything above a certain frequency (which is definitely less that the 4.5 Hz eigen frequency of the geophone— we will face this problem later on).

The low frequencies are necessary to analyze surface wave propagation (MASW, ReMi, ESAC/SPAC, HS, MAAM etc.) while it is usually said that for refraction studies it is necessary to deal with higher frequencies (say between 10 and 200 Hz—above this frequency any information would be completely irrelevant from the seismic/geological point of view and, furthermore, the signal attenuation completely cancel higher frequencies).

When we record a dataset by using a set of 4.5 Hz geophones, we can easily remove the low frequencies by applying a simple high-pass filter (a filter that removes the low frequencies and, consequently, keep only the high frequencies). This way it is possible to use the same dataset both for surface wave analysis (in this case we just focus on the low frequencies) and for refraction studies (by removing the low frequencies with a high-pass filter).

2. But: is it really necessary to remove the low frequencies to analyze the first arrivals associated with body-wave refraction?

Let us consider a vertical-component dataset that contains both Rayleigh waves and P-wave refraction (first arrivals). Of course Rayleigh waves are much slower than refracted P waves. This means that Rayleigh waves arrive much later than the first arrivals related to the refracted P waves and, and if we overlook the very near-source traces (where wave phenomena are horribly complex), there is no "interference" between the two events (see for instance the data shown in Fig. 1.24): refracted P waves and Rayleigh waves are easily identified and there is no possible misunderstanding or interference.

Therefore, in a dataset recorded with a set of 4.5 Hz vertical geophones, we can analyze both the P-wave refraction and the Rayleigh-wave dispersion. There is no problem whatsoever in having (in the same dataset) both the Rayleigh waves and the refracted P-wave refraction.

In case, for any reason, we want to give more emphasis to the early (refracted) arrivals, we can simply apply a high-pass filter and/or an AGC (Automatic Gain Control) to the data recorded with 4.5 Hz geophones (see data presented in Fig. 1.24).

Buying a set of 10 or 14 Hz vertical geophones for refraction surveys and a set of 4.5 Hz geophones for surface-wave acquisitions is quite meaningless from the technical/scientific/practical point of view.

The 4.5 Hz geophones also "contain" the 10–14 Hz geophones. Topologically speaking, the latter (the high-frequency geophones) are in fact a subset of the 4.5 Hz geophones.

Probably, at very high frequencies (several hundreds of Hz and more) the high-frequency geophones provide clearer data but, for common near-surface applications, those frequencies are completely meaningless.

Instead of wasting money in useless sets of geophones, it is often wiser to get a set of 4.5 Hz horizontal geophones useful for the acquisition of Love waves, radial component of Rayleigh waves and, of course, SH-wave refraction surveys.

Are vertical geophones necessary for P-wave refraction surveys? Two synthetic points:

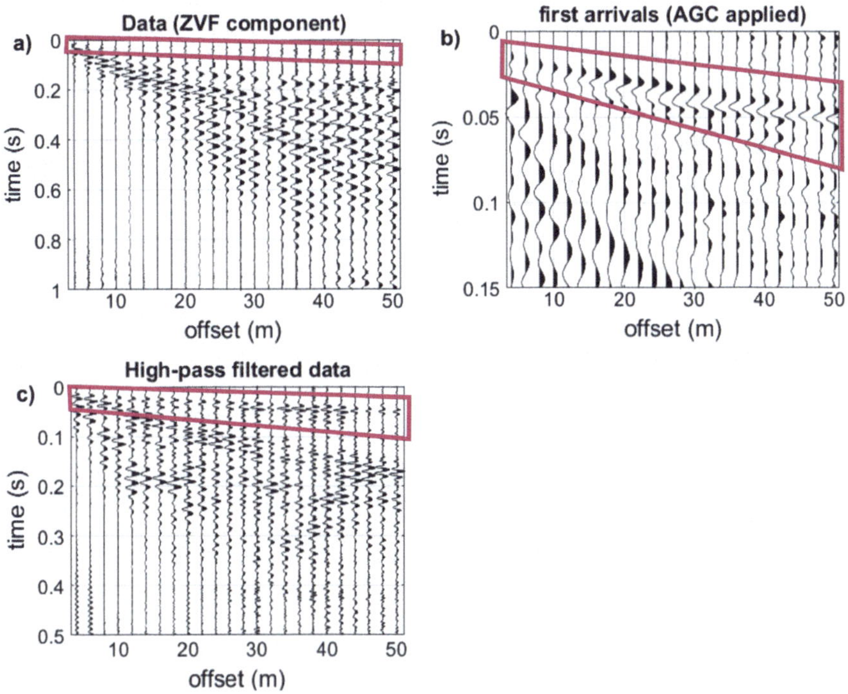

Fig. 1.24 A seismic dataset recorded while using 4.5 Hz vertical geophones: **a** raw data (traces are normalized but no gain or filter is applied—inside the polygon are highlighted the low-amplitude first arrivals associated to P-wave refraction); **b** close up of the first arrivals with the application of a simple AGC (Automatic Gain Control) aimed at increasing the amplitude of the P-wave arrivals; **c** the original traces after the application of a high-pass filter (no AGC) to highlight the P-wave refraction (low-frequency Rayleigh waves are inevitably attenuated). The time axis is different for the three plots so to better highlight the different signals. As can be seen, the refracted P waves arrive well before the (slower) Rayleigh waves. This means that, between refracted P waves and Rayleigh waves there is no "interference" and the two phenomena can be analyzed without any problem from the same dataset (recorded with 4.5 Hz geophones)

(1) first of all it is necessary to remember and underline that P waves are strongly influenced by the saturation conditions and therefore vary their velocity depending on the amount of water in the sediments (this is particularly problematic for the unconsolidated sediments, but to some extent also for rocks—see for example Nováková et al. 2011). On the other hand, S waves are significantly less sensitive to saturation conditions. In case we wish to perform a refraction survey aimed at characterizing saturated sediments, it is much better to do it by means of SH waves and not P waves (of course, surface waves would be a very efficient way to investigate saturated sediments).

(2) refracted P waves do not propagate vertically but obliquely with a critical angle that depends on the ratio between $V_{P(n)}$ and $V_{P(n+1)}$ (Snell's law). Therefore,

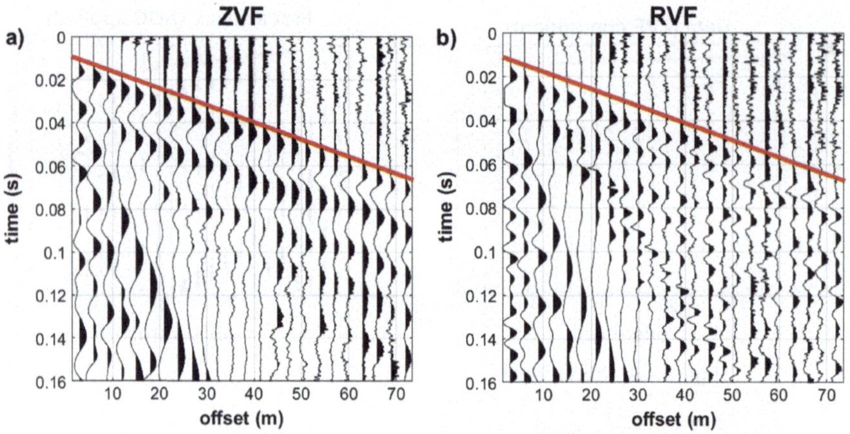

Fig. 1.25 P-wave first arrivals (refraction) and geophones: seismic traces for the ZVF (**a**) and RVF (**b**) components after the application of a simple AGC (close up of the first 0.16 s). First arrivals (associated to P-wave refraction due to the shallow water table in unconsolidated silty sediments) are the same for both the Z and the R components. Data recorded with 4.5 Hz geophones

they have a horizontal (or, to be more precise, radial) component which can be recorded by mean of horizontal geophones set radially with respect to the source (see Fig. 1.7b and related text). Consequence: horizontal geophones can be used to detect the radial component of refracted P waves (see for instance the data shown in Fig. 1.25).

In short (with a focus on refraction): for P-wave refraction surveys, the vertical component is in general necessary but, in some cases, P-wave refraction can be analyzed while considering its radial component (thus by using horizontal geophones set radially with respect the source). The same geophones can then be rotated by 90° so to record the data about the SH-wave refraction (and Love waves—Fig. 1.7c).

1.9 Three-Component (3C) Geophones

First of all we can mention the fact that *3-component geophones* are sometimes called *3D geophones* (the reason is quite obvious). In any case we are just speaking of a "box" that contains two horizontal and one vertical geophones placed next to each other (being the two horizontal sensors mutually perpendicular—see Fig. 1.26).

What is a 3C sensor good for? Unfortunately, too many fellows imagine that it can be used just for the computation of the *Horizontal-to-Vertical Spectral Ratio* (the so-called Nakamura's technique—see Chap. 3) but, as a matter of facts, the possible applications are countless.

Fig. 1.26 Three geophones (one vertical—the yellow one—and two horizontal perpendicular to each other). A 3C geophone is fundamentally a *box* that contains three geophones positioned like in the picture. The computation of the HVSR (see Chap. 3) is just one of the many things that can be done with the (active or passive) data recorded by a 3C geophone (see Chaps. 4 and 5)

It can be useful to clarify the difference between *passive* and *active* geophones. A *passive* geophone is a device that must be connected to a seismograph (standard single-component geophones like those shown in Fig. 1.26 are passive geophones as well as the 3C geophones shown in Fig. 1.23).

Active geophones are those that, inside their box, also have an A/D unit (Analog-to-Digital conversion unit) that converts the electrical signals from the sensors into numbers that are then saved in the laptop that is generally connected to the geophone itself (and that is used to manage the acquisition procedures and store the data files). In practice, they are a sort of mini (3-channel) seismograph.

It is impossible to state in absolute terms what is the *best* solution (active or passive 3C geophones) since the most appropriate solution depends on a huge number of practical considerations. The only recommendation is to opt for this or that solution only after having fully understood all the pros and cons of one solution with respect to all the possible applications of a 3C geophone (see Chaps. 4 and 5). So, be aware that, as the well-known American motto underlines, *the devil is in the details*. Differences that on paper might sound "small" can in fact reflect in colossal differences from the practical point of view.

1.10 Seismic Components and Observables

The scheme in Fig. 1.6 shows the seismic components according to the classical seismological terminology. It is a powerful and useful lingo because the terms *radial* and *transversal* (and *vertical*) precisely describe the mutual relationships between source and receivers. Terms such as x, y and z (that are now and then used in some documents) have no meaning because do not describe the source-receiver relationships and are therefore unable to provide any real information.

The *radial* axis (or component) is defined as the direction that connects the source and the observation point while the *transversal* axis is the one perpendicular to it.

It should be clear that a 3C geophone should be oriented so to obtain a dataset that meets the seismological standards.

But a vector is defined by a *direction* and its *verse*. Let us then consider the two options depicted in Fig. 1.27: is there any difference?

The answer is yes and no. For some kinds of analysis (e.g. the HVSR—see next section) the two options in Fig. 1.27 are equivalent while for some more sophisticate analysis (e.g. the determination of the RPM frequency curve—see Chap. 4) field operations must be accomplished in a more rigorous way and considering each detail of the whole *acquisition system* (the two options in Fig. 1.27 would in fact provide different data—see Fig. 1.18 and related text).

While considering the Rayleigh-wave acquisition, we must also consider that the Z and R components have, in general, difference velocity spectra that provide complementary information. Let us consider the data in Fig. 1.28. If we compare the Z and R phase-velocity spectra we can see that, in this specific case, the R component is the only one that, in the 15–25 Hz frequency range, provides some evidence of the fundamental mode. This does not mean that the R component is (always) *better* than the Z component but simply that is different and that, by jointly analyze two or more components, it is possible to solve issues and ambiguities otherwise inevitable in case of analyses based on just one component.

It also important to underline the difference between *component* and *observable*. The seismic *components* are the three Z, R and T directions (i.e. the seismic traces that represent the particle motion along these three axes) while the *observables* are the "objects" that we can "build" out of these three traces.

The RVSR and RPM curves are two examples (that will be illustrated in detail in Chap. 4): from the Z and R traces (i.e. *components*) it is possible to derive the two respective velocity spectra as well as the RVSR and the RPM curves. In other words, the number of observables is larger than the number of components.

Same *direction*, opposite *verse*:
is there any consequence?

Fig. 1.27 Orientation of a 3C geophone. For the determination of the ordinary HVSR the two configurations are equivalent (i.e. the *verse* does not matter). On the other side, in case we are dealing with active seismics and intend to analyze the polarity of the data, it is essential to know the details of the whole *acquisition system*. If we imagine a seismic source on the right, one configuration is correct while the other is wrong and, nota bene, the North direction often indicated on the 3C geophones has nothing to do with the correct orientation (polarity) to consider in case of active seismics

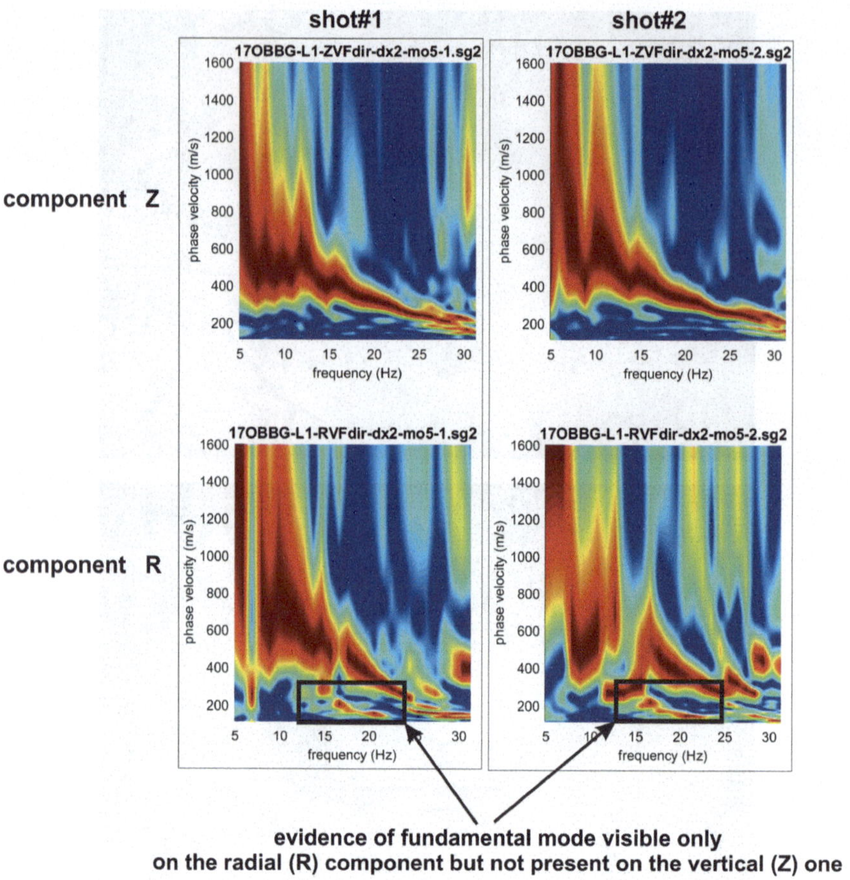

Fig. 1.28 Two examples of Z + R datasets for two consecutive shots (classical *Roll Along* approach for the reconstruction of a 2D profile). For both the datasets we can see that the Rayleigh-wave fundamental mode is visible on the R component (between about 15 and 24 Hz) but is completely missing along the Z component. The consequence is that, in case we would have considered just the Z component, we could have misunderstood the data and obtained an erroneous subsurface model. More: having recorded both the Z and R components, it is possible to compute the RPM curves to include in the joint inversion process (see Chap. 4 and Appendices B and G)

1.11 HVSR: North-South, South-North or What?

How should we orientate our 3C geophone for the acquisition of microtremor data useful for the computation of the HVSR?

The goal is, in general, to verify that the directionality of the H/V spectral ratio is not excessive which, in simple terms, means that, ideally, the H/V ratio should be more or less the same all over the azimuths.

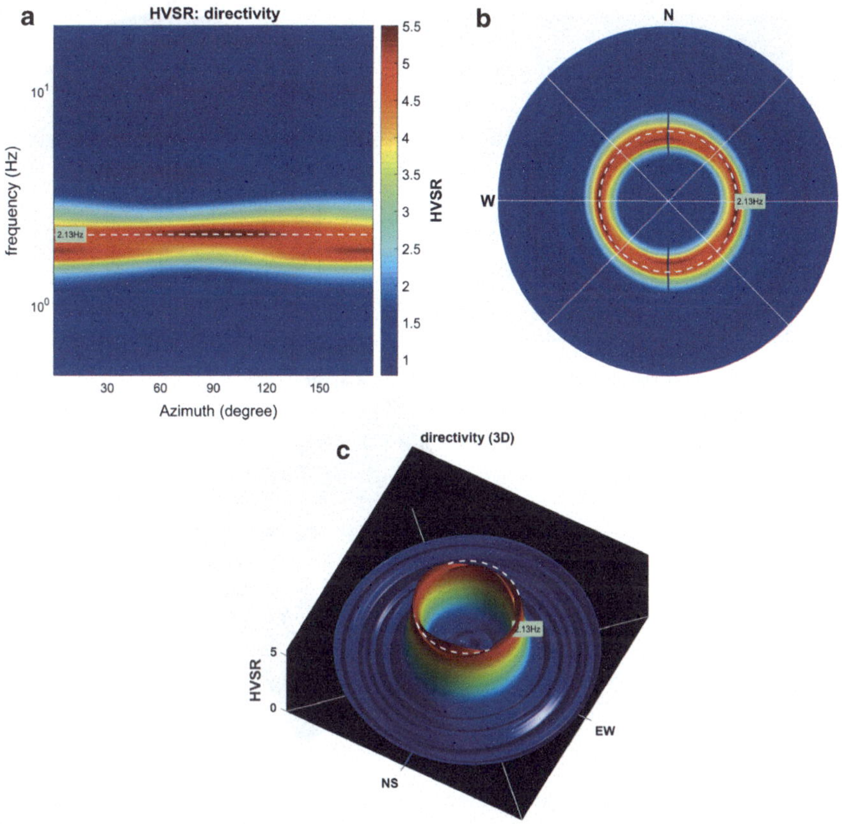

Fig. 1.29 HVSR: example of directivity graphs according to three different data representations: **a** 2D azimuth-frequency graph; **b** same kind of 2D representation but with polar coordinates (the frequency increases from the center towards the external circumference—the frequency limits are the same as in the previous graph [the central point represents the minimum considered frequency while the circumference the maxium frequency]); **c** 3D graph (same meaning as in the previous plot). In this case there is no significant directivity since the HVSR is more or less the same for all the azimuths

Figures 1.29 and 1.30 present two examples of directivity graphs for the HVSR: in the first case (Fig. 1.29) the H/V spectral ratio is highly homogeneous (i.e. the H/V value is more or less the same independently on the azimuth), while in the second case (Fig. 1.30) we can see that the H/V peak at 1.5 Hz shows a maximum value along the N45E-S45W and a minimum value perpendicularly to it (in this case this is due to the fact that such a H/V peak is related to an industrial signal—see Chap. 3).

It is anyway important to underline that, for the standard HVSR, we just deal with *directions*. This is particularly clear while assessing the data in Fig. 1.30 (where there is an apparent directivity). We can say that the HVSR has a maximum

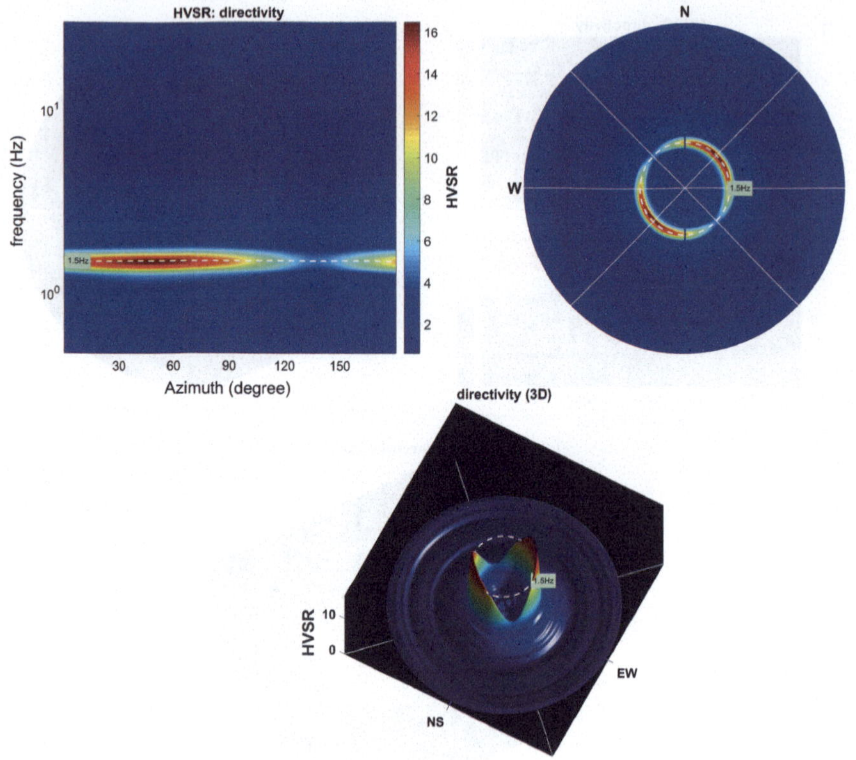

Fig. 1.30 HVSR directivity graphs as in Fig. 1.29. In this case a remarkable directivity of the H/V peak is clearly visible (maximum value along the N45E-S45W axis). See text for further comments

peak along the N45E-S45W direction but there is no way to distinguish the N45E from the S45W since only the axis/direction matters.

This means that in case we rotate by 180° the 3C geophone (i.e. we swap the North and South) the HVSR directivity plots will not change.

On the other side, the situation is totally different in case we record and analyze active data and need to define some observable based on the actual data polarity (we will discuss this issue while presenting in detail the RPM curve).

In that case a 180° rotation of the geophone will produce different (specular) seismic traces (compare Figs. 1.17, 1.18 and 1.27) and we must therefore be sure about the correct orientation to consider. Since most of the 3C geophones are tested only for the computation of the HVSR, we must underline that their "North" has no actual meaning (they should rather draw a North-South line without any North arrow).

What can influence the HVSR directivity? The reasons can be numerous and should be evaluated case by case but let us recall that, from time to time, during this kind of passive acquisitions something can go wrong and the data might be meaningless. This is why it is always recommended to record at least two datasets at two different positions (few meters one from the other) (see Chap. 6 and Appendix A).

In case there are no industrial signals (see Chap. 3) and the observed HVSR depends just on the natural microtremor field, the two most important features that can influence the directivity are the presence of inhomogeneous geological (subsurface) conditions and the topographic characteristics of the area.

A further crucial point is about the orientation of the 3C geophone during the acquisition of microtremor data for the computation of the HVSR. The main point to consider is that the magnetic/geographic North is totally irrelevant. In case we are working in the middle of an Alpine valley with a N45E-S45W axis, it is possible that such a structural element creates some directivity in the HVSR. For this reason it would be quite meaningless to orientate the 3C geophone considering the geographical North (the Polar Star) and, in order to have a straightforward evidence of possible directivities related to the geological structures, the 3C geophone could be conveniently deployed with its NS direction parallel to the valley.

The same applies in case we are working over the side of a mountain: the best option is to orientate the 3C geophone so to have its NS direction along the maximum slope (or perpendicular to it).

Of course, in case there are no apparent structural elements that could influence the HVSR we can consider the geographical north but otherwise we should carefully evaluate all the environmental (or urban/industrial) elements that can influence the data and fix the geophone orientation accordingly.

1.12 Working in a Rational and Productive Way: Components and (Vertical) Stack

In this section, are provided a series of simple suggestions about the organization of the acquisition files in a rational and effective way, so that the all the information are clear and unambiguous. This is particularly important in case we need to exchange the data with some colleague or when, for some reason, we wish to re-process some old dataset and intend to easly recall the acquisition parameters and all the related facts.

Although seemingly trivial, the following recommendations are extremely important: they help in clarifying what we are doing and are a good practice in the professional management of our work and data.

First of all, the file names to be given to the field data should provide clear and immediate hints about their meaning.

Detailed recommendations are provided in Dal Moro (2014) and summarized in Appendix A. File names such as "First_MASWbis.dat", "second_HS.seg2" or "Vienna12.segy" are completely meaningless cause do not provide any information about the kind of data and geometry and should be therefore avoided. Appendix A reports a series of suggestions on how properly naming (during the data acquisition) the data files so to clarify (already from the file name) the geometry and the kind of geophone-source combination.

Figure 1.30 reports the snapshot of a computer folder were are stored the data collected during a seismic survey (in this case the site is called "Torre_Bronzo"). As you can see, six additional sub-folders are created so to organize the different data: "HS" (active seismic data for the HS methodology—Chap. 4), "MAAM" (passive data recorded according to the MAAM approach—Chap. 4), "Multicomponent_MASW" [MASW data about two or more components—for instance Z + R (i.e. the vertical and radial components of Rayleigh waves)], "ESAC" (passive data for the ESAC—Chap. 3), "photos_and_infos" and "HVSR". In turn, in this latter sub-folder we might have two more sub-folders with the data collected at two (or more) different points of the investigated area (it is not advisable to carry out a single HVSR acquisition—see Chap. 3 and Appendix A).

Figure 1.31 shows the snapshot of the "MASW" sub-folder, with the active multi-component and multi-offset data.

As we should know from the introductory courses on Seismology or Applied Geophysics, the stack operation is a fundamental and mandatory operation for any kind of active seismics. While recording you field data, organizing the overall procedures in a rational and efficient way is vital.

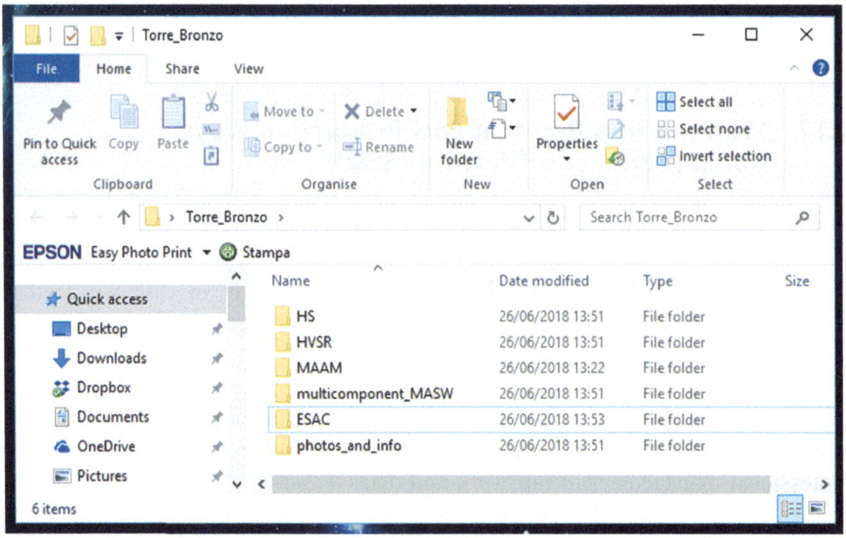

Fig. 1.31 Organizing your own data: snapshot of the subfolders organized within the main folder. Shown the subfolders where all the information and data are accurately, unambiguously and rationally stored. For comments see text

The data/files shown in Fig. 1.32 clearly represent the operation performed on the field during the data acquisition: through the acquisition software, we defined the file names (ZVFdx3mo4 and RVFdx3mo4—please, see Appendix A) and the stack (in this case 10). We then shot 10 times and the seismograph saved all the 10 shots (see suffix *_shot1*, *_shot2* and so on) also adding the average (stacked) data.

In case, for any reason, we want to eliminate one of the shots (for instance because of its poor quality), once we are in our office we can select the nine good shots and re-perform the stack operation without the bad shot/dataset.

For the data used to define the HVSR, in order to avoid misunderstandings, it is advisable to clearly indicate the sequence of the three recorded traces.

For example, if the sequence of the three channels is UD (i.e. vertical), NS and EW, a suitable file name could be:

HV_SouthLondon1_UD_NS_EW.seg2

If the sequence that the geophone-cable combination produces is instead NS, EW and UD, the file name should be something like:

HV_ SouthLondon1_NS_EW_UD.seg2

What is really important is to neatly organize the whole thing and avoid messy or unclear procedures and file names (see Figs. 1.31 and 1.32).

1.13 Paranoia #1: The Asphalt Cover

Very often, the need to record data over an asphalt cover generates anxiety (that can become a sort of paranoia) since we are afraid about the data quality. This section intends to reassure about the problems that might occur while working on this type of surface.

While working on asphalt (or any kind of stiff surface such as the crushed gravels of a city parks), the geophones (including the trigger geophone) are usually supported by simple but effective tripods (we must remove the spikes from the geophone and secure it to the tripod—see Fig. 1.33). Other solutions might be not sufficiently stable to guarantee a good coupling and the absence of unwanted vibrations.

The rest of the procedure is fundamentally the same. Let us see an old field dataset collected during a workshop. The acquisition was carried out considering 12 horizontal 4.5 Hz geophones used to record the radial component of Rayleigh waves (RVF) and Love waves (THF). Figure 1.34 shows the array. We can see all the common elements: the 8 kg sledgehammer, the trigger geophone (and the 12 horizontal geophones), the plate for the vertical-impact shooting and the wooden beam used as shear source to generate Love waves.

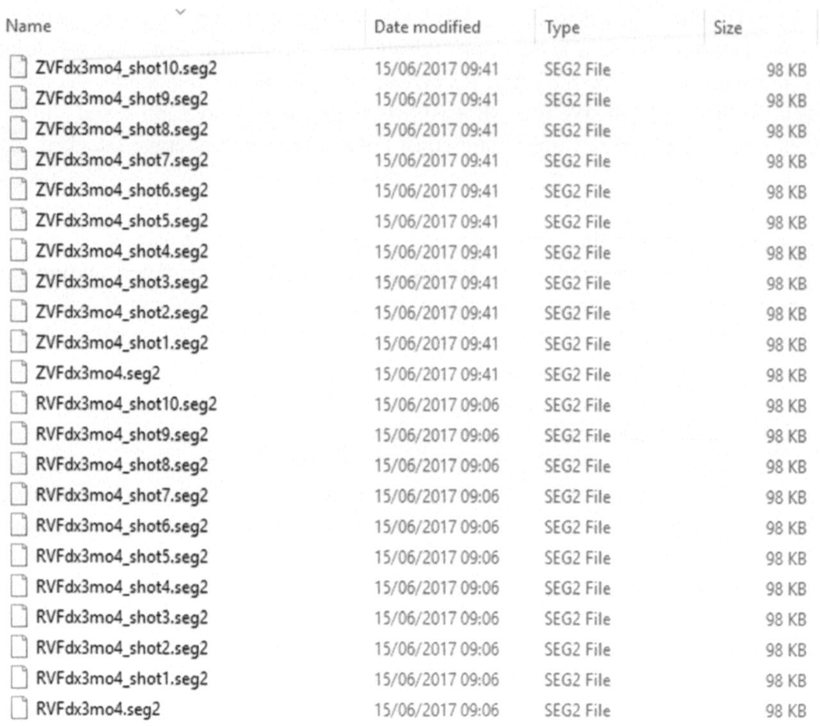

Name	Date modified	Type	Size
ZVFdx3mo4_shot10.seg2	15/06/2017 09:41	SEG2 File	98 KB
ZVFdx3mo4_shot9.seg2	15/06/2017 09:41	SEG2 File	98 KB
ZVFdx3mo4_shot8.seg2	15/06/2017 09:41	SEG2 File	98 KB
ZVFdx3mo4_shot7.seg2	15/06/2017 09:41	SEG2 File	98 KB
ZVFdx3mo4_shot6.seg2	15/06/2017 09:41	SEG2 File	98 KB
ZVFdx3mo4_shot5.seg2	15/06/2017 09:41	SEG2 File	98 KB
ZVFdx3mo4_shot4.seg2	15/06/2017 09:41	SEG2 File	98 KB
ZVFdx3mo4_shot3.seg2	15/06/2017 09:41	SEG2 File	98 KB
ZVFdx3mo4_shot2.seg2	15/06/2017 09:41	SEG2 File	98 KB
ZVFdx3mo4_shot1.seg2	15/06/2017 09:41	SEG2 File	98 KB
ZVFdx3mo4.seg2	15/06/2017 09:41	SEG2 File	98 KB
RVFdx3mo4_shot10.seg2	15/06/2017 09:06	SEG2 File	98 KB
RVFdx3mo4_shot9.seg2	15/06/2017 09:06	SEG2 File	98 KB
RVFdx3mo4_shot8.seg2	15/06/2017 09:06	SEG2 File	98 KB
RVFdx3mo4_shot7.seg2	15/06/2017 09:06	SEG2 File	98 KB
RVFdx3mo4_shot6.seg2	15/06/2017 09:06	SEG2 File	98 KB
RVFdx3mo4_shot5.seg2	15/06/2017 09:06	SEG2 File	98 KB
RVFdx3mo4_shot4.seg2	15/06/2017 09:06	SEG2 File	98 KB
RVFdx3mo4_shot3.seg2	15/06/2017 09:06	SEG2 File	98 KB
RVFdx3mo4_shot2.seg2	15/06/2017 09:06	SEG2 File	98 KB
RVFdx3mo4_shot1.seg2	15/06/2017 09:06	SEG2 File	98 KB
RVFdx3mo4.seg2	15/06/2017 09:06	SEG2 File	98 KB

Fig. 1.32 Files in the multicomponent_MASW subfolder. Two facts are immediately clear: in this case we collected the vertical (Z) and radial (R) component of Rayleigh waves using the classical vertical hammering (VF) as source. Geophone distance was 3 m and the distance between the source and the first geophone 4 m (see Appendix A and/or Dal Moro 2014). During the acquisition we stacked 10 shots (and saved both the stacked data and the single shots). Please, notice that the file names provide all the necessary information about the geometry and the type of source-receiver combination. The data can be sent to anyone without having to explain the meaning of each file (everything is perfectly clear just from the filenames)

In those times (the workshop took place in 2012), 24-bit seismographs were still not very common and we used a 16-bit seismograph with a limited dynamic range (recent state-of-the-art seismographs would surely permit a general higher quality).

The raw data and the respective phase-velocity spectra for the two collected components (RVF and THF) are shown in Fig. 1.35, while Fig. 1.36 presents the seismic traces after a simple low-pass filtering and a quick trace cleaning (also shown the respective phase velocity spectra).

This simple data cleaning considerably affects the data in the time domain (seismic traces), but it is clear that the phase-velocity spectra of the raw and cleaned data are fundamentally identical.

Fig. 1.33 Correct (on the left) and incorrect (on the right) ways to use the tripods while working on asphalt, urban parks etc

Fig. 1.34 Joint RVF + THF acquisition on asphalt. Twelve horizontal geophones (geophone distance 4 m) are used to record the radial component of Rayleigh waves (VF source) and Love waves (HF source). See data presented in Figs. 1.35 and 1.36 as well as the analysis reported later on in the section about the FVS (Full Velocity Spectrum) approach to dispersion analysis (Chap. 2)

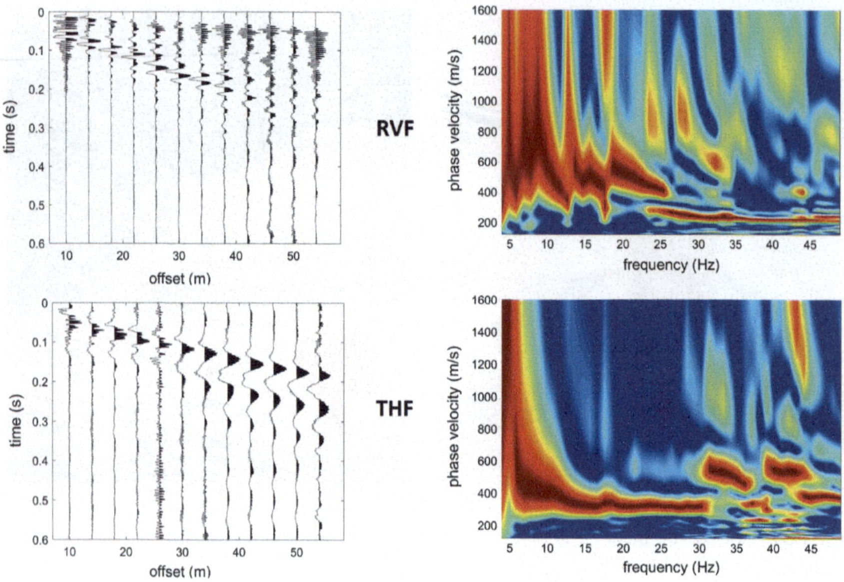

Fig. 1.35 Raw traces and phase-velocity spectra for the RVF and THF components recorded on asphalt (see Fig. 1.34)

The quality of the Love waves (THF component) is remarkably good in spite of the very simple equipment and procedures: a simple and very cheap wooden beam with the "hammer fellow" standing on it and hitting the wood laterally (see also next section, just focused on shear sources).

So, in these situations, the problem is not the asphalt. Some problems can occur because of the several underground utilities and accidents that are often present in urban areas and that can produce a complex series of scattering phenomena that can pollute the data.

In addition, since these areas are often quite noisy because of the traffic, it is often useful (or necessary) to increase the stack so to increase the overall signal-to-noise ratio. The stack value should be fixed by considering the characteristics of the site and the overall noise level and we cannot apply the same value regardless of the noise conditions, the offsets (the longer the array, the higher the attenuation) and the overall general specific conditions.

In very general terms the minimum value can be around five (for very quiet areas) and up to 10–15 in messy urban zones.

In the next chapter (focused on the *Full Velocity Spectrum* analysis) will be shown the joint FVS analysis of the velocity spectra shown in Fig. 1.36.

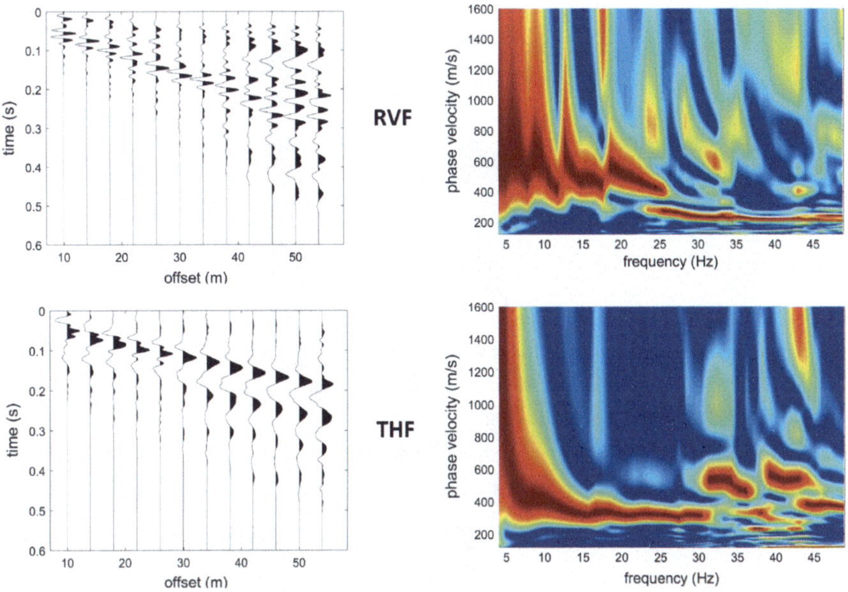

Fig. 1.36 Seismic traces after the application of a low-pass filter and a quick trace cleaning. On the right, the phase-velocity spectra to compare with those shown in Fig. 1.35 (computed while considering the raw traces). The joint inversion of the two velocity spectra (RVF + THF) is presented in the next chapter while illustrating the FVS (Full Velocity Spectrum) approach to dispersion analysis

A further example of data collected in urban areas are reported in the Appendices B and G (acquisition of the Z and R components and joint analysis of the Z and R phase-velocity spectra together with the RPM frequency-offset surface).

1.14 Paranoia #2: Shear Sources, Nails and P Waves

In case we intend to generate and record data for the analysis of SH-wave refraction and/or Love waves (the acquisition is fundamentally identical) we need a shear source and we should try to maximize the amount of shear energy transmitted to the soil.

The most common solution is a simple wooden beam and a sledgehammer as the ones shown in Fig. 1.37.

Where is the problem? Although the bottom of the beam is made in such a way to maximize the friction (for example by means of some heavy rubber surface), because of the hammer impact the beam slips to the side and, consequently, we are losing a certain amount of energy (which should be ideally fully transmitted to the ground).

Fig. 1.37 Simple shear sources (for SH refraction and Love waves). Upper panel: the classical solution (the impact surface is perpendicular to the surface); lower panel: a solution designed to optimize the acquisition process. From the ergonomic point of view, the inclined surface allows an easier hammering and, consequently, the amount of energy input as shear wave is higher. Note that, in both cases, the wooden beam is fixed to the ground thanks to the weight of the person standing on it and, to a small extent, to the use of a couple of big nails that avoid the slip of the beam (an alternative solution is shown in Fig. 1.38)

In order to reduce this problem, to block the beam we can use some big nails (see carefully the beams shown in Fig. 1.37).

In some cases, to further improve the coupling, it is possible to exploit the weight of a vehicle and place one of the car wheels over the beam (see Fig. 1.4a) while a simple and effective alternative to this classical way of generating shear waves is shown in Fig. 1.38.

Fig. 1.38 A further efficient manner to generate Love waves: dig a small hole in the ground and place your plate vertically. This way you do not have to bring a heavy beam with you (see Fig. 1.37) and the source-ground coupling is extremely good (the energy is fully transferred to the ground)

An urban legend insinuates that the nails used to fix the beam to the ground (see set up shown in Fig. 1.37) generate P waves which would contaminate the data and make difficult their interpretation in terms of SH refraction.

What is the problem with this kind of arguments?

Point #1: how much energy propagates as P wave when we hit the surface with a vertical force? About 7% (the rest, 67%, propagate as Rayleigh waves and 26% as shear waves—Miller et al. 1955);
Point #2: and in case we use a shear source? Although the author of this book was unable to find any article about it, it is clear that the percentage of P waves must be significantly less than 7% (the value obtained using a vertical-impact force);
Point #3: over which plane are the P waves travelling? Clearly along the Z-R plane;
Point #4: what is the orientation of the horizontal geophones while recording the SH (or Love) waves? Transversal (i.e. the axis perpendicular to the Z-R plane).

How much P-wave energy can be found along the T component while considering a shear source? Considering the four mentioned points practically none (although some very minor amount is surely possible due to scattering phenomena).

But that is not all. We know that Love waves have a very large amplitude and, since SH waves are fundamentally the "top" of Love waves, in case we record a sufficiently-long dataset (that clearly shows the Love-wave arrivals) the identification of the SH waves will become quite simple.

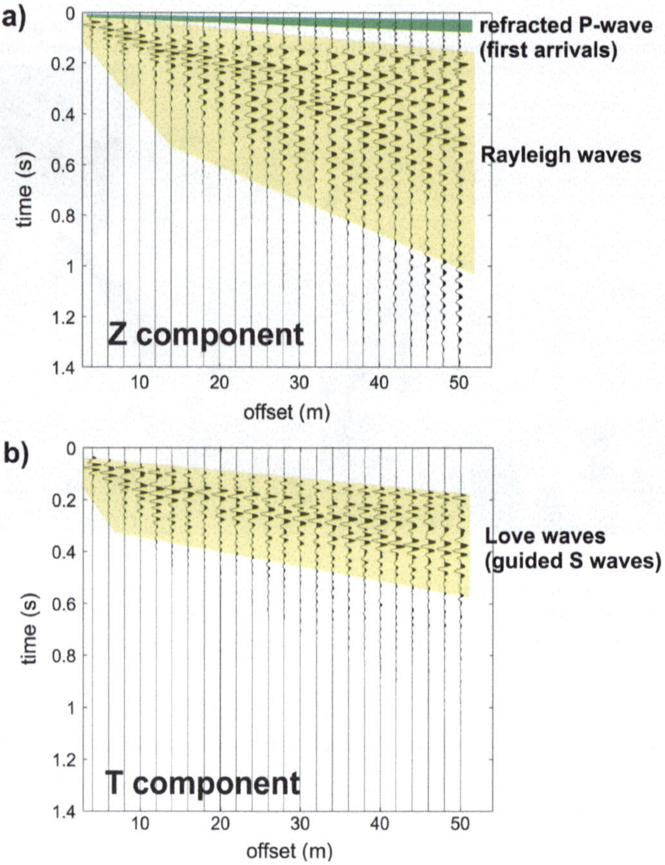

Fig. 1.39 Raw data recorded in a (very tricky) site in the Friuli (NE Italy) alluvial plain (Dal Moro 2011): **a** ZVF component (for the Rayleigh-wave and P-wave refraction analysis); **b** THF component (for Love waves and SH-wave refraction). See also Fig. 1.40

Figure 1.39 presents the data for the Z and T components (the site is the one presented in Dal Moro 2011). While the Z component clearly shows both Rayleigh (high amplitude and low frequencies) and P waves (low amplitude, faster and with higher frequencies), the T-component data clearly shows Love waves. Figure 1.40 presents a close up of the first 0.3 s together with the theoretical arrival times for all the stratigraphic contacts of the model (which is the one discussed in Dal Moro 2011).

It is surely necessary to understand the meaning of all those red lines. The subsoil model is clearly composed of several layers (considered the complexity of the site, in order to obtain a good agreement with the field surface-wave data, it was necessary to define an 11-layer model). Each stratigraphic contact generates a refracted wave (when V_{n+1} is larger than V_n) but, clearly, from the field data we can only define the first arrivals (later arrivals are lost in a complex mixture of different

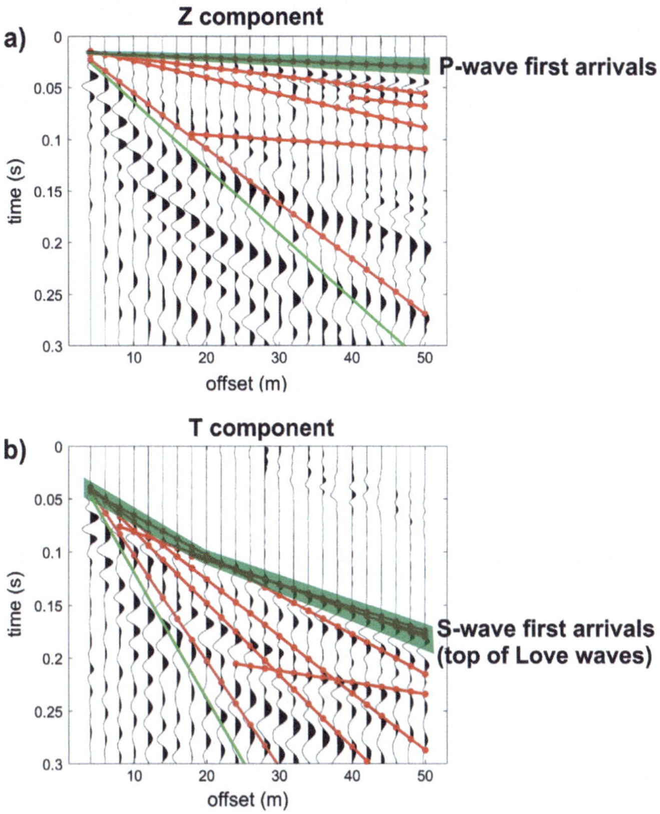

Fig. 1.40 Close up of the data presented in Fig. 1.39 after the application of the AGC (Automatic Gain Control). Shown the arrival times of the refracted P and SH waves (the subsurface model is the one described in Dal Moro 2011). Refracted SH waves are the "top" of the Love waves. For further comments see text

waves and, usually, cannot be identified). Therefore, from the practical point of view, we are interested only in the "fastest events" which, depending on the offsets and velocity model, can refer to different refractors (the thin light-green lines shown in Fig. 1.40 relate to the direct body wave travelling along the surface but detectable only very close to the source).

From the data shown in Figs. 1.39 and 1.40 we can clearly see that:

(1) the P-wave first arrivals are visibly separated from the Rayleigh-wave arrivals (in this case P-wave refraction is associated to the shallow water table);
(2) refracted SH waves are in fact the "top" of the Love waves (Fig. 1.40b).

Therefore, together with the dispersion analysis, it is often useful to verify that our subsurface model is in agreement also with respect to the refracted first arrivals (e.g. case study #2 and #5 in Dal Moro 2014 or Dal Moro 2008).

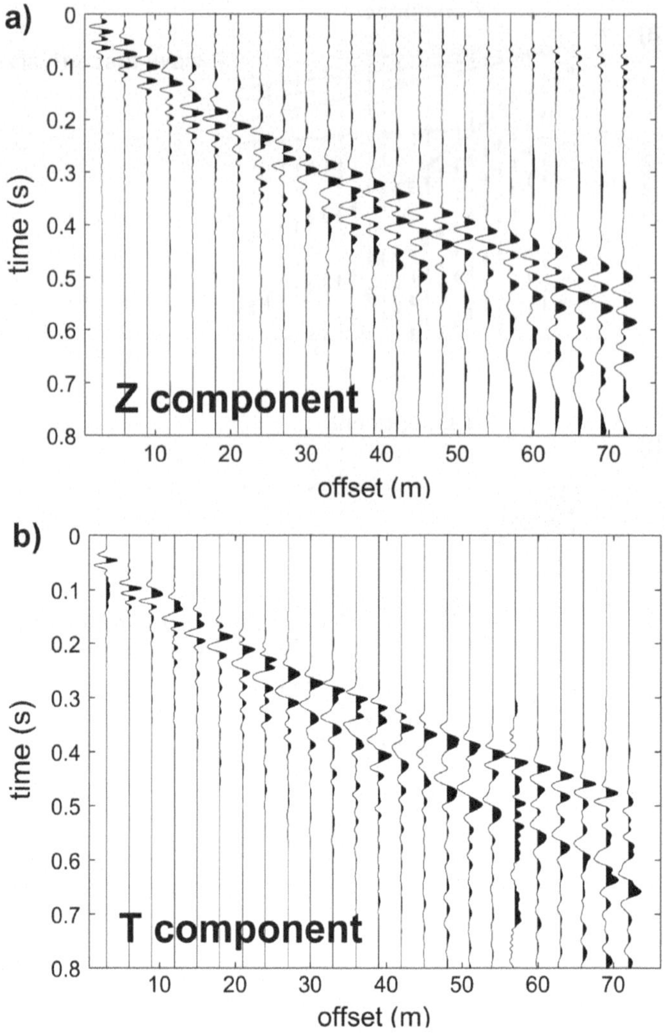

Fig. 1.41 Close up of the ZVF and THF components for a NE Italy site (see case study #2 in Dal Moro 2014). Traces are dominated by the Rayleigh and Love waves but, in the upper panel (Z component), some early P-wave refraction is also apparent (see also Fig. 1.42)

A further and last example is briefly presented in Figs. 1.41 and 1.42 (more details about this dataset are presented in Dal Moro 2010, 2014—case study #2). Fundamentally, the site is a small basin with about 25 m of soft sediments that cover a calcarenitic (fractured) bedrock.

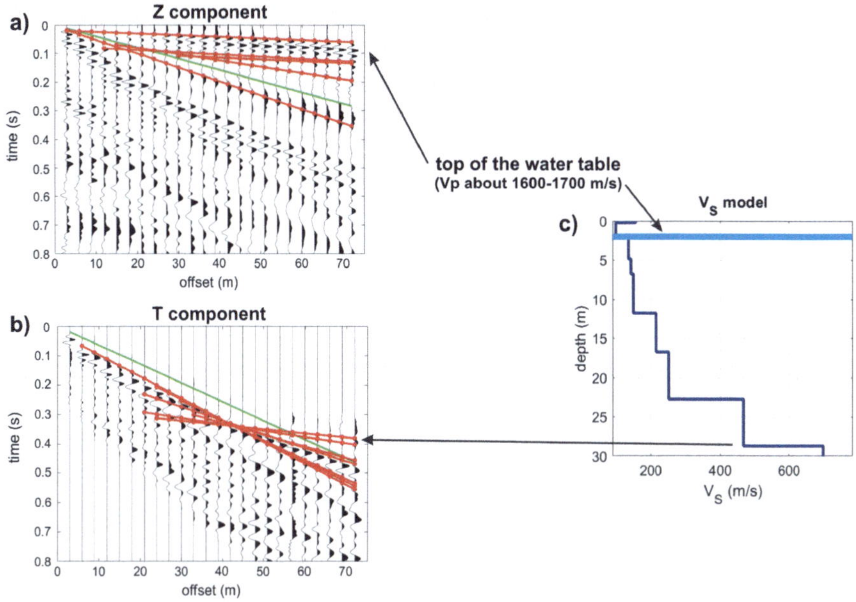

Fig. 1.42 AGC (Automatic Gain Control) applied to the ZVF and THF data shown in Fig. 1.41: **a** the P-wave refraction due to the shallow water table; **b** the SH-wave refraction gives just a vague and inadequate hint of the deep "bedrock". Only the joint analysis of surface waves can provide clear information about the V_S profile (see Dal Moro 2014—case study #2 and, later on in this book, Fig. 4.21 with the related text, where the same site is investigated according to an innovative procedure based on the analysis of the surface waves recorded by a single 3C geophone)

The presence of the water table at a depth of about 1.5 m is visible in particular along the Z component: the first arrivals (Fig. 1.42a) represent the P-wave refraction due to the contact between the superficial unsaturated sediments and the saturated sediments (the velocity is the one typical of saturated soft sediments—1600–1700 m/s) but, due to such a very shallow and remarkable increase in the P-wave velocity, we cannot say anything about deeper layers.

About the THF component we can highlight two points:

(1) the first arrivals of the SH refraction represent the "top" of Love waves;
(2) although at the very far offsets we could maybe spot out some fast event possibly related to the bedrock (at a depth of about 25 m), in spite of the length of the array (about 70 m), it would not be possible to unambiguously identify the bedrock through the analysis of the SH-wave refraction.

Refraction surveys need remarkable offsets (the standard rule of thumb says that we can investigate the subsurface conditions down to a depth of about 1/3 of the considered offsets), a large number of channels (minimum 24) and high-energy sources (body-wave amplitude is by far smaller compared to surface waves).

Furthermore, in order to investigate areas where the water table is shallow, P waves cannot be used and we should necessarily consider SH waves.

On the other side, surface-wave acquisition can be done with an extremely-light equipment and, if properly accomplished, the collected data can provide information about (very) deep layers.

In the following chapters, we will both clarify a series of misunderstandings about some very popular techniques such as the HVSR (*Horizontal-to-Vertical Spectral Ratio*), MASW (*Multichannel Analysis of Surface Waves*) and ESAC (*Extended Spatial AutoCorrelation*), both introduce some state-of-the-art techniques that require a very limited field effort (and equipment) and can also significantly improve the robustness of the retrieved subsurface V_S model.

The software applications used for the analyses shown throughout the book are *winMASW® Academy* and *HoliSurface®*.

References

Arai H, Tokimatsu K (2005) S-wave velocity profiling by joint inversion of microtremor dispersion curve and horizontal-to-vertical (H/V) spectrum. Bull Seismol Soc Am 95:1766–1778

Bard P-Y, Bouchon M (1980a) The seismic response of sediment-filled valleys. Part 1. The case of incident SH waves. Bull Seism Soc Am 70:1263–1286

Bard P-Y, Bouchon M (1980b) The seismic response of sediment-filled valleys. Part 2. The case of incident P and SV waves. Bull Seism Soc Am 70:1921–1941

Bowden DC, Tsai VC (2017) Earthquake ground motion amplification for surface waves. Geophys Res Lett 44:121–127

Carcione JM (1992) Modeling anelastic singular Surface waves in the Earth. Geophysics 57:781–792

Dal Moro G (2008) V_S and V_P vertical profiling via joint inversion of Rayleigh waves and refraction travel times by means of bi-objective evolutionary algorithm. J Appl Geophys 66:15–24

Dal Moro G (2010) Insights on surface-wave dispersion curves and HVSR: joint analysis via Pareto optimality. J Appl Geophys 72:29–140

Dal Moro G (2011) Some aspects about surface wave and HVSR analyses: a short overview and a case study. BGTA—Bollettino Geofisica Teorica e Applicata 52:241–259

Dal Moro G (2014) Surface wave analysis for near surface applications. Elsevier, Amsterdam, The Netherlands, 252 pp. ISBN 978-0-12-800770-9

Dal Moro G (2015) Joint analysis of Rayleigh-wave dispersion and HVSR of lunar seismic data from the Apollo 14 and 16 sites. Icarus 254:338–349

Dal Moro G (2019a) Surface wave analysis: improving the accuracy of the shear-wave velocity profile through the efficient joint acquisition and Full Velocity Spectrum (FVS) analysis of Rayleigh and Love waves. Explor Geophys. https://doi.org/10.1080/08123985.2019.1606202

Dal Moro G (2019b) Effective active and passive seismics for the characterization of urban and remote areas: four channels for seven objective functions. Pure Appl Geophys 176:1445–1465

Dal Moro G (2020) The magnifying effect of a thin shallow stiff layer on Love waves as revealed by multi-component analysis of surface waves. Scientific Reports. https://doi.org/10.1038/s41598-020-66070-1; http://www.nature.com/articles/s41598-020-66070-1 (open access)

Dal Moro G, Ferigo F (2011) Joint inversion of Rayleigh and love wave dispersion curves for near-surface studies: criteria and improvements. J Appl Geophys 75:573–589

Dal Moro G, Pipan M, Forte E, Finetti I (2003) Determination of Rayleigh wave dispersion curves for near surface applications in unconsolidated sediments. In: Proceedings SEG of the 73rd annual meeting, Dallas, Texas, 26–31 Oct 2003. Society of Exploration Geophysicists, pp 1247–1250

Dal Moro G, Al-Arifi N, Moustafa SR (2017a) Analysis of Rayleigh-wave particle motion from active seismics. Bull Seismol Soc Am 107:51–62

Dal Moro G, Al-Arifi N, Moustafa SR (2017b) Improved holistic analysis of Rayleigh waves for single- and multi-offset data: joint inversion of Rayleigh-wave particle motion and vertical- and radial-component velocity spectra. Pure Applied Geophys 175:67–88. https://doi.org/10.1007/s00024-017-1694-8 (open access)

Dal Moro G, Weber T, Keller L (2018) Gaussian-filtered Horizontal Motion (GHM) plots of non-synchronous ambient microtremors for the identification of flexural and torsional modes of a building. Soil Dyn Earthq Eng 112:243–245

Dal Moro G, Al-Arifi N, Moustafa SR (2019a) On the efficient acquisition and holistic analysis of Rayleigh waves: technical aspects and two comparative case studies. Soil Dyn Earthq Eng 125. https://www.sciencedirect.com/science/article/pii/S0267726118310613

Galperin EI (1985) Vertical seismic profiling and its exploration potential. Springer, Netherlands, 442pp

Gao L, Xia J, Pan Y, Xu Y (2016) Reason and condition for mode kissing in MASW method. Pure Appl Geophys 173:1627–1638

Hardage BA (1983) Vertical seismic profiling part A: principles. Geophysical Press, London, 450pp

Hays WW (1980) Procedures for estimating earthquake ground motions. Geological Survey, United States Department of the Interior, Washington, 77pp

Ivanov J, Miller RD, Xia J, Steeples D, Park CB (2006) Joint analysis of refractions with surface waves: an inverse refraction-traveltime solution. Geophysics 71:R131–R138

Jongmans D, Campillo M (1993) The response of the Ubaye Valley (France) for incident SH and SV waves: comparison between measurements and modelling. Bull Seismol Soc Am 83:907–924

Miller GF, Pursey H, Bullard EC (1955) On the partition of energy between elastic waves in a semi-infinite solid. Proceedings of the Royal Society of London. Series A. Math Phys Sci 233 (1192):55–69

Nanda NC (2016) Borehole seismic techniques. In: Seismic data interpretation and evaluation for hydrocarbon exploration and production. Springer, Cham. Print ISBN: 978-3-319-26489-9

Nováková L, Sosna K, Brož M, Najser J, Novák P (2011) Geomechanical parametres of the Podlesí granites and their relationship to seismic velocities. Acta Geodyn Geomater 8(3):353–369

Park CB, Xia J, Miller RD (1998) Imaging dispersion curves of surface waves on multichannel record. In: Proceedings SEG (Society of Exploration Geophysicists), 68th annual meeting, New Orleans, Louisiana, 13–18 Sept 1998, pp 1377–1380

Safani J, O'Neill A, Matsuoka T, Sanada Y (2005) Applications of love wave dispersion for improved shear-wave velocity imaging. J Environ Eng Geophys 10(2):135–150

Scales JA, Smith ML, Treitel S (2001) Introductory geophysical inverse theory. open file. Samizdat Press, Golden, CO, 193pp

Schnabel PB, Lysmer J, Seed HB (1972) SHAKE—a computer program for earthquake analysis of horizontally layered sites. Report no EERC 72-12, Earthquake Engineering Research Center, University of California, Berkeley

Seed HB, Idriss IM (1971) Influence of soil conditions on building damage potential during earthquakes. J Struct Eng ASCE 97(2):639–663

Tokimatsu K, Tamura S, Kojima H (1992) Effects of multiple modes on Rayleigh wave dispersion characteristics. J Geotech Eng ASCE 118:1529–1543

West M, Menke W (2001) Fluid-induced changes in shear velocity from surface waves. In: Symposium on the application of geophysics to engineering and environmental problems (SAGEEP), pp 21–28

2 Surface-Wave Analysis Beyond the Dispersion Curves: FVS

I should not like my writing to spare other people the trouble of thinking. But, if possible, to stimulate someone to thoughts of his own.

Ludwig Wittgenstein (Philosophical Investigations)

G. Dal Moro, *Efficient Joint Analysis of Surface Waves and Introduction to Vibration Analysis: Beyond the Clichés*,
https://doi.org/10.1007/978-3-030-46303-8_2

Abstract

The FVS (*Full Velocity Spectrum*) approach is a way to analyze the dispersion of surface waves recorded according to active methodologies. The goal is go beyond the classical and problematic approach based on the (interpreted) modal dispersion curves that, in some cases, can be difficult or impossible to be defined. In the classical approach, the velocity spectrum is interpreted in terms of modes (fundamental and/or higher modes). Therefore, we do not invert something "objective" but a subjective *interpretation*. The same velocity spectrum can be interpreted in different ways and, consequently, lead to different solutions. The FVS approach considers the whole frequency-velocity matrix without any interpretation in terms of modal curves. A series of single- and multi-component examples are presented. In this Chapter we will focus on the phase velocities obtained from multi-offset data but in the Chap. 4 we will see how to efficiently exploit the FVS approach in case of single-offset data (HoliSurface approach).

2.1 A Brief but Important Introduction

The power of language by which the *Homo sapiens* developed in the ways we know, is that of being able to identify, distinguish and describe things and processes in a timely manner, going beyond the indistinct *chaos* that represented the world in which we lived when the language was limited to very few forms of representation of elementary phenomena.

The advances of these last decades in terms of surface wave analysis have made acronyms like MASW completely unable of expressing something clear and precise so that we are now back in that original *chaos*. Technological progresses are in fact frenetic not only in mobile communications but in geophysics as well.

What does MASW nowadays mean? As a matter of fact, nothing:

(1) What kind of *surface waves* are we talking about?
(2) What kind of geophones do we connect to our *multi-channel* acquisition system?
(3) What kind of *analysis* do we talk about?

Since all these questions have multiple answers that can be combined in different ways, that nowadays the acronym MASW does not mean anything clear and univocal.

Are we going to consider a single component or different components?

Are we going to analyze phase or group velocities?

What kind of geophones are used and what observables are considered when we do a multi-offset seismic (classical MASW)?

Is the dispersion analyzed through the modal curves or by means of some more "sophisticated" approach?

Is the ESAC/SPAC/ReMi/MAAM (see Chaps. 3 and 4) a kind of MASW? Yes, of course (we are in fact recording the data while using a multi-channel system and analyzing the surface wave propagation).

Is the HS method (see Chap. 4) a kind of MASW? Yes, of course (we are analyzing surface wave propagation while considering multi-channel data [in that case we deal with multi-component, single-offset data]).

In short: nowadays MASW is an acronym unable to define anything clear. And there is a further (related) point that needs to be clearly highlighted and understood.

Surface-wave analysis is carried out through the correct execution of two distinct operations:

(1) determination of the dispersive properties of the medium/site;
(2) analysis/inversion of the previously-retrieved properties.

These two steps must be kept clearly separate because different choices are possible and, consequently, it should be always clearly stated what is required (or done) during both the first and the second step.

There are in fact several ways to determine the dispersive properties of the medium/site and several ways to analyze/invert them.

The idea that "MASW" means to deploy 24 (vertical) geophones and analyze the dispersion according to the modal curves (as often done) is nowadays utterly anachronistic. For all the reasons summarized for instance in Dal Moro (2014, 2019a, b) and briefly recalled in the first chapter of this volume, none of the examples considered in the present book is solved by following this problematic and inefficient approach.

While referring to surface wave analysis we should be therefore very precise about *what* and *how* (so not to leave any room to personal interpretations in the meaning of the used words/expressions). Here three examples of descriptions sufficiently clear not to be misunderstood:

(1) Determination of the V_S profile by joint analysis of the phase velocity of the Love waves (according to the modal curves or the FVS approach) and the HVSR curve obtained as the average of the HVSR curves collected in, at least, two points along the array.
(2) Determination of the V_S profile through the joint analysis of the phase velocities obtained from ESAC (vertical component) and modeled as *effective* dispersion curve together with the HVSR obtained as mean curve of the HVSR curves of, at least, three points along the ESAC array.
(3) Determination of the V_S profile by joint analysis of the group velocities of the vertical and radial components of Rayleigh wave (analyzed according to the FVS approach) together with the RPM (Rayleigh-wave Particle Motion) curve (see Chap. 4).

In these three descriptions, each single word has a very precise (technical) meaning that leaves no room for personal interpretations (the book is aimed at providing the background necessary to understand all the details).

These are just three examples but the observables can be actually combined in several different ways.

We should once again underline all these three "methodologies" can be listed under the term "MASW" since all of them refer to the analysis of surface waves collected through a multi-channel acquisition system.

It should be emphasized that the expression *multi-channel* and *multi-offset* indicate very clear and very different things: *multi-channel* means that we are dealing with at least two channels (but it says nothing about what kind of geophones we connect to the channels and in what kind of combination); on the other side, *multi-offset* indicates the presence of various geophones at different offsets.

The HVSR, for instance, is a multi-channel (but not multi-offset) methodology (since it requires three channels). ESAC (Chap. 3) and HS (Chap. 4) are also *multi-channel* methodologies: ESAC is multi-offset while the HS is single-offset (like the HVSR).

2.2 Introduction to the FVS Analysis

The FVS (*Full Velocity Spectrum*) approach is a way to analyze the dispersion of surface waves recorded according to active methods. The goal is go beyond the classical and problematic approach based on the (interpreted) modal dispersion curves that, as we saw in the previous Chapter, can be quite difficult (or impossible) to define.

In the classical approach (based on the modal dispersion curves), the velocity spectrum is interpreted in terms of modes (fundamental and/or higher modes). Therefore, we do not invert something "objective" but a subjective data *interpretation*. This is a key fact: the same velocity spectrum can in fact be interpreted in different ways and, consequently, lead to different solutions (the substantial difference between *velocity spectrum* and *dispersion curve* has been widely emphasized in Dal Moro 2014).

To solve these problems, an approach based on the analysis of the full velocity spectrum has been introduced (e.g. Dal Moro 2014, 2019a, b, 2020; Dal Moro et al. 2019a). This methodology is based on the analysis (inversion) of the entire velocity spectrum (or spectra in case of multi-component data) without any subjective interpretation in terms of modal dispersion curves.

Such a method is actually not dissimilar from the inversion of the wave field as described in Dou and Ajo-Franklin (2014).

In its automatic implementation (inversion), the FVS approach consists of three steps (details in Dal Moro 2019a; Dal Moro et al. 2014, 2015c, 2019a):

(1) computation of the synthetic traces of the considered components (for example the radial component of Rayleigh waves and/or Love waves) for a tentative model;

(2) computation of the velocity spectrum/a (phase or group velocities depending on the considered approach) of the previously-computed synthetic traces;
(3) computation of the misfit between the velocity spectrum/a of the field and synthetic traces.

These three steps are implemented within a heuristic optimization scheme aimed at minimizing the misfit (i.e., identifying a subsurface model whose velocity spectrum is as similar as possible to the velocity spectrum of the field data). This way, we deal with the whole velocity spectrum (i.e. the entire frequency-velocity matrix) and not with a dispersion curve (i.e. a frequency-velocity curve) that represents a *subjective* interpretation of the velocity spectrum.

Figure 2.1 shows how, during a FVS inversion, we aim to identify a subsurface model whose velocity spectrum is as close as possible to that of the field data (the caption provides all the elements necessary to understand the fundamental points).

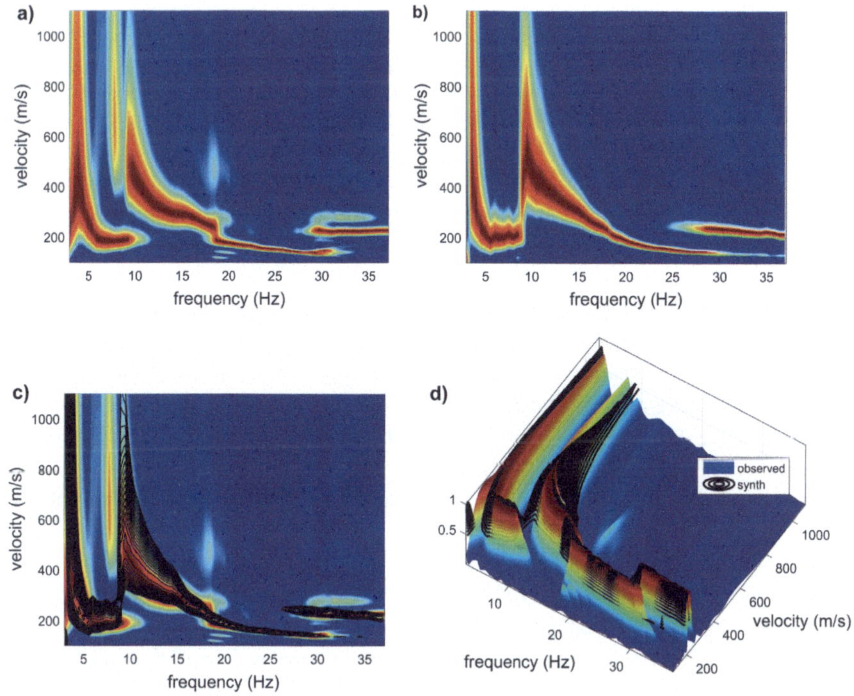

Fig. 2.1 Example of single-component FVS analysis: **a** phase velocity spectrum of the field data; **b** phase velocity spectrum of the model identified through the FVS approach; **c** the two spectra in a single graph (the black contour lines refer to the velocity spectrum of the model while the colors in the background pertain to the field data); **d** same data as in the previous graph but according to a 3D representation (the velocity spectrum of the model envelops the velocity spectrum of the field data)

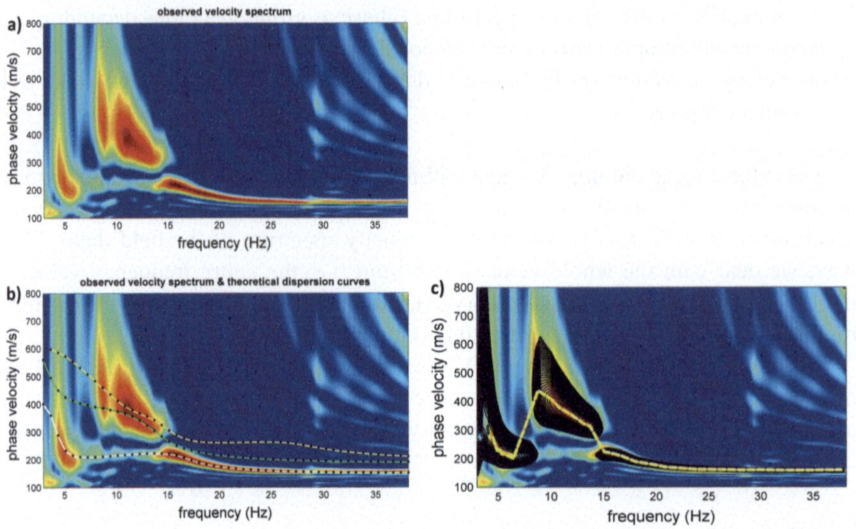

Fig. 2.2 The importance of the FVS analysis: **a** phase velocity spectrum of a field dataset; **b** classical analysis according to the modal dispersion curves; **c** FVS approach: it is shown that, between 8 and 15 Hz, the tentative model actually excites the higher modes (the dashed curve is called *effective* curve). Details in the text. After Dal Moro (2014)

One of the most important aspects of this approach is that, this way, it is possible to demonstrate that a certain subsurface model actually excites certain modes (or not).

Figure 2.2 should help clarify this crucial point. At the top left (plot *a*) is shown the phase velocity spectrum of a field dataset. At the bottom left (plot *b*), together with the velocity spectrum are shown the modal dispersion curves of the first three modes of a tentative model. The agreement is quite good but the question is: are we sure that, in the 8–15 Hz frequency range, the considered model actually excites those higher modes (second and/or third)?

The modal curves cannot provide any answer in this regard. In fact, the modal curves do not say why a certain frequency range shows evidence of a given mode while in another frequency range other modes dominate.

Looking at Fig. 2.2b, we could wonder: why does the fundamental mode dominate at 25 Hz, while at 10 Hz dominate the first higher modes?

Let us underline once again this point: the modal curves do not give any answer in this regard. On the other side, if we use the FVS approach (Fig. 2.2c) we can demonstrate (not simply supposing, assuming or hoping) that in the 8–15 Hz range the considered (tentative) model actually excites the higher modes (which outside that frequency interval are not relevant). In fact, note that the black contour lines perfectly overlap with the field velocity spectrum. This means that, this way, we can demonstrate that a certain model actually excites certain modes.

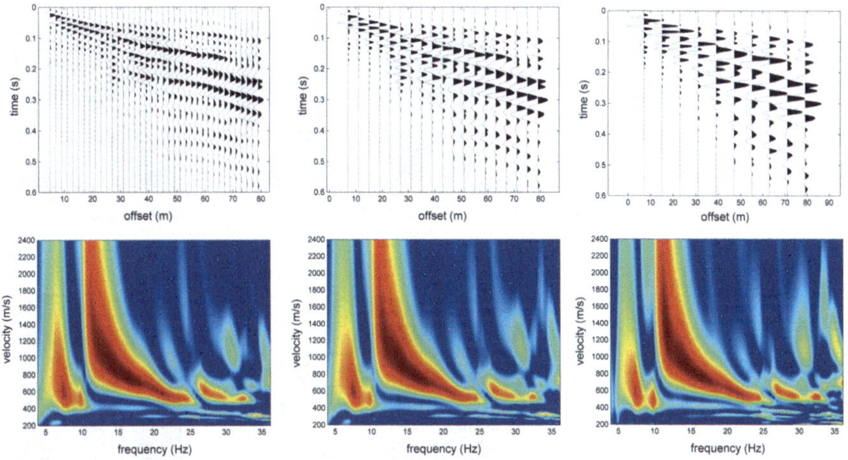

Fig. 2.3 Trace decimation (from left to right traces are reduced from 38 to 19 and, eventually, 10). Is there any significant difference in the velocity spectra? Is there any significant difference if we analyze surface wave dispersion (phase velocities) while dealing with 38, 19 or 10 traces? See text for further comments

A further point is crucial and is linked, as always, to some simple "theoretical" facts that need to be fully understood. Since from the computational point of view the FVS approach is rather heavy, it is important to address an old issue: how many traces are necessary to analyze surface-wave dispersion (while dealing with phase velocities)?

We already addressed the "problem" in the first chapter but it is useful to recall the point with a further (concrete) example.

Figure 2.3 shows a dataset originally with 38 channels/geophones (array length about 80 m). From left to right we reduced the number of traces so to obtain a dataset with 19 and, finally, 10 traces. We then computed the phase velocity spectra for all the three datasets (lower plots): is there any significant difference between the obtained velocity spectra?

Why should we work with many traces (buying more geophones and wasting a lot of time in useless field operations) when the same result is obtained with fewer traces?

It is probably useful to recall some technical facts, not before having recalled that we are here considering the case of active multi-offset data (a, so-to-speak, standard "MASW" approach) used to determine the phase velocities.

In Fig. 2.4 we present a classical 24-trace dataset and its phase velocity spectrum computed according to the phase shift method (see Chap. 1) and the *f-k* transform. Those who are using this latter method to analyze surface-wave dispersion, in order to obtain the phase velocities must then use the well-known $v = f/k$ equation.

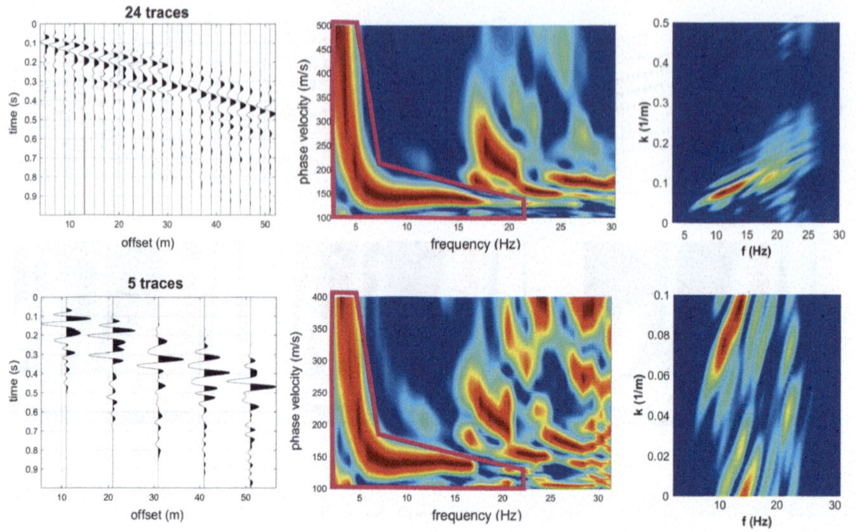

Fig. 2.4 Determination of the dispersive properties via phase shift and *f-k* transform. In the upper part the original dataset (24 traces) with, in the center, the velocity spectrum (*f-v* domain) (highlighted inside the magenta polygon the signal pertinent to the fundamental mode). On the right is presented the *f-k* spectrum (from which it is possible to determine the velocity by means of the $v = f/k$ equation). In the lower panel are reported the same data while considering a decimated dataset (only 5 traces maintained). The phase velocity spectrum obtained via phase shift is still clear (with just some "noise" at the higher frequencies) while the *f-k* spectrum inevitably suffers from spatial aliasing (thus compromising its exploitation in terms of phase velocity determination)

In the lower part of the figure, are reported the data after having reduced the dataset to just 5 traces (it must be considered that, as a matter of fact, the total length of the array is more or less maintained).

From the assessment of the 5-trace data we can see that while the velocity spectrum determined via phase shift is fundamentally the same (see signal within the magenta polygon), the *f-k* spectrum is suffering from major spatial aliasing that jeopardizes the possibility of accurately determining the propagation velocities.

Moral: by using the phase shift we are safe from significant aliasing problems and we can afford to use a limited number of traces saving both time and money.

Does the FVS represent the final solution in surface wave analysis? Certainly not.

The intrinsic ambiguity of the velocity spectra (see Chap. 1 and Dal Moro 2014) prevents from simplistic answers (we should remember that all the surface methodologies are somehow intrinsically non-unique).

Through the joint analysis of several observables (all the examples and case studies presented in this volume are considering different possible types of joint inversions often considering the dispersion analysis via FVS).

In order to slowly getting familiar with FVS, we now begin by considering two single-component examples and then move on to two examples of joint analyses (RVF + THF).

The synthetic traces presented in the following (and used to perform the FVS analysis) are computed according to the modal summation technique (Panza 1985; Herrmann 2013).

2.3 Two Examples of Single-Component FVS Analysis

In this section we present two simple (single component) examples regarding a couple of case studies were the goal was not the Vs30 but the geotechnical characterization of the very first meters of sediments (surface-wave analysis is not useful only for the "infamous" Vs30 parameter).

Example #1 The acquisition, carried out with 24 vertical geophones, was performed with a very short array (about 20 m) which, quite clearly, does not allow us to identify Rayleigh wave velocities at very low frequencies. On the other hand, the low frequencies were not relevant with respect to the problem to address, which was the definition of the V_S values in the first 5–10 m along a 2D section (the shot presented here is just one of the several datasets acquired by moving the array and the source along the section to investigate—see for instance the case study reported in the Appendix G).

Since the FVS inversion is computationally heavy, the original 24-trace dataset was decimated to 12 traces (Fig. 2.5) and the obtained phase-velocity spectrum was then inverted with the results shown in Figs. 2.6 and 2.7. ◀

Example #2 For this second FVS example (ZVF single component), the acquisition was carried out with just 12 vertical geophones (array length 33 m Fig. 2.8) (the array length is the distance between the first and the last geophones). The result of the FVS inversion is shown in Fig. 2.9 where the agreement between the two phase velocity spectra (the one of the field traces and the one of the retrieved model) is apparent (they perfectly overlap).

In order to discuss (and understand) a bit more the data, in Fig. 2.10, together with the phase velocity spectrum of the field data (background colors) and the synthetic one (contour black lines), are also reported the modal dispersion curves of the first seven modes of the identified V_S model (show in Fig. 2.9).

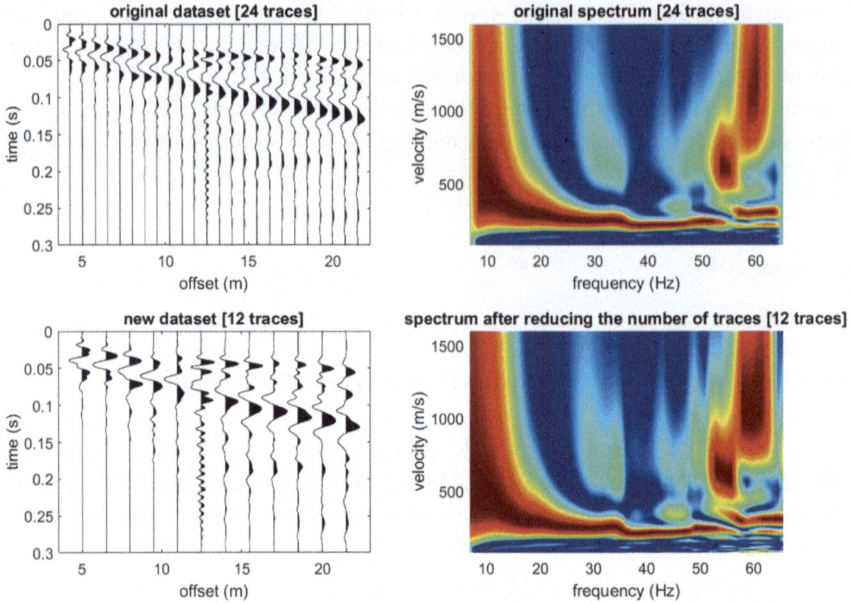

Fig. 2.5 Field data for the FVS example #1: in the upper panel the original 24-trace dataset while, in the lower panel the data after trace decimation (we kept only the even traces). Since the FVS procedure is computationally heavy, we can proficiently deal with the 12-trace dataset (using 24 traces is in fact clearly useless)

We can see that, while below 30 Hz the data are dominated by the fundamental mode, at frequencies higher than 30 Hz the energy is due to the sixth higher mode.

Especially for Rayleigh waves, mode jumps are not necessarily progressive and things can be extremely complicated (here at 30 Hz there is a jump from the mode #0—the fundamental one—to the sixth higher mode).

To get to further fine details, we could also observe that the energy at very low frequencies (less the 4 Hz), to a large extent is due to the first higher mode, which means that at about 4 Hz there is a mode jump between the mode#1 and the mode#0 (compare velocity spectrum in Fig. 2.8 and data reported in Fig. 2.10). ◀

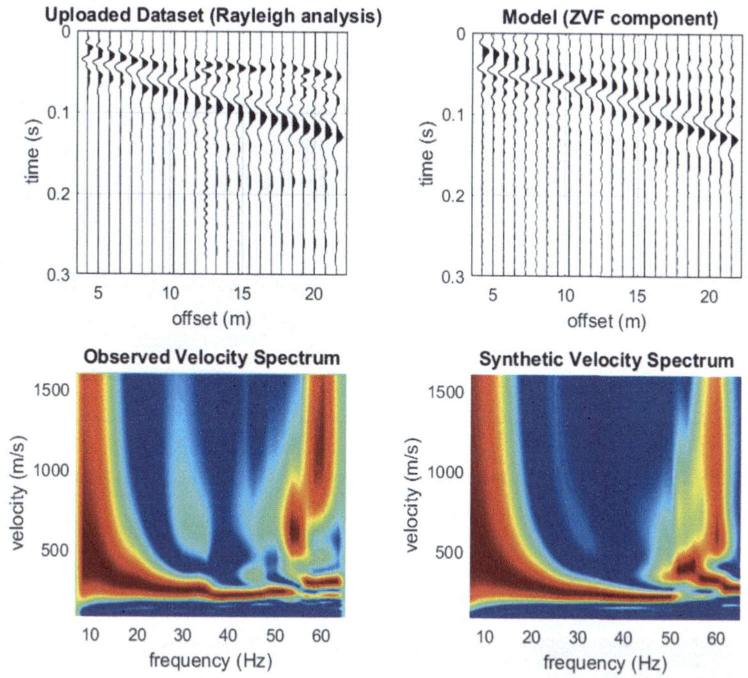

Fig. 2.6 FVS example #1: at the end of an FVS inversion, it is possible to compare the traces and the phase velocity spectrum of the field data (on the left) with the synthetic traces and the phase velocity spectrum of the identified model (on the right). The V_S model is shown in Fig. 2.7

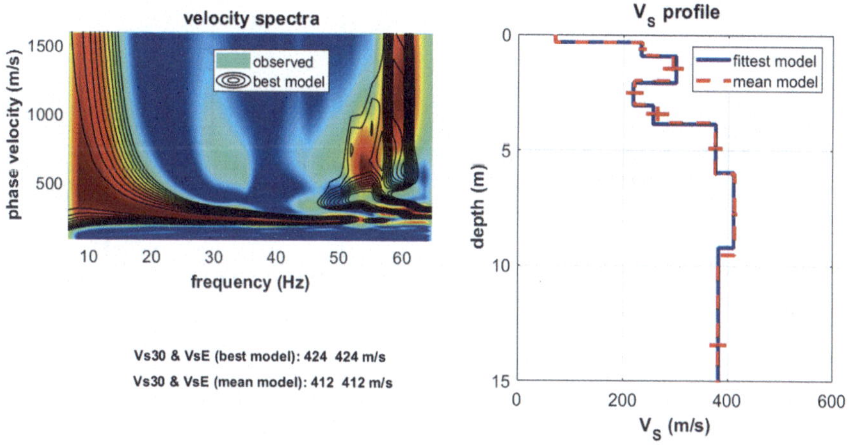

Fig. 2.7 FVS example #1: on the right the V_S model identified via FVS inversion and, on the left, the phase velocity spectrum of the field data (background colors) and of the synthetic data (black contour lines). These velocity spectra are the same as those reported in Fig. 2.6 and are represented in the same graph so to give a clearer and more immediate evidence of the goodness of the accomplished inversion

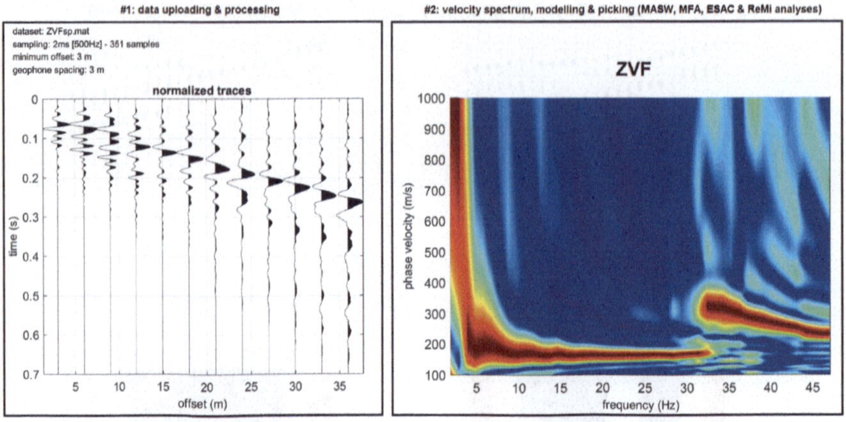

Fig. 2.8 FVS example #2: field traces (ZVF component) and, on the right, the phase velocity spectrum

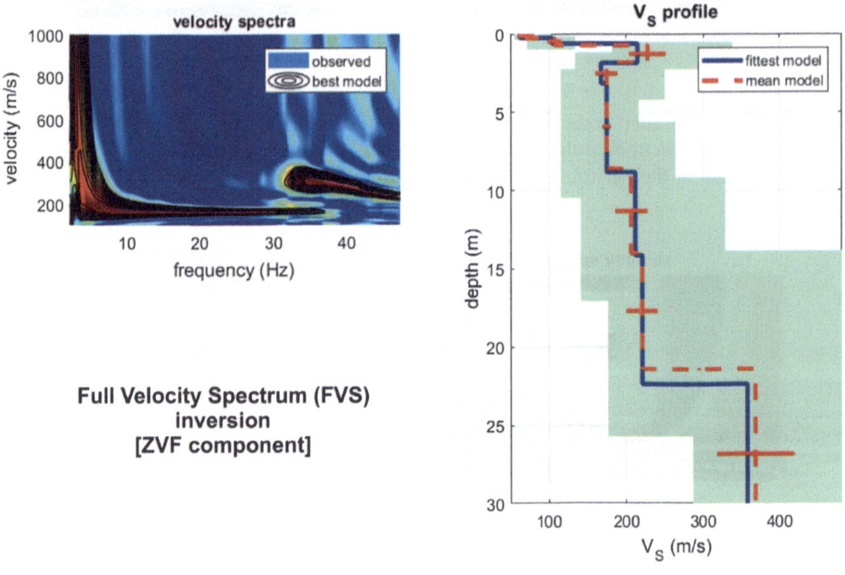

Fig. 2.9 FVS example #2: on the right, the V_S model identified by means of the FVS inversion (the green area reports the adopted search space) and, on the left, the velocity spectra of the field data (colors in the background) and of the identified model (overlapping contour lines)

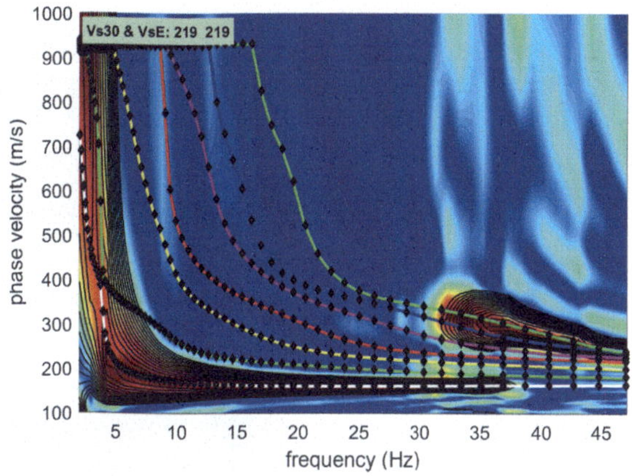

Fig. 2.10 FVS example #2: non-progressive mode jumps. Together with the phase velocity spectrum of field data (background colors) and the synthetic spectrum (black contour lines) are now superimposed the modal curves of the first seven modes of the identified subsurface model. We can see that below 30 Hz data are dominated by the fundamental mode while at higher frequencies the energy is fundamentally due to the sixth higher mode. A careful analysis of the data also shows that below 4 Hz the energy belongs to the first higher mode (compare with field data in Fig. 2.8)

2.4 Joint FVS Analysis of the Phase-Velocity Spectra of the RVF and THF Components: Example #1

In case two or more components are jointly analyzed, the inversion scheme we use is based on the Pareto criterion (MOEA—Multi-Objective Evolutionary Algorithm), an approach that allow the determination of the models that represent the best compromise with respect to the two (or more) considered objective functions (for details see Van Veldhuizen and Lamont 2000; Ramík and Vlach 2002; Dal Moro and Pipan 2007; Pardalos et al. 2008; Dal Moro 2008, 2010, 2019a, 2020; Dal Moro et al. 2015c, 2019a).

We can here only mention the fact that, in the multi-objective approach, the two most important models are: 1) the mean model obtained from all the final Pareto front models; 2) the model with the minimum geometric distance from the utopian point (that we often call the "best model" or "minimum-distance model"—see Dal Moro et al. 2019a). It must be emphasized that, if everything is done properly, these models are (and must be) very similar while in case they are significantly different, the reason is that the inversion process was not properly set up.

The data used for this RVF (radial component of Rayleigh waves) + THF (Love waves) joint analysis were recorded over a quaternary fluvial terrace dominated by unconsolidated sediments (silt, clay and sand), alternated with gravel deposits that are occasionally cemented (conglomerates).

Data were acquired by means of 22 horizontal (4.5 Hz) geophones and using a standard 8 kg sledgehammer but, during the processing, traces were immediately decimated so to obtain the velocity data/spectra shown in Fig. 2.11 (for details about this case study, see Dal Moro 2019a).

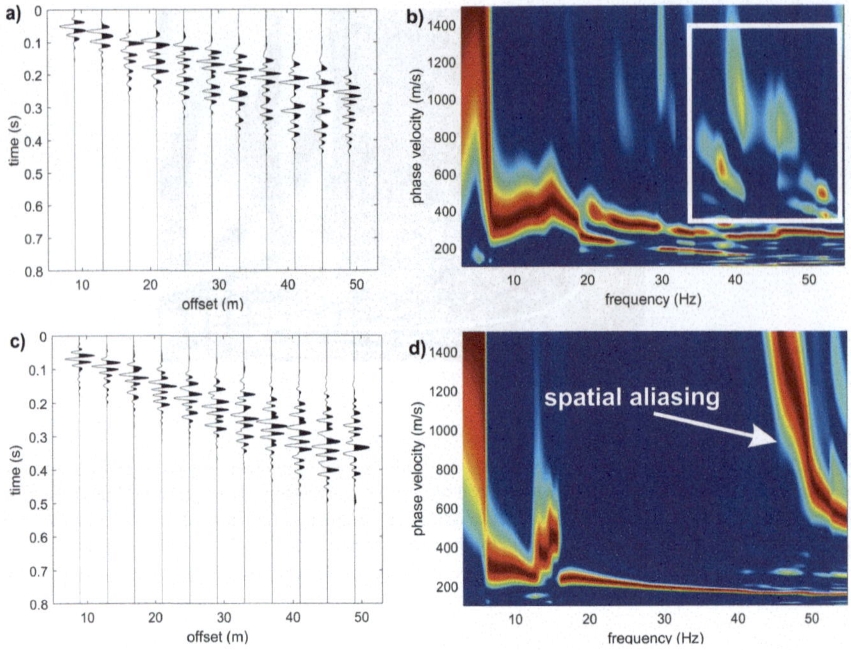

Fig. 2.11 A first example of joint RVF+THF analysis. Decimated field traces (the original 22-trace data were decimated so to deal with lighter 11-trace datasets) and phase velocity spectra for the RVF (above) and THF (below) components. In the Love wave velocity spectrum (THF component), the signal at frequencies higher than 45 Hz is due to spatial aliasing that does not influence the analysis. For details see Dal Moro (2019a)

The spatial aliasing apparent in the THF phase-velocity spectrum (Fig. 2.11d) does not represent a problem (further examples of spatial aliasing in the phase-velocity spectra are presented and commented in Dal Moro 2014), but, in order to avoid it, we considered the velocity spectra just up to 45 Hz.

In this regard, it should be considered that, from the geological/engineering point of view, 45 Hz are an extremely-high frequency that refers to the first decimeters of soil and are therefore completely irrelevant for our purposes.

Figure 2.12 shows the results of the accomplished joint FVS inversion. The velocity spectra of the field and synthetic data for both the components are shown and the overall consistency between field and synthetic data is apparent (field and synthetic velocity spectra overlap very well for both the RVF and THF components).

In order to highlight the importance of the joint FVS approach (Rayleigh + Love), Rayleigh-wave data (velocity spectrum) were processed according to the standard approach based on the picked modal dispersion curve(s). After picking the (interpreted) modal dispersion curve (see Fig. 2.13a), we inverted it and obtained the V_S profile shown in Fig. 2.13b. This is the classical and common approach that inevitably provides a seemingly good result since the picked and synthetic/inverted curves are apparently in good agreement (see curves in Fig. 2.13a).

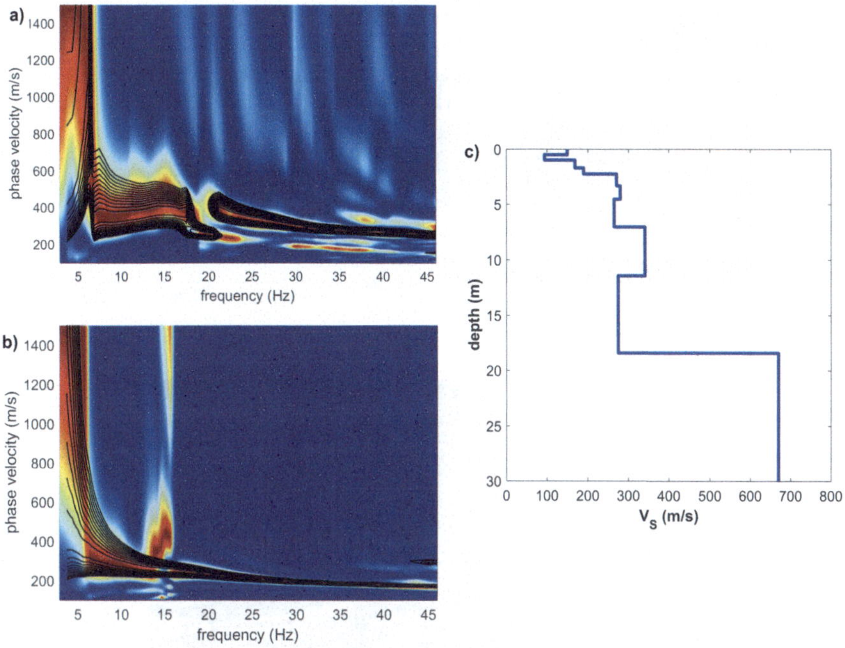

Fig. 2.12 Result of the joint inversion of Rayleigh (RVF—radial component) (a) and Love (THF) waves (b). Note the excellent agreement between the field velocity spectra (background colors) and the synthetic velocity spectra (black contour lines) of the identified V_S model (c). From Dal Moro (2019a)

It is only with the joint approach that we can see that, as a matter of fact, something is wrong.

In fact, if we now consider the identified V_S model (Fig. 2.13b) and compute the modal dispersion curve (fundamental mode) for Love waves, we realize the bad agreement between the field velocity spectrum and the dispersion curve obtained from the analysis of Rayleigh waves only (Fig. 2.13c). It is clear that, for frequencies lower than 15 Hz, the model identified through the inversion of the Rayleigh-wave modal dispersion curve is not consistent with the Love-wave dispersion. This means that the retrieved subsurface model is wrong (although seemed to be in pretty good agreement with the picked Rayleigh-wave dispersion curve—see Fig. 2.13a).

Such inconsistency should be compared with the outcome of the joint (RVF + THF) FVS inversion presented in Fig. 2.12, where the velocity spectra of the identified model and those of the field data match very well for both components (Rayleigh and Love waves).

In summary, the joint FVS approach aims to avoid ambiguities and errors associated with the use of single-component data analyzed in terms of modal dispersion curves.

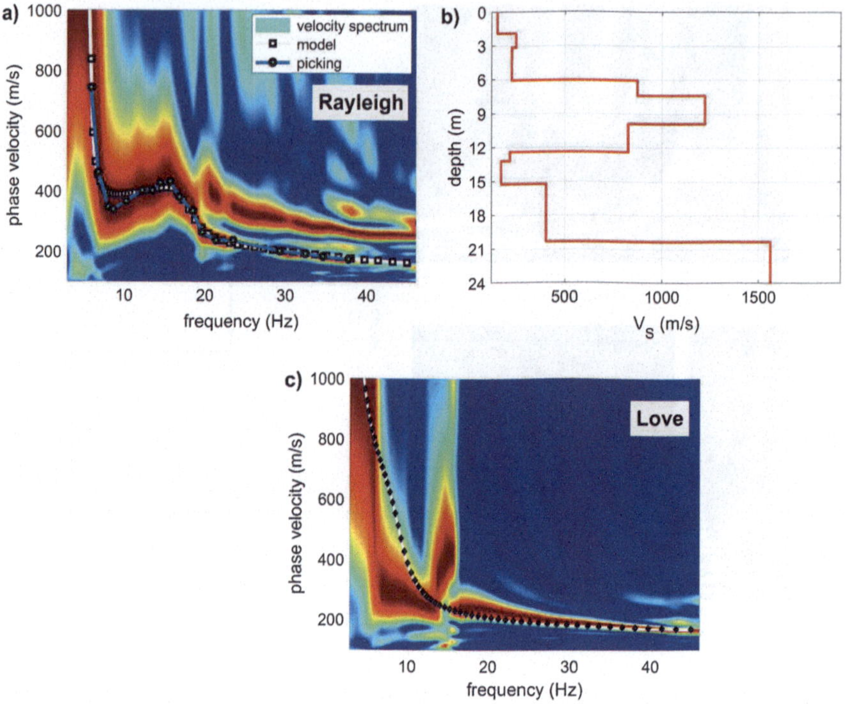

Fig. 2.13 Usefulness of a second component (in this case Love waves) to verify the correctness of the Rayleigh wave analysis accomplished according to the standard modal dispersion curves: **a** Rayleigh-wave velocity spectrum (from the field traces) with the picked (interpreted) modal dispersion curve and the curve retrieved from its inversion (classical approach based on the inversion of the picked dispersion curve); **b** V_S model obtained from the inversion of the interpreted Rayleigh-wave modal dispersion curve (fundamental mode); **c** Love-wave velocity spectrum (background colors) with the Love-wave fundamental mode of the model identified through the Rayleigh-wave inversion (the mismatch is apparent). For details see text and Dal Moro (2019a)

2.5 A Further Example of Joint Analysis (RVF + THF): The Asphalt Dataset

In the first chapter, while speaking about the asphalt paranoia (Sect. 1.13), we saw an example of a joint RVF + THF acquisition and showed that, although we were working with an old and moderate-quality equipment, the data had no particular problem.

Figure 2.14 presents the result of the joint RVF+THF analysis carried out according to the FVS approach.

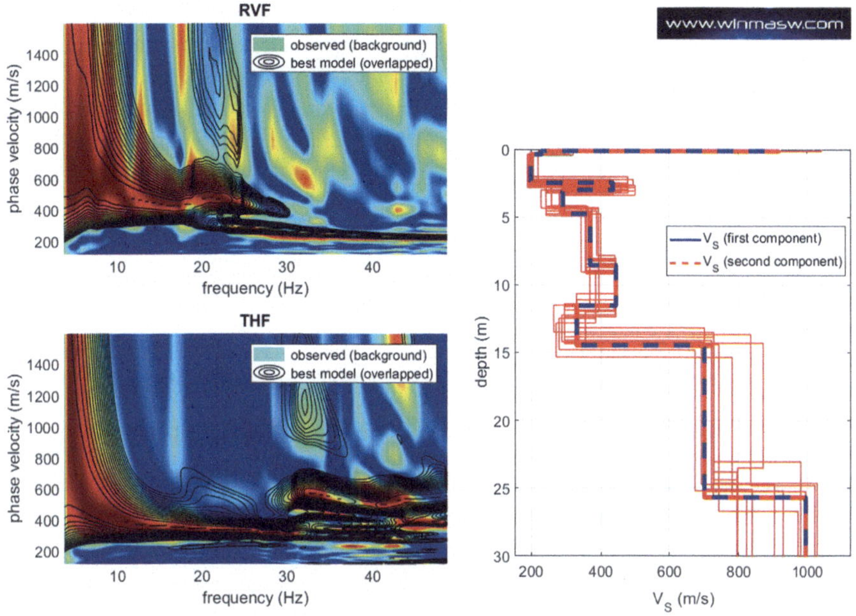

Fig. 2.14 Example of joint FVS analysis of the RVF (Rayleigh waves) and THF (Love waves) components. On the left are shown the phase velocity spectra of field data (background colors) and of the best model (minimum-distance model) (black contour lines) identified through the joint FVS analysis. On the right are shown all the Pareto front models. Field data are the same shown in Sect. 1.13 (Paranoia #1: the asphalt cover)

While presenting the HS (HoliSurface) methodology (Chap. 4) we will compare the V_S profile obtained from this multi-component and multi-offset R + T MASW (Fig. 2.14) with the profile identified via HS, i.e. through the analysis of the active data recorded by means of a single 3C geophone.

The suggested approach for MASW (multi-offset) analyses

As explained and suggested for instance in Dal Moro (2014, 2019a), the joint RVF + THF approach requires just a dozen of 4.5 Hz horizontal geophones and the analysis can be efficiently carried out by following the FVS approach.

This way, by adopting a pretty simple and straightforward acquisition procedure (we set the axis of the horizontal geophones perpendicular to the array for recording Love waves and parallel for recording the radial-component of Rayleigh waves—see Sect. 1.2 and Appendix A), one avoids ambiguities and issues that would raise in case we use a single component and that would eventually mirror in erroneous solutions. Furthermore, it was recently shown that in case a shallow stiff layer is present, the analysis of the Love-wave higher modes can provide valuable information about the deep layers even at high frequencies (Dal Moro 2020).

In case it is necessary to investigate deeper strata, we could simply add the HVSR (see the several examples in Dal Moro 2014 as well as in this book).

References

Dal Moro G (2008) V_S and V_P vertical profiling via joint inversion of Rayleigh waves and refraction travel times by means of bi-objective evolutionary algorithm. J Appl Geophys 66:15–24

Dal Moro G (2010) Insights on Surface-wave dispersion curves and HVSR: joint analysis via Pareto optimality. J Appl Geophys 72:29–140

Dal Moro G (2014) Surface wave analysis for near surface applications. Elsevier, Amsterdam, The Netherlands, 252 pp. ISBN 978-0-12-800770-9

Dal Moro G (2019a) Surface wave analysis: improving the accuracy of the shear-wave velocity profile through the efficient joint acquisition and Full Velocity Spectrum (FVS) analysis of Rayleigh and Love waves. Explor Geophys. https://doi.org/10.1080/08123985.2019.1606202

Dal Moro G (2019b) Effective active and passive seismics for the characterization of urban and remote areas: four channels for seven objective functions. Pure Appl Geophys 176:1445–1465

Dal Moro G (2020) The magnifying effect of a thin shallow stiff layer on Love waves as revealed by multi-component analysis of surface waves. Scientific Reports. https://doi.org/10.1038/s41598-020-66070-1; https://www.nature.com/articles/s41598-020-66070-1 (open access)

Dal Moro G, Pipan M (2007) Joint inversion of surface wave dispersion curves and reflection travel times via multi-objective evolutionary algorithms. J Appl Geophys 61:56–81

Dal Moro G, Coviello V, Del Carlo G (2014) Shear-wave velocity reconstruction via unconventional joint analysis of seismic data: a case study in the light of some theoretical aspects. In: Engineering geology for society and territory, vol 5. Springer, Dordrecht, pp 1177–1182

Dal Moro G, Moura RM, Moustafa SR (2015) Multi-component joint analysis of surface waves. J Appl Geophys 119:128–138

Dal Moro G, Al-Arifi N, Moustafa SR (2019a) On the efficient acquisition and holistic analysis of Rayleigh waves: technical aspects and two comparative case studies. Soil Dyn Earthq Eng 125. https://www.sciencedirect.com/science/article/pii/S0267726118310613

Dou S, Ajo-Franklin JB (2014) Full-wavefield inversion of surface waves for mapping embedded low-velocity zones in permafrost. Geophysics 79:EN107–EN124

Herrmann RB (2013) Computer programs in seismology: an evolving tool for instruction and research. Seismol Res Lett 84:1081–1088

Panza GF (1985) Synthetic seismograms: the Rayleigh waves modal summation. J Geophys 58:125–145

Pardalos PM, Migdalas A, Pitsoulis L (eds) (2008) Pareto optimality, game theory and equilibria. Springer, New York. ISBN 978-0-387-77247-9

Ramík J, Vlach M (2002) Pareto-optimality of compromise decisions. Fuzzy Sets Syst 129:119–127

Van Veldhuizen DA, Lamont GB (2000) Multiobjective evolutionary algorithms: analyzing the state-of-the-art. Evol Comput 8:125–147

HVSR, Amplifications and ESAC: Some Clarifications

3

Mirrors should reflect a little before throwing back images.

Jean Cocteau

G. Dal Moro, *Efficient Joint Analysis of Surface Waves and Introduction to Vibration Analysis: Beyond the Clichés*,
https://doi.org/10.1007/978-3-030-46303-8_3

Abstract

This chapter discusses some points regarding two classical passive techniques more and more popular in seismic micro-zonation studies: the *Horizontal-to-Vertical Spectral Ratio* (HVSR) and the ESAC (*Extended Spatial AutoCorrelation*), a sort of "generalized" SPAC (*SPatial AutoCorrelation*). The influence of industrial components on the HVSR, the role of Love waves in the HVSR modelling and few more important issues are illustrated. Furthermore, the idea that the HVSR curve represents the so-called site amplification is discussed also in the light of experimental data based on the computation of the SSR (*Standard Spectral Ratio*). A series of recommendations about the analysis and modelling of the *effective* dispersion curve retrieved from ESAC (a methodology that provides a dispersion curve much clearer with respect to the ReMi (*Refraction Microtremor*) technique) are then reported together with some concrete examples.

This chapter discusses some crucial (but often neglected) points regarding two classic methodologies that are more and more popular especially in seismic micro-zonation studies: the *Horizontal-to-Vertical Spectral Ratio* (HVSR) and the ESAC (*Extended Spatial AutoCorrelation*).

Both belong to the family of passive methodologies since for both of them we are just *listening* to the background microtremor field in order to extract information about the relative amplitude of the horizontal and vertical particle motion (HVSR) and the dispersive properties of the medium (i.e. the propagation velocities) (ESAC).

Both these techniques provide complementary information about the subsurface conditions; in addition, the HVSR was also claimed to be a sort of estimate of the so-called site amplification although, as we will see, this is highly disputed and experimental data seem to disprove such a simplistic assumption.

Since the ESAC is a sort of "generalized" SPAC (SPatial AutoCorrelation) and the deployment of the geophones can be done according to any (arbitrary) bi-dimensional geometry, we will just focus on this technique (SPAC requires circular geometries and therefore more complex field operations but, in general terms, the performances are quite similar).

Since the data we can obtain from ESAC (and SPAC) are by far clearer with respect to the ReMi (*Refraction Microtremor*) performances, this latter methodology will not be considered.

3.1 HVSR: Few Initial Remarks

The *Horizontal-to-Vertical Spectral Ratio* (HVSR) is something that was "created" when most of the readers of this book had not even been born yet (it is in fact something inherent in the phenomenology of surface waves) but, because of several recent national and international regulations about the seismic-hazard assessment, in the last decade it became extremely popular even among geologists without a specific and robust seismological background.

Because of the simple data acquisition and processing, its popularity has because extremely vast and, inevitably, increased proportionally to a series of related over-simplifications about its actual meaning.

Here we cannot (and do not want) to provide a comprehensive essays about HVSR but it is important to underline few points and try to avoid some annoying misunderstandings.

Nowadays, HVSR has two main applications: on one side it can be used to obtain some information about the vertical shear-wave velocity profile (e.g. Arai and Tokimatsu 2004, 2005), on the other side the HVSR curve is also interpreted as expression of the so-called site amplification (this second application is often referred to as the Nakamura's technique—e.g. Nakamura 1989, 1996, 2000, 2008, 2019; Lanchet and Bard 1994).

These two applications should be kept and commented separately because they suffer from different problems.

HVSR was used to obtain information about the V_S profile already in the 70s. For instance, Mark and Sutton (1975) used HVSR recorded during the Apollo missions to obtain information about the lunar regolith (see also Dal Moro 2015). Since the HVSR curve suffer from a severe non-uniqueness of the solution, it is in general recommended to use only together with quantitative information about dispersion (e.g. Arai and Tokimatsu 2005; Dal Moro 2010).

Main problems are about the way HVSR is modelled. Although the most important parameters are the shear-wave velocity and thickness of the layers, P-wave velocities and quality factors also influence the HVSR curve. Furthermore, the amount of Love waves in the background microtremor field significantly alter the HVSR (see α factor in Sect. 3.1.1). In simple terms: the amount of variables is so large that is quite difficult to use the HVSR to get detailed information about the shear-wave velocities. It can surely provide valuable information about the overall structure but should be used only together with stringent data about surface-wave dispersion.

On the other side, the idea that the HVSR curve represents the so-called amplification curve (Nakamura's technique) is based on non-verified assumptions and when we compare the spectral ratio between the motion at a reference (rocky) site and the motion over a basin we can see that HVSR fails to represent the real (empirical) amplification (see Sect. 3.1.5).

There is a further reason that should be mentioned while critically discussing the HVSR and the excessive emphasis given to the determination of the amplification curve. The HVSR is a "sexy thing" (just to mention an expression often used by Prof. Panza—University of Trieste, Italy) but has a limited and relative meaning both because the actual amplification is not a persistent characteristic of the site and depends on the characteristics of each quake (e.g. Olsen 2000), both because the actual ground accelerations depend first of all on the earthquake spectra (i.e. on the precise and detailed modelling of the seismic source and propagation phenomena—e.g. Panza et al. 2012, 2014; Fasan et al. 2016).

Due to its popularity (the necessary equipment is cheap and data acquisition and analysis is relatively simple), HVSR is too often seen as a panacea and a straightforward expression of a non-well-defined "seismic hazard" and microtremor analysis for the definition of the HVSR is therefore often accomplished more as an apotropaic ritual rather than as a scientific experiment.

Writing down on our field note book the weather conditions during the data acquisition is fine, but what really matters is also, for instance, the correct alignment of the 3C geophone with respect the structural and topographical features of the area (and/or with respect to industrial facilities that can produce spurious signals that can alter the HVSR itself).

The geographical North (i.e. the Polar star) has no particular meaning for our HVSR. In case we are working in a valley with a N45W-S45E axis, it makes much more sense to position the 3C geophone with its NS axis parallel to such a structural axis which could potentially produce some directivity in the H/V spectral ratio.

Covering the 3C geophone with a box in order to protect it from the wind is surely not a mistake but is it really useful? In case of wind, the problem is not only the effect of the wind directly over the geophone but the effects of the wind on the trees and building and structures. These latter can transmit to the soil (and alter the HVSR) but cannot be avoided by simply covering the 3C geophone. Therefore, when possible, windy days should be simply avoided.

A typical issue (often little understood and considered) is about the minimum frequency that can be soundly analyzed. According to the SESAME guidelines, it is recommended that the minimum number of cycles within a window is ten. What does that mean from the practical point of view? That the lower the minimum frequency we want to verify, the larger the window we need to fix. In case we want to compute the HVSR down to 0.5 Hz (i.e. 2 s), we should fix a 20 s window. If we want to explore the HVSR down to 0.1 Hz, a 100 s window is recommended.

And about the recording time? The general rule of thumb is similar suggested by the SESAME criteria is similar: the lower the frequency we want to deal with, the longer the time series we need. For some reason, most of the people think that the recommended length is 20 min but it actually depends on the lowest frequency we are interested in and, also according to AA. VV.—SESAME(2005), in case we are not interested in frequencies lower than 2 Hz, the recording time can be reduced to 5 min (a small table is provided in the mentioned document).

Of course, all these values should be considered as mere general suggestions and not as the Tables of the Law since only a well-balanced combination between a robust theoretical background and a well-digested field practice can guide us during the field operations.

And about the characteristics of the 3C geophone?

The HVSR and the eigenfrequency of the 3C geophone: legends

There is quite a lot of confusion about the characteristics of the geophones needed to determine accurate HVSR curves in the low-frequency range.

In order to correctly determine the H/V ratio at a certain frequency, a single condition is required: that the response curves of the three sensors are (at that frequency) identical (regardless of their absolute value).

Does this have anything to do with the eigenfrequencies of geophones? Not necessarily.

Consider the case of three geophones (two horizontal and one vertical) with a natural frequency of 2 Hz. One of the two horizontal sensors has a response curve which (for example between 0.2 and 2 Hz) is different (lower or higher) than that of the other two geophones. Along that frequency range the H/V ratio will be erroneous.

Consider instead a 3C geophone with 3 sensors with a natural frequency (eigenfrequency) of 4.5 Hz but with identical response curves (for all the three sensors) for instance in the 0.2–100 Hz frequency range.

In this case, the H/V spectral ratio will be corrected throughout all that frequency range.

There is no relationship between the eigenfrequency (f_0) of the sensors and the reliability of the spectral ratio.

The point can be summarized through an elemental logical sequence:

(1) the HVSR is a mere spectral *ratio*
(2) on what depends the *ratio*? On the response curves of the sensors (the vertical and the two horizontal geophones)
(3) if the response curves of the three sensors are identical (*independently* on their shape/trend), the ratio is correct.

Question to address: are the response curves of the three sensors identical in the f_1-f_2 frequency range?

If the answer is positive, we can compute the spectral ratio in the f_1-f_2 frequency range (even if $f_1 < f_0$).

In case we know the response curve of the sensors we can also compute the absolute value of the microtremors (i.e. the amplitude spectra for the three components expressed in physical units), but this is not necessary for the computation of the HVSR.

The eigenfrequency of the sensors should not be confused with the fact that the three sensors have (or not) the same response curve.

There are low-quality 2 Hz geophones (with response curves different for the three sensors) and excellent 4.5 Hz 3C geophones equipped with sensors having exactly the same response curve over a very wide frequency range, which allow the determination of accurate HVSR curve well-below the 4.5 Hz eigenfrequency. We can have a look at the data presented in Figs. 3.1 and 3.2.

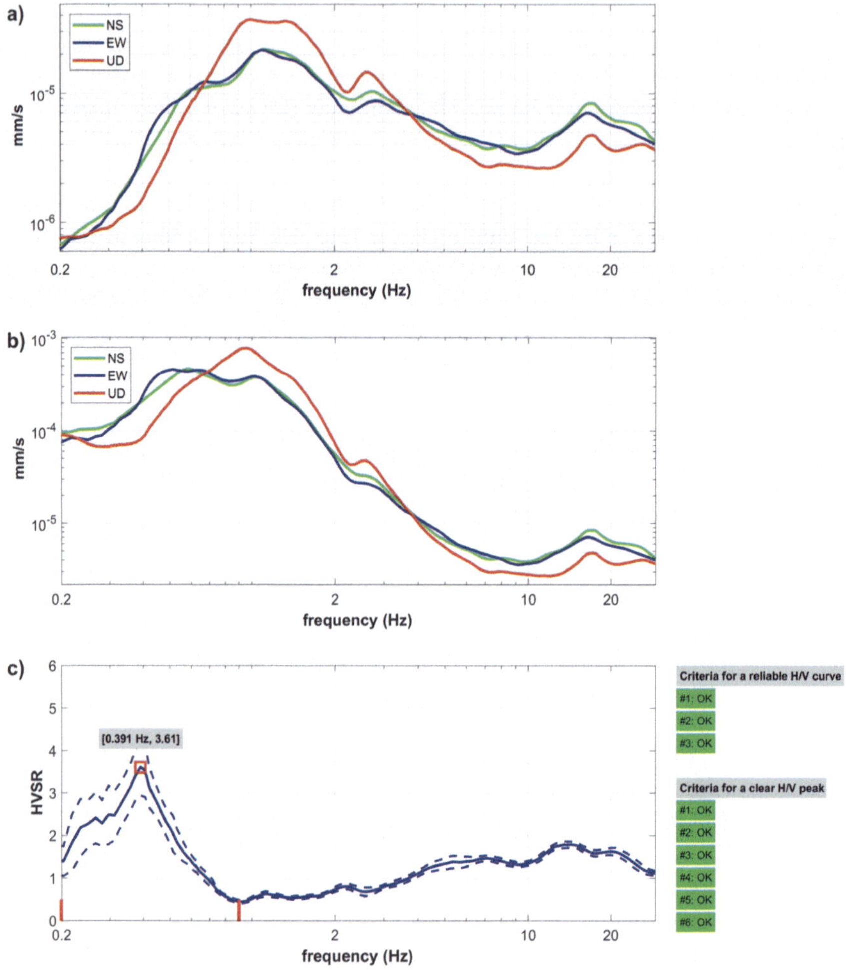

Fig. 3.1 HVSR curve obtained through the analysis of the data recorded by a 4.5 Hz 3C geophone having the three sensors with the same response curve down to about 0.2 Hz: the 0.39 Hz peak is consistent with the depth of the local bedrock (which is known from VSP and reflection seismic data): **a** amplitude spectra before the data equalization; **b** amplitude spectra after the data equalization; **c** HVSR curve (since the three sensors have exactly the same response curve, the HVSR does not change). In this case, all the six SESAME criteria for a reliable peak are fulfilled

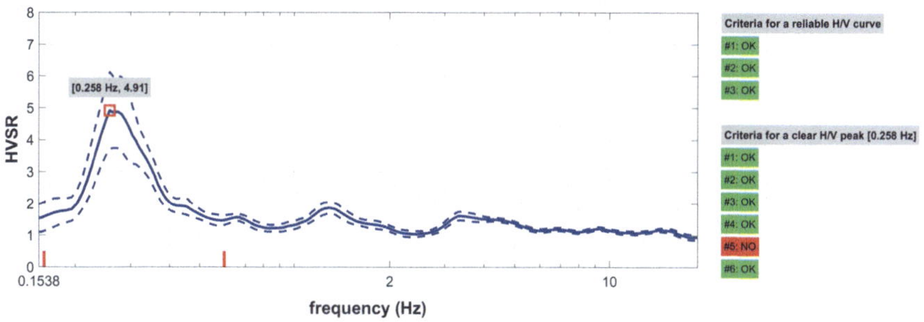

Fig. 3.2 HVSR obtained with the same 4.5 Hz 3C geophone used for the data presented in Fig. 3.1 (the response curves are identical in a very wide frequency range, well below the 4.5 Hz eigenfrequency): the peak around 0.258 Hz is fully consistent with the known depth of the bedrock. In order to compute the HVSR in such a low-frequency range, the window was fixed to 65 s (see the 10-cycle rule recommended by the SESAME guidelines)

There are surely many other aspects that should be considered for a correct and accurate data acquisition (e.g. the *dynamic range* of the digitizer or the units of the output data) but we prefer to focus on the geophysical aspects of the data (assuming that they were properly recorded).

3.1.1 The H/V Spectral Ratio and the Contribution of Love Waves (the α Factor)

The correct modeling of the HVSR is a bit more problematic than commonly believed. Leaving aside the possible contribution of body waves (which can occur in special cases), we here briefly focus on the effects of Love waves on the observed HVSR from microtremor analysis.

The experimental HVSR is not representative of the Rayleigh-wave ellipticity (as often stated). It is the result of the combined effect of Rayleigh and Love waves:

$$\mathrm{HVSR}\ (f) = \sqrt{\frac{\alpha \mathrm{H_L}(f) + \mathrm{H_R}(f)}{\mathrm{V_R}(f)}} \tag{3.1}$$

being $\mathrm{H_R}$ and $\mathrm{V_R}$ the Rayleigh-wave contribution (in terms of power spectra—see Arai and Tokimatsu 2004) on the horizontal (H) and vertical (V) components, and $\mathrm{H_L}$ the contribution of Love waves (the α parameter can therefore be considered as the contribution of Love waves on the observed HVSR).

In order to model the HVSR, the α value should be properly considered.

Of course, when we fix the α parameter to 0, we are just considering the Rayleigh waves, while if we fix it to 1 we are simulating a microtremor field where the amount of Rayleigh and Love waves is the same.

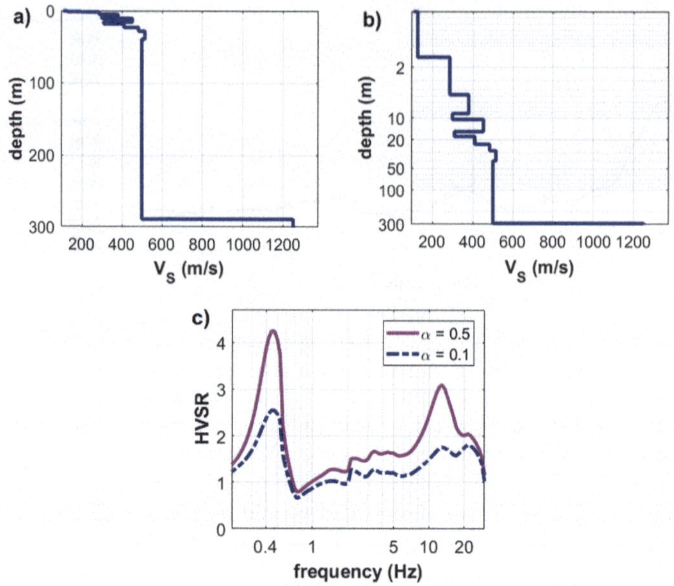

Fig. 3.3 The effect of Love waves on the HVSR. A simple modelling: **a** V_S profile; **b** V_S profile graphed with a logarithmic scale so to emphasize the shallow layers; **c** HVSR curves obtained by considering two different amounts of Love waves (the α parameter is 0.1 and 0.5—i.e. 10 or 50% with respect to the amount of Rayleigh waves). Since the amount of Love waves actually present in the background microtremor field is unknown, the HV modelling is necessarily ambiguous (see text for further comments)

The modeling presented in Fig. 3.3 shows an example of the effect of Love waves on the HVSR curve: the same V_S profile is used to compute the HVSR while considering two different values for the α parameter. Obviously, the larger the amount of Love waves, the larger the motion along the horizontal plane (Love waves move only on the horizontal plane). In other terms, if we consider a larger amount of Love waves, the HVSR curve we obtain has higher values.

Two straightforward consequences:

1. The amount of Love waves should be considered as one of the several variables in the HVSR inversion/modeling (experience shows that this value generally assumes a value between 0.1 and 0.5);
2. HVSR is insufficient to constrain a reliable V_S profile even when geological information are available (we always need some quantitative constrain provided by dispersion data). In fact, in case we intend to fit an experimental HVSR curve, in order to obtain a good agreement we can modify both the V_S and thickness of the strata, both the α parameter (so the modelling itself is intrinsically ambiguous).

The quality factors (Q_S and Q_P) can also influence the HVSR but their influence is in general minor (for an example of major effects during the analysis of the seismic data collected during the Apollo missions on the Moon, see Dal Moro 2015).

3.1.2 Multi-modal Populations and SESAME Criteria

An issue often poorly considered during the assessment of the HVSR, is the correct computation of the SESAME criteria for the definition of the reliability of a peak. Such criteria are represented by a series of statistical tools used to define whether certain "stability conditions" are (or not) met (AA.VV.—SESAME 2005).

The main point we intend to address in this section can be expressed as follows: what happens to the SESAME criteria in case the HVSR has two (or more) peaks?

From the statistical point of view, a multi-modal population can be a problem since such a condition breaks a bottom-line assumption made in those sort of statistical assessments. We can easily understand the problem (and the solution) through a concrete example.

Figure 3.4a presents an HVSR curve with two peaks (one at about 0.9 and the other at about 6.8 Hz). If we try to compute the SESAME criteria for the definition of a clear peak while considering the data over the entire frequency range (for example 0.3–13 Hz), a series of problems caused by the presence of two peaks (multi-modal population) will produce meaningless results.

In Fig. 3.4b are shown the peak frequency values for each of the 36 windows considered during the analysis. In most of the windows the 0.9 Hz peak wins over the 6.7 Hz peak (i.e. it is larger) but in some windows the higher-frequency (6.7 Hz) peak has a larger amplitude and "wins" over the 0.9 Hz peak.

The consequence is simple: the standard deviation (σ_f) of the peak frequency values increases considerably and the fifth SESAME criterion (which is based on such a standard deviation—i.e. on the stability of the peak frequency) is not fulfilled (for details see AA.VV.—SESAME 2005).

This does not mean that the two peaks do not meet the six SESAME criteria. The problem is just due to the fact that this kind of statistical analyses cannot be performed on a multi-modal population.

To solve the problem we simply need to limit the analysis in the frequency range around each peak (the assessment of the robustness and stability of a peak must be clearly defined for each single peak).

In our case, we must therefore limit the analyses in the intervals around the two peaks. Figures 3.4c and d present the results obtained while considering the 0.3–1.3 Hz and 3–13 Hz frequency ranges (separately). Shown data (see standard deviations σ_f) demonstrate that, once we focus the analyses around each specific peak, both the peaks fulfill the fifth SESAME criterion (as well as the other criteria, which are also influenced by the malicious effects of multi-modal populations).

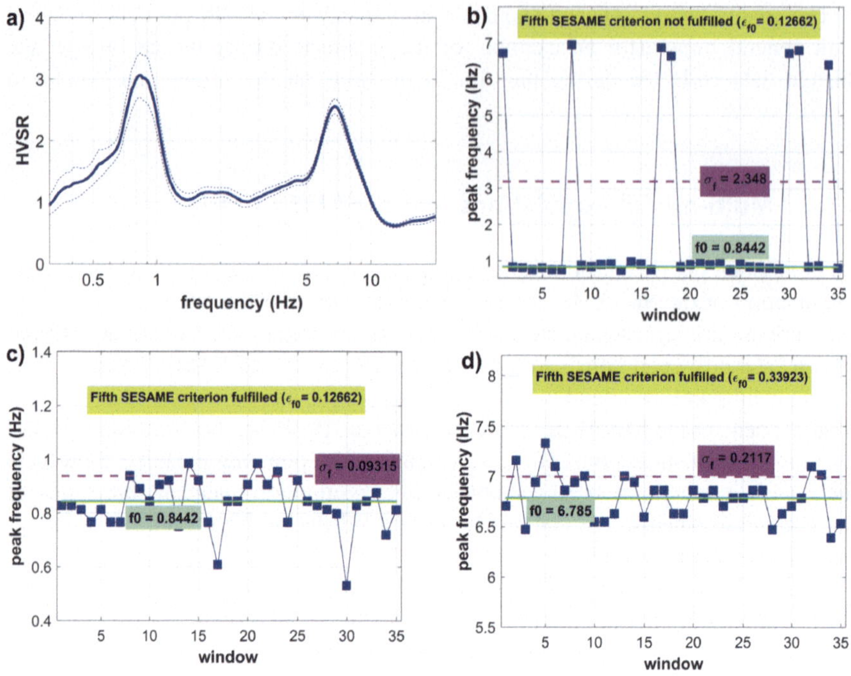

Fig. 3.4 Computations of the fifth criterion SESAME (for the definition of the statistical validity of a H/V peak) for a dataset characterized by two distinct peaks: **a** HVSR average curve with its standard deviations; **b** determination of the peak frequency values for all 36 windows while considering the entire 0.3–13 Hz frequency range (the value of the standard deviation σ_f is equal to 2.3); **c** determination of the peak frequency by considering the data in the 0.3–1.3 Hz frequency range (centered around the 0.85 Hz peak) (σ_f = 0.09315); **d** determination of the peak frequency by considering the data in the 3–13 Hz (centered around the 6.8 Hz peak) (σ_f = 0.2117). For comments see text

Few remarks

The six SESAME criteria for the definition of the statistical robustness of a peak refer (this is pure tautology) to a peak.

If the analyzes are carried out while considering curves with two or more peaks, the multi-modal nature of the population will necessarily produce meaningless results.

Often it is required to provide the so-called fundamental frequency as revealed from the HVSR curve. In the following we will see that the HVSR cannot be considered as the expression of the site amplification but, other than that (and pretending that the simplistic idea that an HVSR peak represents a resonance frequency), we might wonder whether the "*fundamental period*" is that of the lower-frequency peak or the one with the greater amplitude. This question is actually ill-posed since it just addresses a merely formal point and assumes a bureaucratic point of view.

If we are dealing with a curve with two (or more) peaks, a meaningful question to address is: what is potentially the most dangerous resonance frequency with respect to the local urban landscape?

In other terms, what kind of buildings are prevalent in the area? Three-floor family houses (sensitive to high frequencies) or tall skyscrapers (sensitive to low frequencies)? In one case the 6.8 Hz peak can be more relevant, while in case of high-raise buildings the 0.85 Hz frequency is definitely more important.

But is the HVSR sufficient to describe the "seismic hazard"? Two quick points:

(1) HVSR is not a straightforward and faithful expression of the amplification (see Sect. 3.1.5);
(2) the amplification (the real one) is just part of the story since the actual ground accelerations depend on the seismic source and a complex series of effects related to the seismic-wave propagation.

Site effects just modify the ground shaking but, first of all, it is fundamental to properly compute/simulate the ground shaking at the bedrock. The precise computation of the effective ground acceleration is therefore based on the correctness of two facts:

(1) The seismic source must be properly modelled (since it represents the input signal) as well as the wave propagation from the source to the site under study;
(2) The local V_S profile must be precise (since it modifies the earthquake signal).

The actual accelerations surely depends on local site effects but, before that, are determined by the specific characteristics of the fault: its distance from the study area, its depth, its precise motion (slip angle and time history), the azimuth between the fault direction (strike angle) and the site and many other parameters that can influence the actual ground shaking.

Several papers are available about this issues and an intense academic dispute between the probabilistic (PSHA—Probabilistic *seismic* hazard analysis) and the neo-deterministic approaches (NDSHA—Neo-Deterministic Seismic Hazard Assessment) is currently going on (e.g. Zuccolo et al. 2011; Panza et al. 2012, 2014; Fasan et al. 2016).

Furthermore, to come to the main focus of the present book, most of the V_S profiles determined by means of simplified (or simplistic) methods that do not properly take into account all the issues related to the non-uniqueness of the solution and the ambiguities of the data, are inaccurate and, as a consequence, cannot be used to correctly define the local site effects.

In short: the HVSR curve surely provides important information that can help in the definition of the V_S profile in the deeper strata but its assessment for the evaluation of the local amplification (the Nakamura's technique) should be taken with caution (see Sect. 3.1.5).

3.1.3 HVSR and Industrial Components

One of the main problems in the HVSR assessment is the presence of possible industrial signals that can be misinterpreted as representative of a resonance frequency.

Two important points should be clarified:

(1) Industrial signals can travel for tens of kilometers and, consequently, can contaminate the HVSR even if no industrial facility is present in the surrounding area;
(2) Traffic (and similar transient events) has nothing to do with this kind of spurious signals: industrial components are mono-chromatic (or quasi mono-chromatic) correlated signals while road traffic generates (uncorrelated) tremors in a wide frequency range (and do not necessarily procude negative effects on the HVSR).

Now let us have a look at few examples so to get familiar with this kind of spurious industrial components widely described in Dal Moro (2020).

Example #1 (*NE Italy*) Figure 3.5 presents the mean amplitude spectra for the three NS, EW and vertical components and the HVSR for a dataset recorded in the NE Italy area (data are available for the download from the link provided in the preface of the book). In order to deal with an easily-interpretable curve and compute the SESAME criteria, a 10–15% smoothing is always applied to the amplitude spectra and the HVSR. ◀

A well-trained eye can immediately identify a spurious (industrial) signal at about 1.5 Hz: the two horizontal components (NS and EW) show very large amplitudes while the vertical (UD) component shows a smaller amplitude. This fact clearly produces a significant increase of the H/V spectral ratio.

It should be underlined that, due to historical reasons, the amplitude spectra are usually reported with a logarithmic scale. Furthermore, spectra are always smoothed by 10–15% so to make the data easily interpreted and to allow the application of the SESAME statistical tools. In some cases, these two facts can unfortunately mask the presence of industrial components. For this reason, together with the HVSR, it can be useful to compute and assess two further observables (Fig. 3.6): the mildly-smoothed (e.g. 3%) amplitude spectra plotted with linear scales and the coherence functions for all the possible channel combinations (UD-NS, UD-EW, NS-EW). It must be underlined that the coherence functions define how similar are the data frequency-by-frequency and in case of correlated signals reach very high values (while the background microtremor field is composed of several uncorrelated sources/events, industrial signals are produced by specific and repetitive sources/facilities and are therefore highly correlated—see Dal Moro 2020).

In these graphs (Fig. 3.6), the presence of industrial components is now extremely clear, in particular for the 1.5 Hz (very large) signal which, incidentally, can be found (almost every day) over the whole NE Italy area.

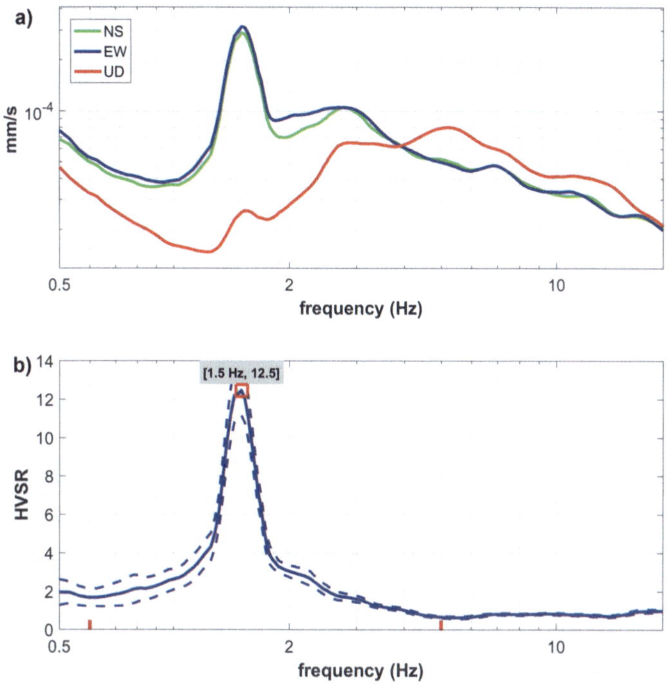

Fig. 3.5 Example #1: mean amplitude spectra for the three components (top) and H/V spectral ratio (bottom). A well-trained eye, can easily identify the complex nature of the HVSR curve around 1.5 Hz: an industrial signal overlap to a real H/V peak (related to the subsurface conditions). Further evidences are provided by the coherence functions and unsmoothed amplitude spectra reported in Fig. 3.6 (see text for comments). After Dal Moro (2020)

Can we somehow remove the contribution (contamination) of the industrial component and obtain a "clean" HVSR curve, expression of the subsurface conditions?

In some cases, we can try to pick the amplitude spectra so to "cut out" spurious (industrial) components and re-compute the HVSR while using the picked amplitude spectra. Figure 3.7 shows the effect of this procedure on the considered data: the peak (now at about 1.3 Hz) has now dropped down to about 4 (compare with the original HVSR curve shown in Fig. 3.5b).

Example #2 (*Central Italy*) The dataset was acquired in Tuscany (Apuan Alps) as part of a seismic micro-zonation study. The HVSR curve reported in Fig. 3.8 shows a major peak at about 1.34 Hz and a secondary one at about 2.7 Hz. The shape of the 1.34 Hz peak risks to be quite misleading but the joint assessment of the coherence functions and the mildly-smoothed amplitude spectra (graphed in Fig. 3.9) clearly reveals that, below 3 Hz, a complex series of industrial components makes the HVSR almost completely indecipherable (i.e. useless). Around

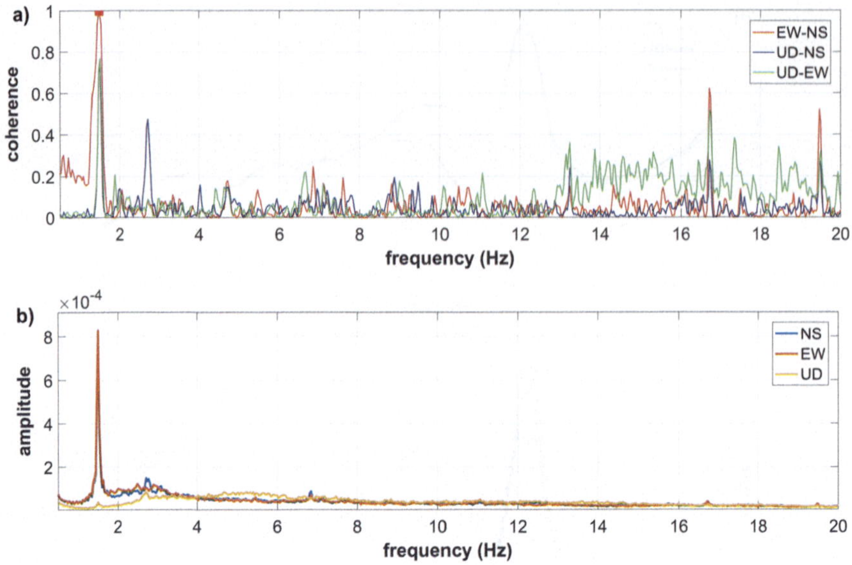

Fig. 3.6 Top panel: coherence functions for all the three possible trace combinations (EW-NS, UD-NS and UD-EW); bottom panel: amplitude spectra (3% spectral smoothing) graphed with a linear scale for both the frequency and the amplitude. The presence of industrial components is obvious at 1.5 Hz (very large) as well as at 16.8 and 19.6 Hz (much smaller). After Dal Moro (2020)

1.3 Hz some "real" peak is probably present but its actual trend is completely altered by complex industrial signals that characterize the area and are probably related to the several well-known marble quarries that can be found all over the area. ◀

Please, notice that, as usual (because of its persistence), the 1.34 Hz industrial peak fulfills all the six SESAME criteria (Fig. 3.8, lower panel).

Example #3 (*NE-Italy*) A final example is summarized by the data presented in Figs. 3.10 and 3.11, where the complexity of the interactions between the natural microtremor field and a series of industrial components is so-to-speak extreme. ◀

At first glance, the HVSR curve (Fig. 3.10b) seems characterized by two clear and distinct peaks but a deeper analysis demonstrates the presence of a complex series of industrial signals whose amplitude and interactions create a complex situation.

In a relatively small frequency range (between about 1 and 5 Hz), the assessment of the coherence functions shown in Fig. 3.11a clearly reveals the massive presence of five major industrial frequencies (components) labelled as A, B, C, D and E (incidentally, the A signal is the same 1.5 Hz component that was identified for the Example #1 and that can be often found in most of the NE italian region).

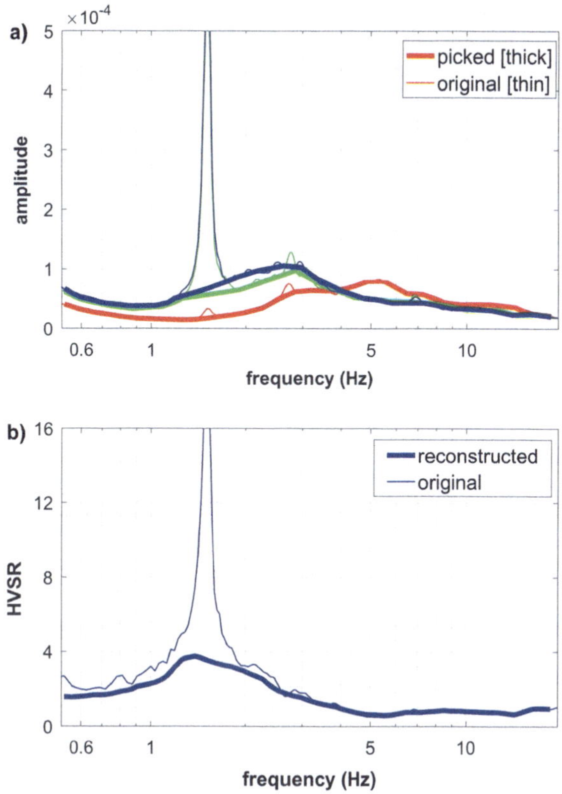

Fig. 3.7 Attempt to remove the industrial signal that dominates the considered dataset: **a** original (thin lines) and picked (thicker lines) amplitude spectra; **b** original (thin line) HVSR curve (contaminated by the 1.5 Hz industrial signal) and re-constructed (thicker lines) HVSR obtained by considering the picked amplitude spectra (the peak is now at 1.3 Hz and its amplitude has decreased to about 4)

For each of the five frequencies, the relationships between the horizontal and vertical components are different and, consequently, the H/V spectral ratio modifies accordingly (further details in Dal Moro 2020).

Three facts to underline about the data collected over the time at this site:

(1) an HVSR curve not too contaminated by industrial components was shown in Dal Moro (2010) where a distinct (and single) peak is present at about 2.1 Hz (further data in Dal Moro 2020);
(2) over the time, the presence, intensity and quantity of industrial components that can be identified is quite variable (from day to day the obtained HVSR can be rather different depending on the industrial components present in that specific day);
(3) the HVSR "hollow" at about 1.95 Hz (see Fig. 3.10b) is also the effect of an industrial component (the signal B in Fig. 3.11): depending on the relationships between the amplitudes along the vertical and horizontal (NS and EW) components, an industrial component can in fact determine a decrease of the H/V spectral ratio. In this case, both the two peaks (at 1.5 and 2.5 Hz) and the "valley" in-between (at 1.95 Hz) are the result of the interaction of the components A, B and C (the D and E components do not significantly alter the

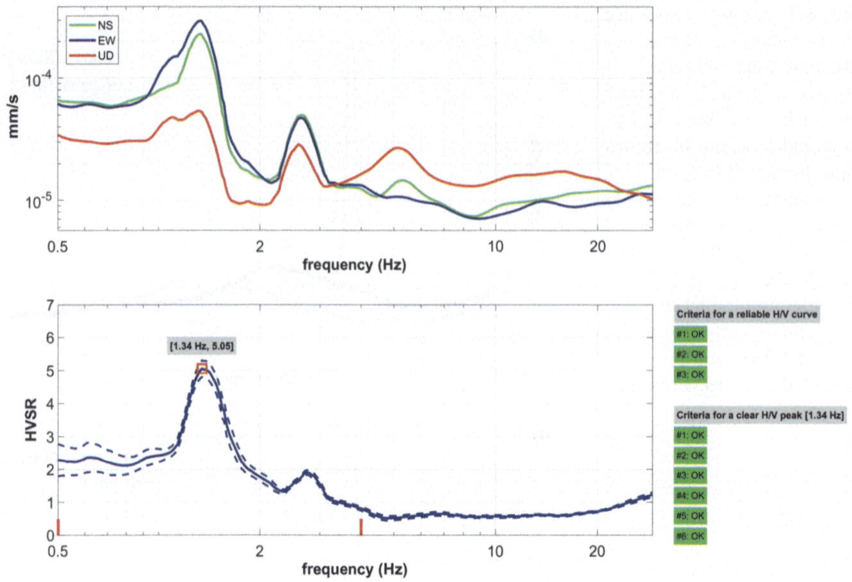

Fig. 3.8 Example #2: mean amplitude spectra of the three components (top) and H/V spectral ratio (bottom). The H/V peaks at 1.34 and 2.7 Hz are dubious. The data reported in Fig. 3.9 (coherence functions and mildly-smoothed amplitude spectra plotted with linear scales) reveal crucial to properly indentify as spurious (industrial) these two peaks

natural HVSR since the ratios between the amplitudes along the horizontal and vertical axes do not modify).

Since in this case the amount and the complexity of the industrial components cannot be easily handled through data processing (see Example #1), the only solution is to record some new dataset in some other day. Although the origin of these signals is unknown (we do not know the facility - or facilities - responsible for them), it could be useful to record the data also during the week end (in the hypothesis that the industrial facilities are closed).

An example of identification and removal of industrial components through an automatic procedure based on the explotation of the coherence functions (and the derivative of the amplitude spectra) is presented in the Appendix H.

3.1.4 What Is H?

For the computation of the HVSR and/or (as we will see later) the SSR (*Standard Spectral Ratio*), the definition of the motion along the horizontal (H) plane from the two perpendicular horizontal sensors (usually referred to as North-South and East-West) is possible according to different "solutions". There are (at least) three

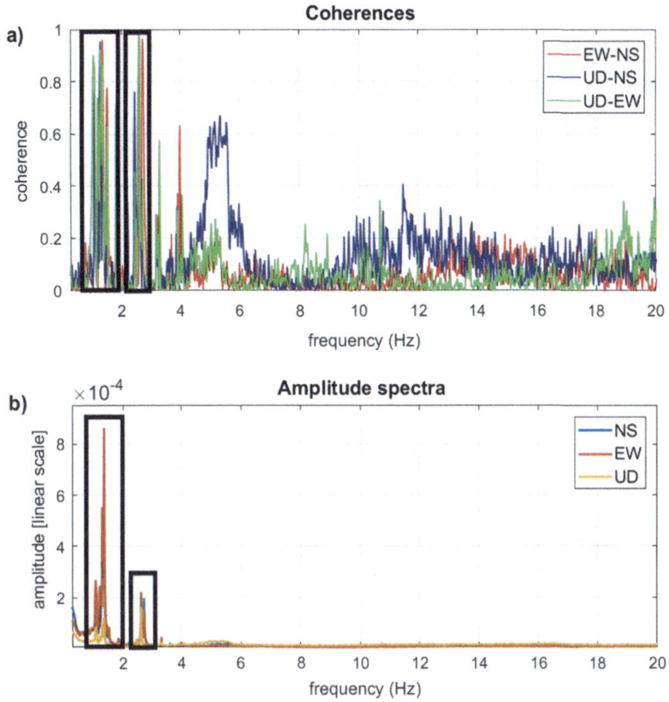

Fig. 3.9 Example #2: **a** coherence functions for the three trace combinations (EW-NS, UD-NS and UD-EW); **b** mildly-smoothed (3%) amplitude spectra (linear scales). The presence of industrial signals (highlighted by the two black rectangles) that heavily alter the H/V ratio is obvious (compare with the curves reported in Fig. 3.8)

possible ways to do it: the quadratic, the geometric and the arithmetic mean (e.g. Albarello and Lunedei 2013). The respective equations are:

$$\mathrm{HVSR} = \frac{\sqrt{(\mathrm{NS}^2 + \mathrm{EW}^2)/2}}{\mathrm{UD}} \tag{3.2}$$

$$\mathrm{HVSR} = \frac{(\mathrm{NS} + \mathrm{EW})/2}{\mathrm{UD}} \tag{3.3}$$

$$\mathrm{HVSR} = \frac{\sqrt{(\mathrm{NS} \cdot \mathrm{EW})}}{\mathrm{UD}} \tag{3.4}$$

being NS, EW and UD the North-South, East-West and vertical (Up-Down) components.

Which one is the best way to define the horizontal component? Now and then this issue is raised while stating that *this* way is better than *that*. But does that really make sense?

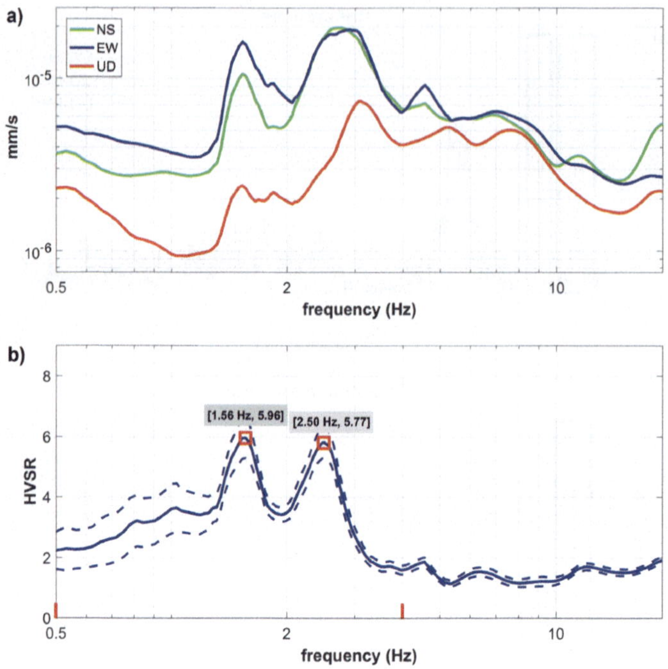

Fig. 3.10 Example #3: **a** mean amplitude spectra for the three components; b) H/V spectral ratio (standard 15% smoothing). The two peaks at 1.56 and 2.5 Hz are generated by the presence and complex interaction of a series of industrial components (see text and Fig. 3.11). Further data are presented in Dal Moro (2010) and further details about the analysis can be found in Dal Moro (2020)

Let us try to face the problem by computing the HVSR of a microtremor dataset according to the three possible approaches. The results are presented in Fig. 3.12 and should reveal how pointless are often this kind of issues (the differences among the three curves are apparently negligible).

In order to further clarify the negligible nature of this kind of disputes, in Fig. 3.13 are shown the directivity plots for all the three possibilities. Once again, the differences are completely insignificant (remember the Saint Matthew's warning quoted in the first chapter?).

Later on (in this chapter), similar evidences will be presented while briefly introducing the SSR or, more precisely, the SSRn (a kind of spectral ratio computed from the microtremor data recorded at a rocky reference site and at a site we are investigating—the *n* stands for *noise*; Perron et al. 2018).

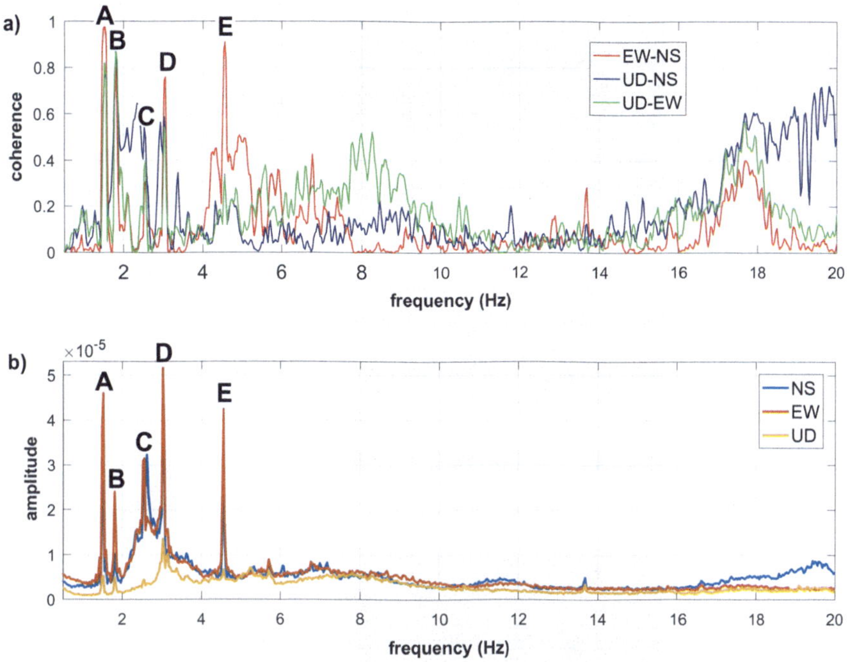

Fig. 3.11 Example #3: **a** coherence functions for the three trace combinations (EW-NS, UD-NS and UD-EW); **b** mildly-smoothed (3%) amplitude spectra (linear scales). The letters A, B, C, D and E identify a series of industrial components that profoundly alter the HVSR curve (signal A is the well-known 1.5 Hz component which can be often found in the NE Italy area—see also Example #1). From Dal Moro (2020)

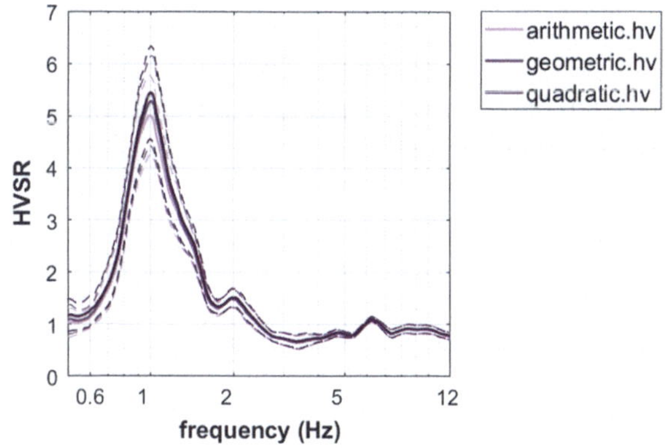

Fig. 3.12 HVSR curves computed from the same microtremor dataset while considering the definition of the H component according to the three possible approaches (arithmetic, geometric and quadratic). See also Fig. 3.13

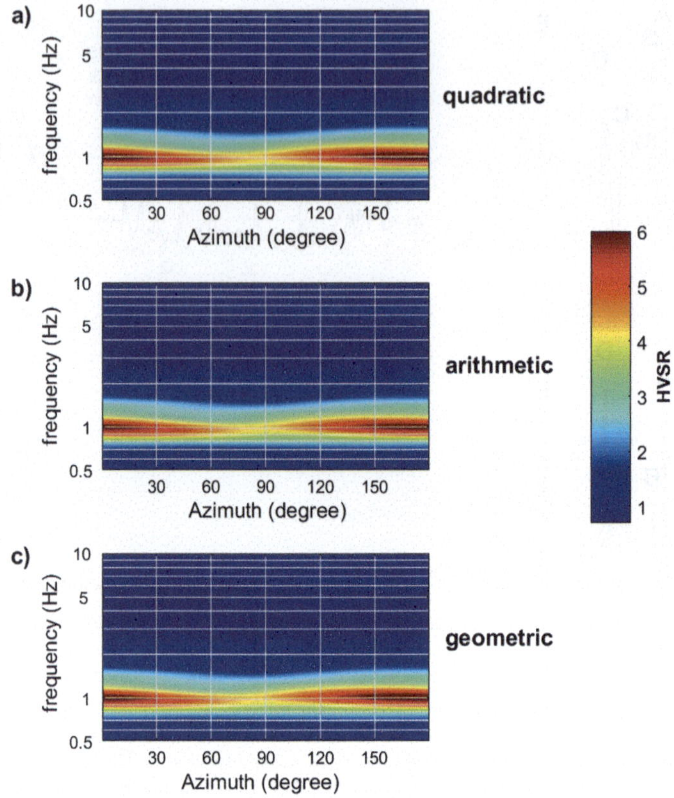

Fig. 3.13 HVSR directivity while computing the H component according to the three possible approaches: **a** quadratic; **b** arithmetic; **c** geometric. No significant difference is apparent. See also Fig. 3.12

3.1.5 Does the HVSR Represent the Actual Site Amplification During an Earthquake?

Among the several issues about the nature and meaning of the HVSR, we should surely mention the simplistic idea that the HVSR curve represents the so called *site amplification*.

The problem is, first of all, linguistic (*cf.* Wittgenstein's *Tractatus*): the HVSR is the HVSR, the amplification is the amplification. If we mix these two things, we end up with serious problems.

On one side, the HVSR curve is not a straightforward expression of the actual site amplification and, on the other side, the amplification is not a straightforward expression of the actual *seismic hazard*.

The ground accelerations recorded during an earthquake depend on the complex combination of several aspects: the characteristics of the seismic source (the fault and its rupture, its depth, distance and azimuth with respect to the study area), the complex wave-propagation phenomena and, at the very end, the possible site effects.

If we just focus on these latter facts, we risk to end up in a pointless scientific *cul de sac*. Since this book deals with the analysis of surface waves and vibration data in very general terms, we cannot get into subtle details about the actual value and meaning of the HVSR in terms of estimation of the site effects in case of earthquake. With the aim of providing some evidence about the fact that the HVSR does not correspond to the amplification function, we just present a couple of brief and simple comparative examples:

(1) a comparison between the HVSR and the analytic SH-wave transfer function;
(2) a comparison between the HVSR and the SSR (*Standard Spectral Ratio*).

Case #1: the SH-wave transfer function
Given the V_S profile down to the *bedrock*, it is possible to compute the SH-wave transfer function (see Schnabel et al. 1972 and all the vast literature that followed the seminal work that led to the SHAKE program). The goal is to simulate the propagation of the emerging SH waves from the bedrock up to the surface, thus verifying how the local stratigraphy modifies the original signal (different frequencies can be amplified or de-amplified). Does this represent the real and actual amplification? Several researchers doubts about it (during an earthquake the wave phenomena are not limited to the vertical propagation of the SH waves but to complex phenomena involving different kinds of waves—e.g. Bowden and Tsai 2017) but it is important to point out that, in general terms, the difference between the experimental HVSR and the SH-wave transfer function can be large. Figure 3.14 shows an example where such a difference is quite apparent (the HVSR significantly underestimate the amplification curve estimated from the SH-wave transfer function—see Fig. 3.14d).

Case #2: experimental SSR (*Standard Spectral Ratio*)
The SSR (*Standard Spectral Ratio*) is the very classical method to assess the local amplification through experimental data recorded during an earthquake at a reference (rocky) site and somewhere else (e.g. Borcherdt 1970; Mittal et al. 2013).

In Fig. 3.15 the very simple experimental setting is shown: we have a 3C sensor at a reference site (usually a bedrock outcrop) and a 3C sensor in a second site (for instance the middle of a sedimentary basin).

If these two stations are permanent stations and we record the data of some actual earthquake, by comparing the amplitude spectra at the two locations we can determine the so-called *Standard Spectral Ratio*. On the other side, if we just record microtremor data we can obtain a curve that can be called SSRn, where the *n* stands for *noise*, i.e. the ambient seismic microtremor field (see Perron et al. 2018 and references therein mentioned).

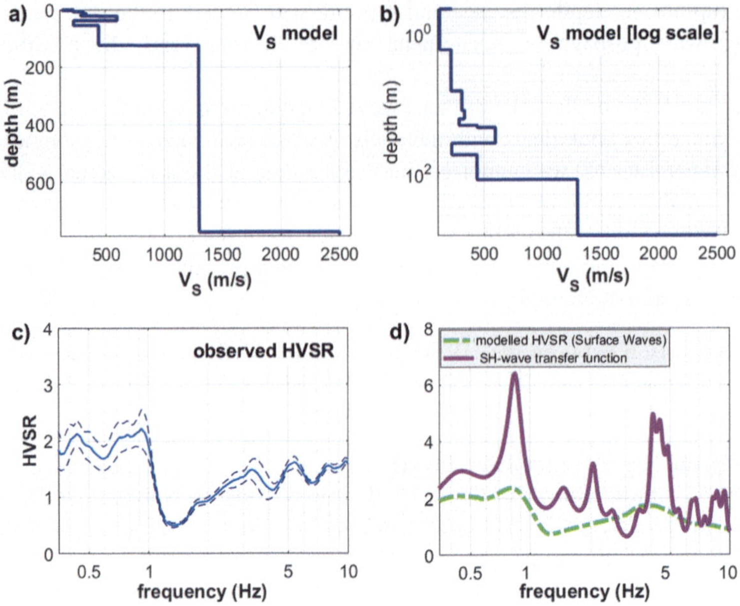

Fig. 3.14 Comparison between the experimental HVSR and the SH-wave transfer function computed according to Schnabel et al. (1972): **a** V_S profile obtained through the joint analysis of the dispersive properties (determined via ESAC and HS) and the HVSR curve; **b** same V_S profile but graphed with a log scale (so to emphasize the shallow layers); **c** experimental HVSR; **d** comparison between the HVSR modelled considering the retrieved V_S profile (plot *a* and *b*) and the SH-wave transfer function. The difference between the two curves is large and apparent

Fig. 3.15 Experimental setting for the recording of the data necessary to define the *spectral ratio* between a reference rocky site (site #1) and the location that we intend to evaluate from the amplification point of view (site #2)

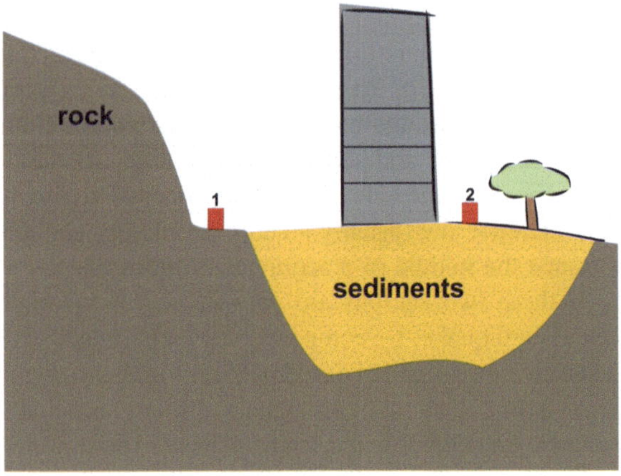

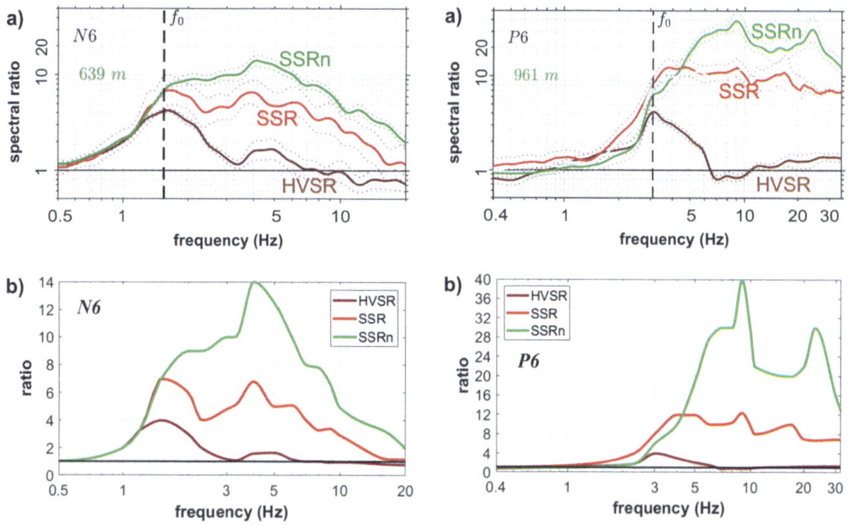

Fig. 3.16 Comparison between HVSR, SSR and SSRn for two sites (N6 on the left and P6 on the right) in the south-eastern France Alpine foreland (after Perron et al. 2018). Data are reported according to both a logarithmic (upper plots) and linear (lower plots) vertical scale. The huge difference between the HVSR and the SSR (i.e. the experimental actual amplification) is apparent for both the sites. For further comments see text

Figure 3.16 presents the data for two of the sites presented in Perron et al. (2018). HVSR, SSR and SSRn curves are reported with the original logarithmic scale (upper plots) and according to a linear scale (lower plots).

For a series of reasons (currently not definitely and fully clarified), the SSRn curve systematically tends to overestimate the SSR, which is the expression of the real (experimental) amplification.

We surely cannot get into finer details about that, but we must clearly highlight that the SSR and the HVSR curves are significantly different. For the point N6 (on the left in Fig. 3.16), at frequencies higher than about 1.2 Hz the HVSR completely fails to estimate the experimental amplification defined by the SSR (the logarithmic scale for the vertical axis compresses the curves and risks to provide a false perspective of the actual differences, which are clearer if a linear scale is used).

For the point P6 (on the right in Fig. 3.16), the difference between the HVSR and SSR curves is dramatic along the entire frequency range. For instance, between 6.5 and 14 Hz, the HVSR is 1 (or less) while the SSR curve exceeds the value of 10.

On the other side, it is also clear that the SSRn (the spectral ratio from the analysis of the microtremors) somehow reflects the SSR (i.e. the “true” experimental amplification) but overestimates the actual amplitudes.

So what is the HVSR about? In near-surface applications, HVSR reveals quite helpful for constraining the V_S of the deep layers that, usually, cannot be sufficiently clear from dispersion data alone (usually poorly defined in the low frequency range). HVSR can surely provide a first very rough idea about possible amplification phenomena but cannot be taken as an valuable estimation of the actual amplification over the entire frequency range.

If we consider the SSR curves reported in Fig. 3.16, it is clear that even the very idea of a "fundamental frequency" (f_0) is quite simplistic and almost meaningless. Of course, we can decide to call "fundamental frequency" the frequency of the lowest peak of the HVSR curve but such a frequency is then hardly related to the actual amplification. The SSR curves show that the actual amplification is very large in a very wide frequency range well beyond the f_0 (the huge difference between the HVSR and SSR curves should be carefully underlined and understood). The HVSR provides just rudimentary information about the frequency where the amplification starts (in the two examples presented in Fig. 3.16 this happens at about 1 and 2.5 Hz) but, at higher frequencies, completely fails to represent the actual amplification.

The SSR (and/or SSRn) is simply the comparison of the amplitude spectra computed for a reference rocky site and for the investigated site. During the data recording, the orientation of the sensors must be clearly the same and it is then possible to compute (and compare) the spectral ratios for the NS and EW directions separately (as well as for the vertical component).

But it is also possible to compute the mean H (horizontal) motion and obtain a "mean" SSR (for the horizontal motion as a whole—curves reported in Fig. 3.16 were computed by considering the quadratic mean).

If we decide to do so, once again the issue of the "proper" way to define the horizontal motion raises (we introduced this problem while considering the HVSR computed according to the arithmetic, geometric and quadratic—see Sect. 3.1.4).

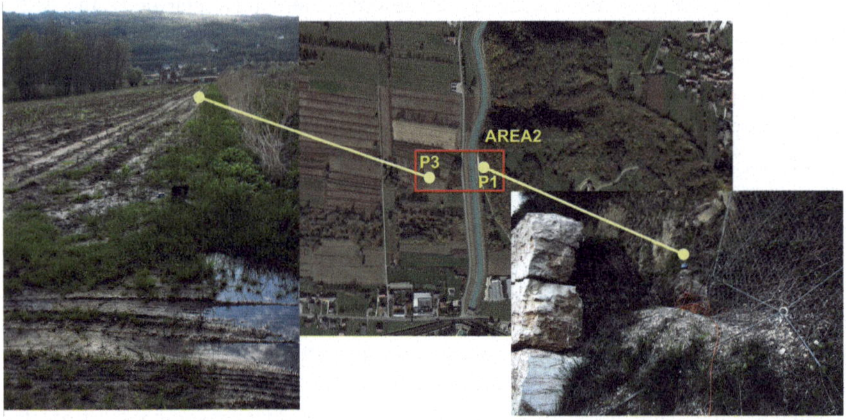

Fig. 3.17 An Alpine valley in NE Italy (Alpago area). The direction of the valley (actually a paleo lake created by a landslide) is approximately NS. Indicated the location of the reference rocky site (P1) and of the investigated point (P3) in the valley. At P1 and P3 we recorded two synchronous 20-min microtremor datasets

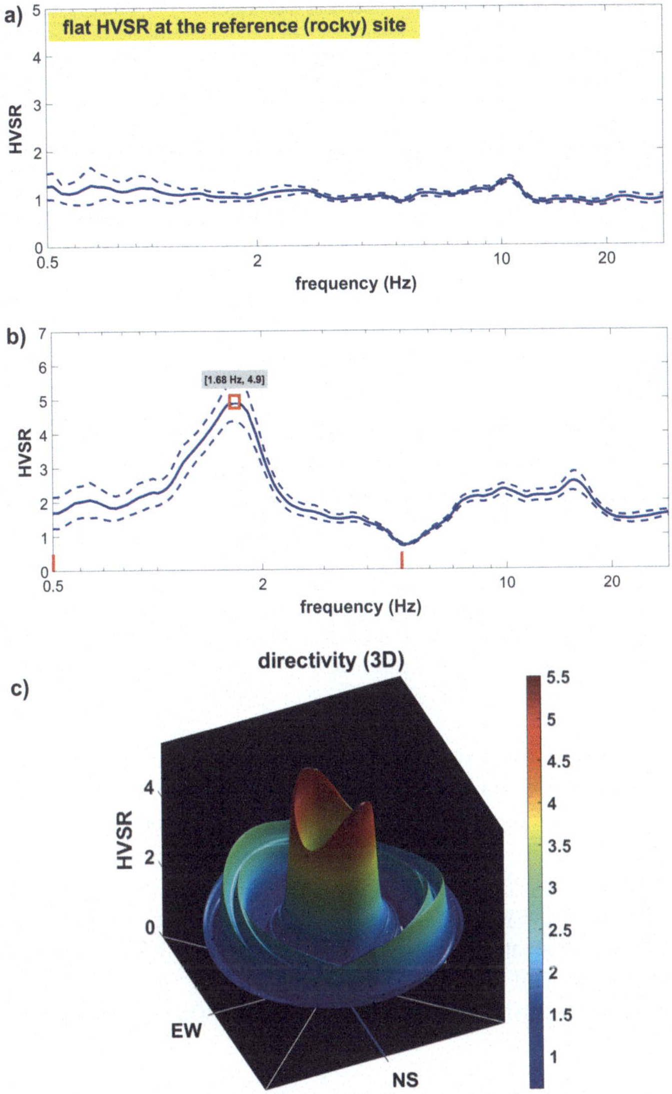

Fig. 3.18 Alpago valley (NE Italy): the HVSR curves (H component computed as quadratic mean) for the reference (rocky) site P1 (**a**) and for the investigated site P3 (**b**) (see location map Fig. 3.17). Also reported the directivity of this latter HVSR (**c**) (the HVSR is larger along a direction approximately NS and smaller perpendicularly to it—compare with the SSRn data presented in Figs. 3.19 and 3.20)

Let us see how this problem applies to the SSRn through a comparative example about the microtremor data recorded in a NE-Italy Alpine valley. In Fig. 3.17 are shown the location of the two points where two 3C geophones recorded a 20-min

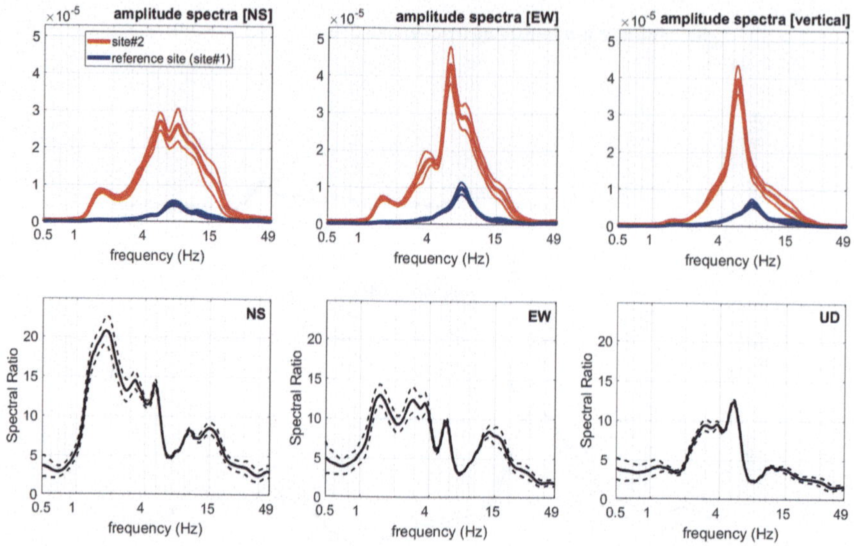

Fig. 3.19 Computation of the SSRn separately for the three components (from left to right: NS, EW and vertical). Upper plots: amplitude spectra at the reference rocky site (blue lines) and in the middle of the valley (site #2, red curves). Lower plots: spectral ratios. Shown the mean values and the standard deviations (window length 20 s, total length of the recording 20 min)

synchronous dataset. Data were then divided into 20 s segments and mean amplitude spectra and spectral ratios computed.

Figure 3.18 presents the HVSR data at the two considered points (the reference rocky site P1 and the site P3 in the middle of the valley) and Fig. 3.19 reports the SSRn separately for the three components.

At the reference (rocky) site P1, the HVSR is (unsurprisingly) nearly-perfectly flat (Fig. 3.18a), while at the investigated site (P3) the HVSR curves shows a well-defined peak around 1.68 Hz (Fig. 3.18b).

The analysis of the SSRn for the three components (Fig. 3.19) shows two clear facts:

(1) compared to the EW component, the spectral ratio of the NS component is larger (directivity of the SSRn);
(2) the amplification of the vertical component is quite significant and this might be one of the reasons why the HVSR fails in the determination of the amplification (see also Mohammadioun 1997; Dimitriu et al. 1999; Diagourtas et al. 2001 as well as the further comparative case study presented in Appendix E).

The SSRn curves computed considering the mean horizontal component and shown in Fig. 3.20 provide the evidence that, as for the HVSR (see Sect. 3.1.4), also for the SSRn the differences between the three approaches for the computation

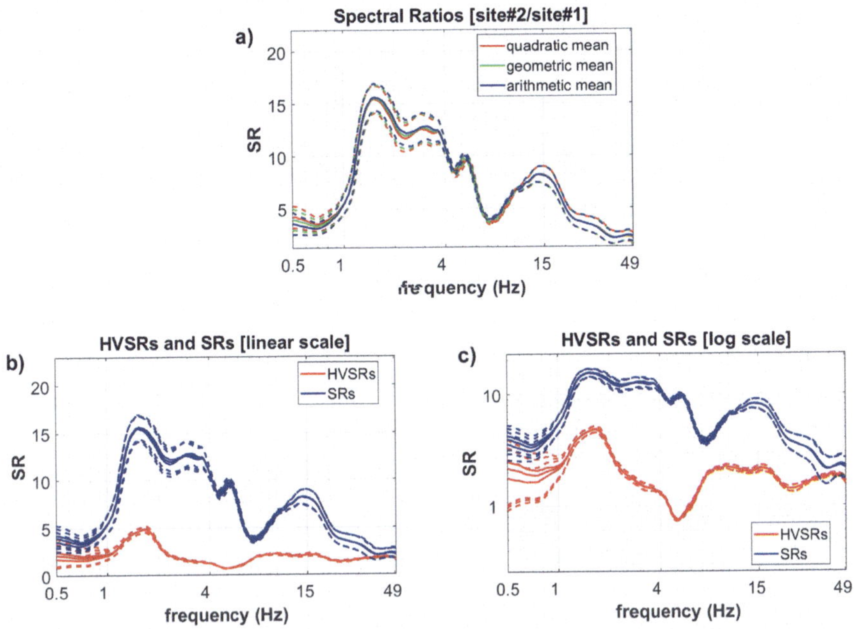

Fig. 3.20 HVSR at site P3 (see Figs. 3.17 and 3.18) and SSRn for the Alpago data (NE Italy): **a** SSRn curves computed according to the three possible ways to define the H component (the differences are negligible); **b** comparison between the HVSR at P3 and the SSRn curves (computed considering the arithmetic, geometric and quadratic mean) with a linear vertical scale; **c** same curves as in the previous graph but with a logarithmic vertical scale

of the mean horizontal motion (arithmetic, geometric and quadratic mean) is insignificant.

To summarize (before moving on):

(1) in general terms, the HVSR does not represent the amplification actually observed during an earthquake;
(2) the SSRn significantly overestimates the SSR but somehow mirrors its general trend (see Perron et al. 2018 and examples in Fig. 3.16).

3.2 ESAC (Extended Spatial AutoCorrelation)

As briefly recalled in the introduction of this chapter, ESAC is the "generalized" version of the SPAC (SPatial AutoCorrelation) methodology. The purpose is the determination of the propagation velocities of the surface waves through the analysis of the microtremors.

This is obtained by deploying a series of geophones according to a bi-dimensional geometry and by analyzing the Bessel's functions.

From the practical point of view, the main difference between the SPAC and ESAC methodologies is that while SPAC is based on the analysis of data recorded according to a circular geometry, ESAC can be performed with any arbitrary geometry (this is possible because the way the Bessel's polynomials are treated is slightly different for the two methodologies).

ESAC is therefore by far more flexible and the acquisition procedures definitely simpler: the deployment of a series of geophones over two perpendicular lines (simple example of ESAC geometry) is in fact definitely simpler with respect to the deployment of the same geophones along a circle (with the geophones regularly spaced).

While here we will just deal with a series of practical aspects, technical details about the computation of the dispersion curves according to these techniques are available for instance in Asten (1978, 2006), Asten and Henstridge (1984), Ohori et al. (2002), Okada (2003, 2006), Asten et al. (2004), Tada et al. (2007), Asten et al. (2014), and Asten and Hayashi (2018).

Although these techniques can be modified so to analyze both Rayleigh (radial and vertical components) and Love wave dispersion (this would require the use of 3C geophones and complex data processing), in most of the common applications, SPAC and ESAC are applied to analyze the vertical component of Rayleigh waves only. The reason is quite simple: while the particle motion along the vertical axis is necessarily due to Rayleigh waves, the motion along a horizontal axis can be attributed to either Rayleigh (radial component) or Love waves. Since the discrimination between these two waves requires the use of 3C geophones and some advanced processing (e.g. Poggi and Fäh 2010), the analysis of the dispersion from passive data is usually limited to the vertical component of Rayleigh waves.

We should also recall that, due to its ambiguity, the ReMi (*Refraction Microtremors*, Louie 2001) approach is nowadays practically abandoned since the linearity of the adopted array (ReMi is a sort of passive MASW) does not allow to properly handle the directivity of the signals which, in a passive experiment, is unknown (see Dal Moro 2014).

Two important notes

Note #1: what is ESAC?

ESAC and SPAC (as well as other passive methodologies) are techniques aimed at determining the dispersive prosperities of the medium and are in use approximatively since the eighties.

ESAC/SPAC does not state anything about the way the retrieved dispersion properties are modelled/inverted: ESAC/SPAC is just a way to determine the dispersive properties through the analysis of passive microtremor data.

Note #2: the *effective* dispersion curve

It is then fundamental to strongly underline that the modelling/inversion of the dispersion curve(s) obtained through the analysis of passive data (ReMi, SPAC, ESAC and MAAM—see next chapter) must be done while considering the mathematics of the *effective* dispersion curve.

In fact, the way the retrieved dispersion curve is modelled is a key fact since the dispersion curve obtained through the analysis of passive data does not have to be modelled according to the fundamental mode (modal approach) of the Rayleigh waves. As a matter of fact, in general, the obtained dispersion curve is not representative of a single mode and it must be modelled according to the mathematics of the *effective curve*, i.e. as the result of the contribution of all the modes that a specific site excites (see Tokimatsu et al. 1992; Ikeda et al. 2012).

If we model/invert a dispersion curve obtained via ESAC (or any other passive methodology) by considering it as the fundamental mode, we commit a big mistake since, even if the retrieved curve is correct, we model it according to the wrong mathematics (see also the case study #12 in Dal Moro 2014).

Figure 3.21 shows an example of ESAC + HV acquisition setting. Since ESAC provides the average dispersive properties over the considered area (in this case a triangle) and lateral variations are possible, it is recommended the acquisition of at least three HVSRs at different positions. By comparing the three HVSR curves we are able to understand whether significant lateral variations occur. In that case, a single HVSR curve would be insufficient and it could be difficult to find a single subsurface model that satisfies both the dispersion curve (which represents the average subsurface conditions over the area) and the HVSR (which instead represents very local conditions—see the HVSR data presented in Figs. 3.22 and 3.23 and the comments reported in the captions).

What is the difference between the ESAC and a standard multi-offset ZVF-MASW? As a matter of fact, the difference is quite small: in both cases we just define the dispersion of the vertical component of Rayleigh waves (in the case of the ESAC, this is accomplished through the computation of the correlations between traces and the optimization of the Bessel's functions—see Figs. 3.24 and 3.25).

This means that an ESAC + ZVF-MASW survey is quite pointless and it cannot be considered as a "joint analysis" since both the observables are fundamentally providing the same information (the dispersive properties of the vertical component of Rayleigh waves). The only difference is that, if the array is sufficiently large (this is a fundamental pre-requisite), the very low frequencies are defined more accurately by the ESAC (while MASW data provide fuzzier information).

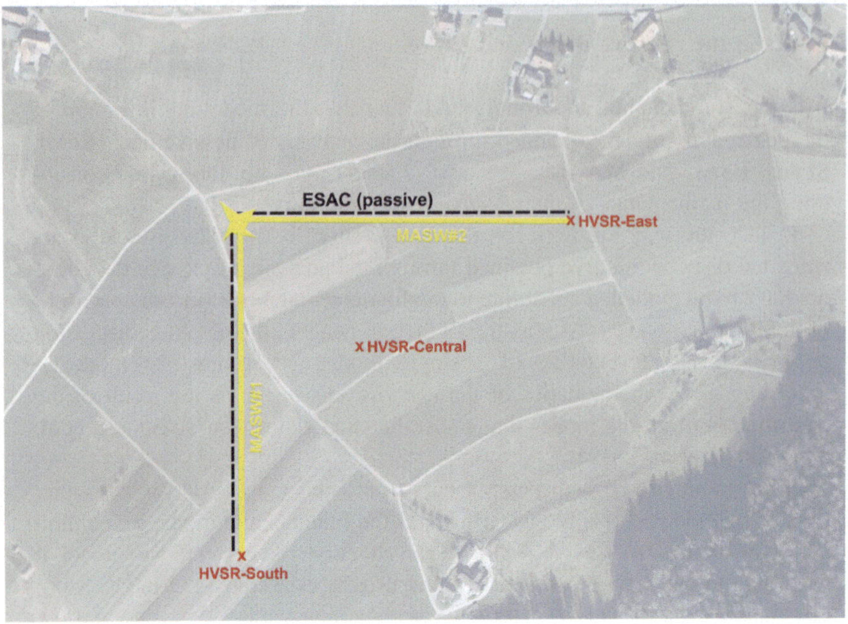

Fig. 3.21 Example of common ESAC array (two arms perpendicular to each other). If we suppose we have placed 12 geophones along one arm and others 12 along the other, it is also possible to acquire two MASW datasets (which will have directions perpendicular to each other). The possible location of three HVSRs (two at the ends of the string and one in the center position) is also shown. From Dal Moro (2014)

But, on the other side, we should also consider that the low frequencies can be easily explored via HVSR.

In case we do not have enough room to deploy our geophones, ESAC is almost useless and a survey based on the joint analysis of the data collected via MASW/HS (see next chapter) + HVSR is usually more than fine (please, notice that in this volume the acronym MASW does not necessarily mean the standard ZVF-MASW accomplished through the analysis of the interpreted modal curves—see Chaps. 1 and 2 as well as the case studies reported in the appendices).

Non-uniqueness

Because of the nature of the *effective* dispersion curve (which is the resultant of all the excited modes), the non-uniqueness is particularly significant. Fig. 3.26a shows five V_S profiles while, on the right (Fig. 3.26b), in the background is shown the velocity spectrum obtained from an ESAC survey and, overlaying, the effective dispersion curves of the five models shown on the left. We can see that the five considered V_S models (quite different in particular below 13 m) have practically the same effective dispersion curve. The consequence is that, although ESAC provides clearer data in comparison to other passive methodologies such as ReMi, it cannot

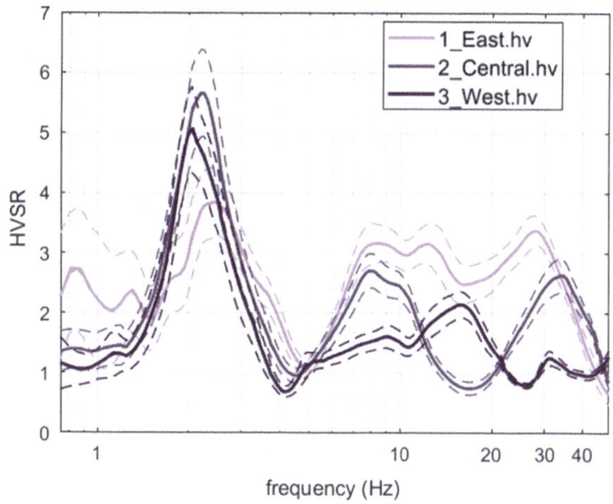

Fig. 3.22 Example of three HVSR curves obtained at three points (along an ESAC array). The considerable differences suggest significant variations in the subsurface conditions at the three points. In the low frequencies, one of the curves (1_East) is very different from the other two and, during the joint analysis with the dispersion curve obtained via ESAC, we might decide to remove it and compute the mean HVSR from the other two curves (see Fig. 3.23)

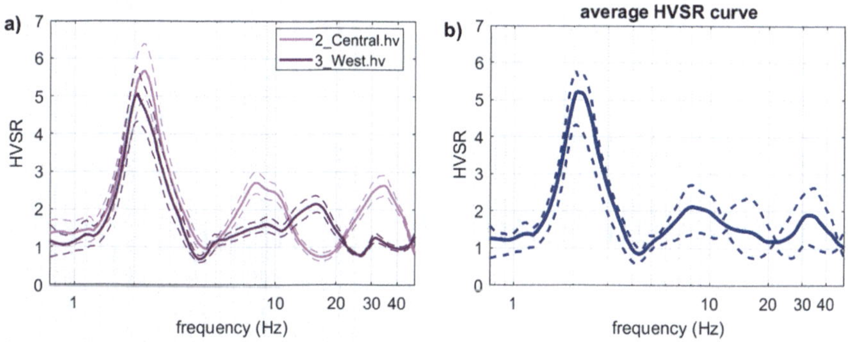

Fig. 3.23 Computation of the mean HVSR curve from two of the curves shown in Fig. 3.22: **a** the two selected curves; **b** the average curve. For frequencies higher than about 6 Hz the two curves significantly differ and, in order to obtain an average subsurface model over the area, we might for instance decide to consider the HVSR only in the 0.8–6 Hz frequency range and perform the joint analysis with the dispersion curve obtained via ESAC (which is necessarily an average curve influenced by the average subsurface conditions over the whole area)

represent *the* solution since, in any case, it provides a single observable (see conceptual scheme in Fig. 1.8 and related text). Also considered the field effort, it is therefore usually more efficient a hybrid active + passive approach represented by the combination of MASW or HS data (see next Chapter) jointly with the HVSR.

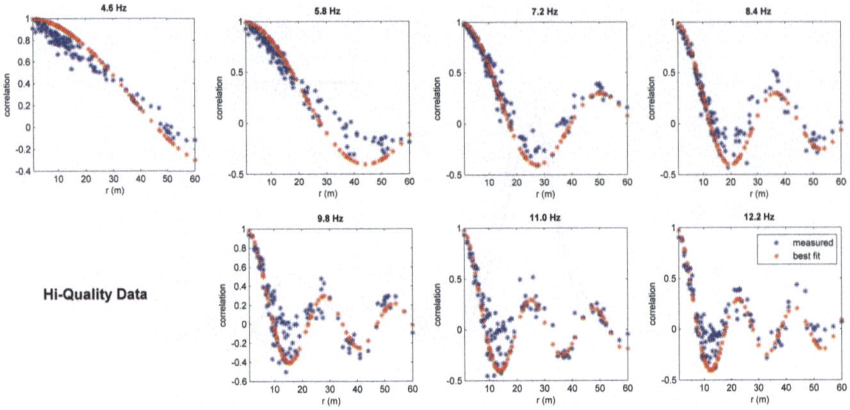

Fig. 3.24 Correlation coefficients as a function of the distance between receivers for seven sample frequencies: the data quality is apparent by the clear "wavy" nature of the data. The propagation velocities is related by the "wavelength" of the data (blue dots), which is modeled (red dots) by means of the Bessel's polynomials so to define the effective phase velocity frequency by frequency

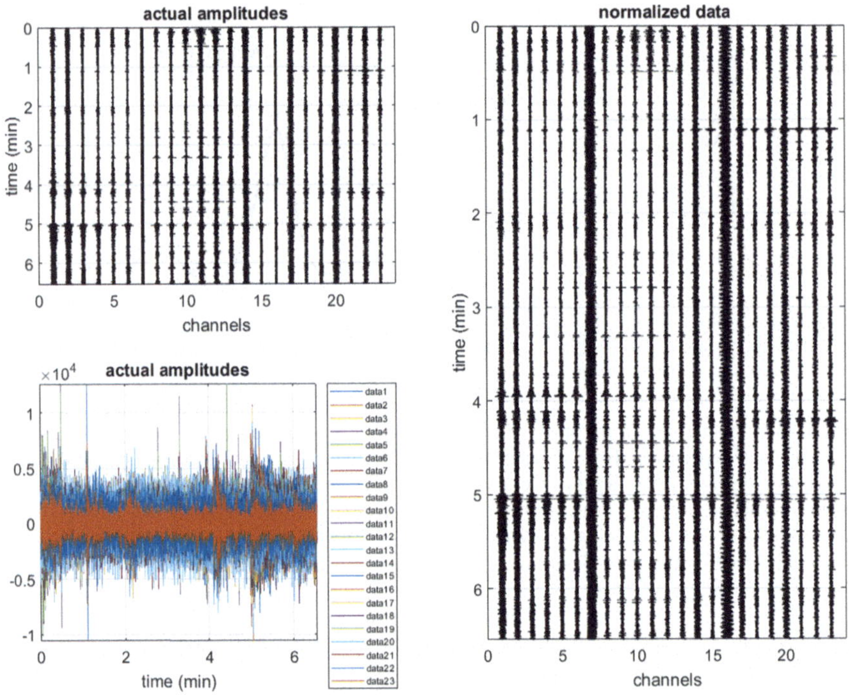

Fig. 3.25 Example of passive data (Z component) according to three different representations: on the left the data with their actual amplitudes, on the right the normalized traces (two dead traces [channels 7 and 16] can be removed so to increase the quality of the ESAC processing)

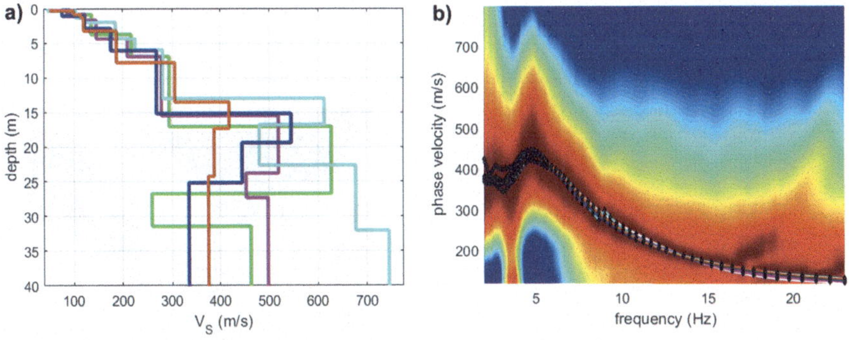

Fig. 3.26 Non-uniqueness of the solution for the *effective* dispersion curve (obtained from any passive methodology): **a** five V_S profiles; **b** effective dispersion curves for the five V_S profiles. As for the standard MASW, the non-uniqueness of the solution is remarkable. While comparing these profiles with the ones presented in Fig. 1.9 (about the standard ZVF-MASW data), it should be considered that for the MASW data the lowest considered frequency was much larger (5 Hz). Since the ESAC data here presented are modelled down to about 2 Hz, we cannot compare the non-uniqueness of the V_S models in exceedingly-strict terms and we should simply consider that non-uniqueness is not about the method used to determine the dispersion properties but is an intrinsic ambiguity of the dispersion curve per se (independently on the way such a curve is retrieved)

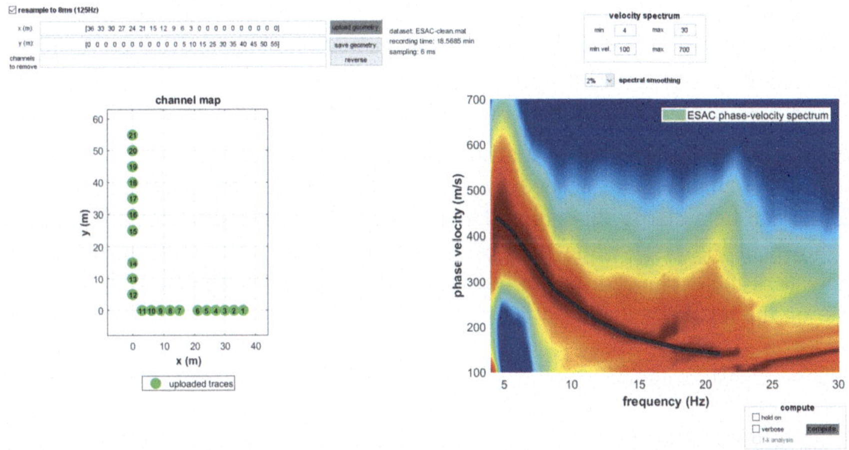

Fig. 3.27 ESAC analysis considering a 22-trace dataset (recording time 18 min). The channel map (on the left) reports the location of the geophones while on the right is shown the retrieved *effective* dispersion curve. Compare with the results obtained while considering a subset of seven traces (Fig. 3.28)

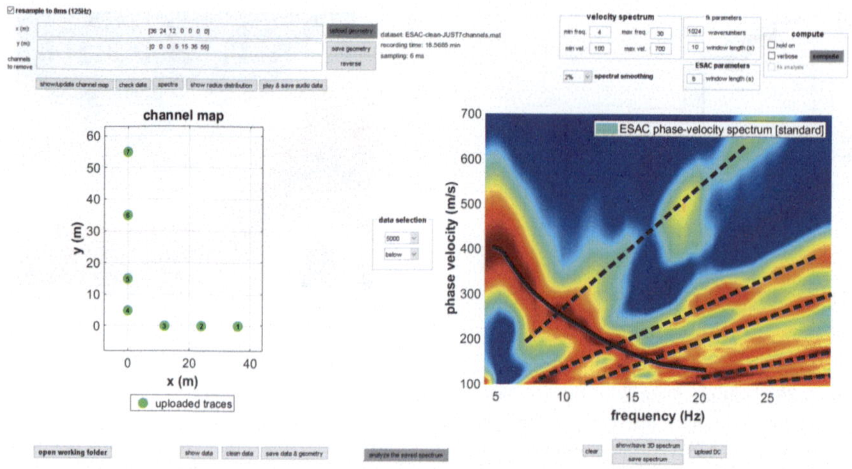

Fig. 3.28 ESAC analysis from the same dataset considered in the previous figure but after having removed 15 traces and reduced the dataset to just 7 traces (shown in the channel map on the left). The spatial aliasing phenomena that inevitably occur (straight dotted lines) do not seriously affect the possibility of identifying the effective dispersion curve which, of course, appears identical to the one obtained while using 22 traces (see black curve in the velocity spectrum on the right and compare it with the curve shown in Fig. 3.27)

On the side, it must be underlined that ESAC is a very robust methodology since it is not significantly infulenced by possible acquisition problems (dead traces and/or noisy environment) and, if the local conditions allow a sufficiently large array, can provide robust information about very low frequencies.

ESAC: number of channels, spatial aliasing and few final notes
How many channels are needed for the ESAC analysis? The analysis of a field dataset can help clarify this point. The analysis of the original dataset (22 traces) is shown in Fig. 3.27 while the next figure (Fig. 3.28) reports the result of the analysis of the decimated dataset composed of just 7 traces (shown in the channel map on the left).

Some considerations:

- the obtained (effective) dispersion curve is the same and, in spite of the spatial aliasing phenomena that inevitably occur, can be easily identified;
- of course, in case we use a limited number of channels, it is important to deploy the geophones so to properly sample different wavelengths in a "continuous" fashion (this means that the distances between adjacent geophones are different —see channel map in Fig. 3.28);
- even if ESAC can be performed while using a limited number of channels, the deployment of a series of non-equally spaced geophones (Fig. 3.28) can be a bit

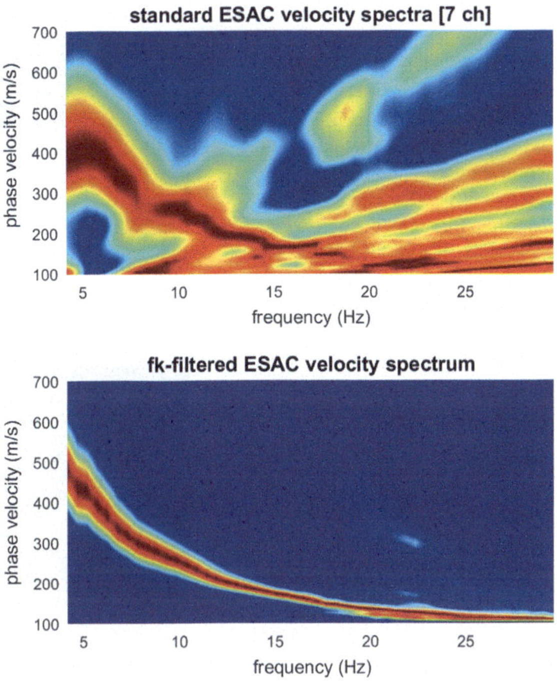

Fig. 3.29 ESAC analysis from the same dataset considered in the previous figure (i.e. by considering just 7 traces) but after the application of a simple *fk* filter that removes the signals with a "meaningless" slope (in this case the signals created by spatial aliasing phenomena). The effective curve is now clearer but is this kind of operation always feasible? See text and next figure

cumbersome (all the positions must be carefully checked) and, in some cases, the use of two simple and classical seismic cables along an L-shaped array (Fig. 3.21) can be easier from the practical point of view.

- while using a limited number of channels, the application of a *fk* filter can improve the obtained velocity spectrum (Fig. 3.29) but caution should be applied since artefacts can araise (see example in Fig. 3.30).

Figures 3.31 and 3.32 report an example of joint analysis of the *effective* dispersion curve from ESAC and the mean HVSR obtained by averaging two HVSR curves from two different points within the ESAC array. Because of shallow lateral inhomogeneities (the explored area was quite large—see channel map in Fig. 3.31a), at high frequencies the two HVSR curves were exceedingly different and we therefore kept the average curve only for frequencies lower than 6 Hz (Fig. 3.31c). Actually, a third HVSR was recorded but because of low data quality (one of the two horizontal geophones did not work properly) was not considered.

It should be understood that this example was chosen not because everything is simple and easy (the ESAC array is very large, the HVSR data are relatively complex and the area is surely not fully homogenous) but because it represents a typical real-world (relatively complex) case.

The results of the joint inversion are shown in Fig. 3.32 and allow to emphasize once more (and in a very concrete way) two points:

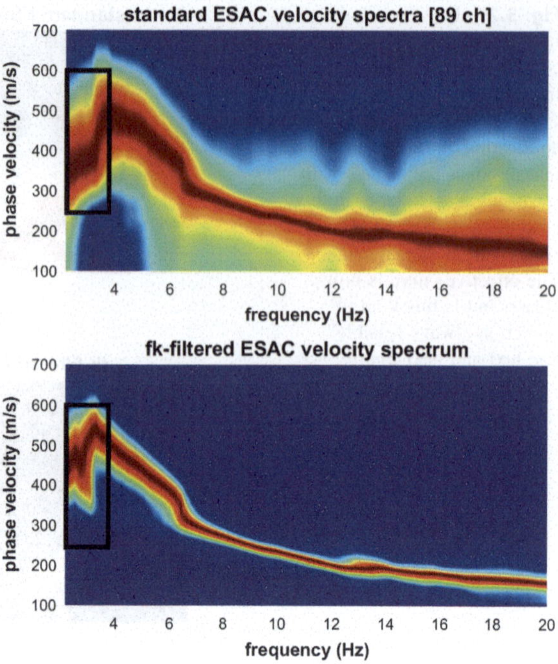

Fig. 3.30 ESAC analysis for a Swiss dataset (courtesy of *roXplore.ch*). The complex trend of the effective dispersion curve between 2 and 5 Hz is not an artefact and it represents the real *effective* dispersion curve: in cases like this one, the application of a *fk* filter risks to alter the real dispersion and create an artefact reproducing an erroneous dispersion curve (compare the curves in the two black rectangles)

(1) The effective dispersion curve retrieved from ESAC must be modeled according to the mathematics of the *effective* dispersion curve (Tokimatsu et al. 1992; Ikeda et al. 2012). Considering the dispersion curve obtained via ESAC/SPAC/ReMi/MAAM (all passive techniques) as representative of the fundamental mode is a huge mistake that can lead to overestimate the V_S values (compare the *effective* and *modal* dispersion curves shown in Fig. 3.32b);
(2) HVSR needs to be modelled also considering the effect of Love waves (see the α factor illustrated in the previous pages and Arai and Tokimatsu 2004).

Of course, the same *effective* (sometimes called *apparent*) dispersion curve can refer to different models that excite different modes (non-uniqueness of the dispersion curve). In some cases it can be attributed just to the fundamental mode, in other cases just to higher modes but, usually, is a complex mixture of different modes. In the present (very common) case it is a mixture of the fundamental and first higher mode (Fig. 3.32b): at frequencies higher than about 9 Hz the *effective* curve practically corresponds to the fundamental mode while for lower frequencies is the result of the combination of the fundamental and first higher modes (between 7 and 8 Hz almost coincides with the first higher mode).

As widely discussed in this book, only the joint analysis of several observables (in this case the *effective* dispersion curve and the HVSR) enables us to solve these ambiguities.

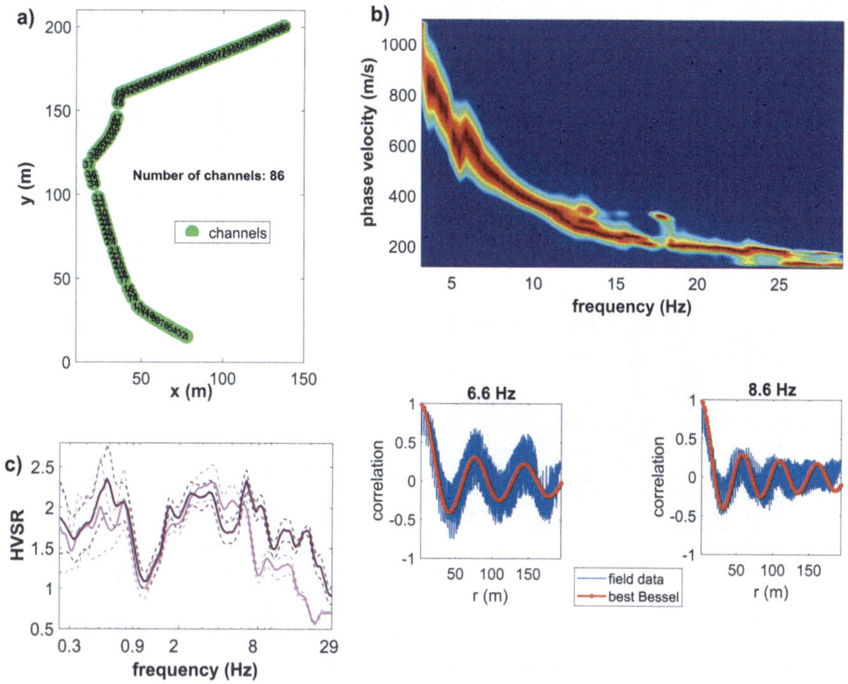

Fig. 3.31 Example of ESAC + HVSR processing for a large array: **a** channel map; **b** obtained phase-velocity spectrum (also shown the experimental correlation values and the best Bessel functions for two sample frequencies—6.6 and 8.6 Hz); **c** two HVSR curves from two points within the ESAC array. The result of the joint analysis is shown in Fig. 3.32

Let us then clearly underline few general points about the ESAC.

ESAC with linear array?

The ESAC technique is a sequence of mathematical operations useful for analyzing the dispersion from passive data.

It can therefore be applied to any kind of array but a key point should be considered. A linear array is not able to provide information on the directivity and, consequently, performing the ESAC on the data recorded by a linear array is fundamentally equivalent to a ReMi (even if the mathematics is completely different).

For an *Extended Spatial AutoCorrelation* analysis it is therefore important to deal with bi-dimensional arrays. The simplest geometry is clearly an L-shaped array (e.g. Figures 3.21 and 3.27) but, because of local logistical reasons, in some cases (Fig. 3.31a) more complex arrays are necessary.

Investigated depth

As usual, it cannot be easily defined. On one side it depends on many (site and data dependent) facts, on the other side, we must always consider that the uncertainty of

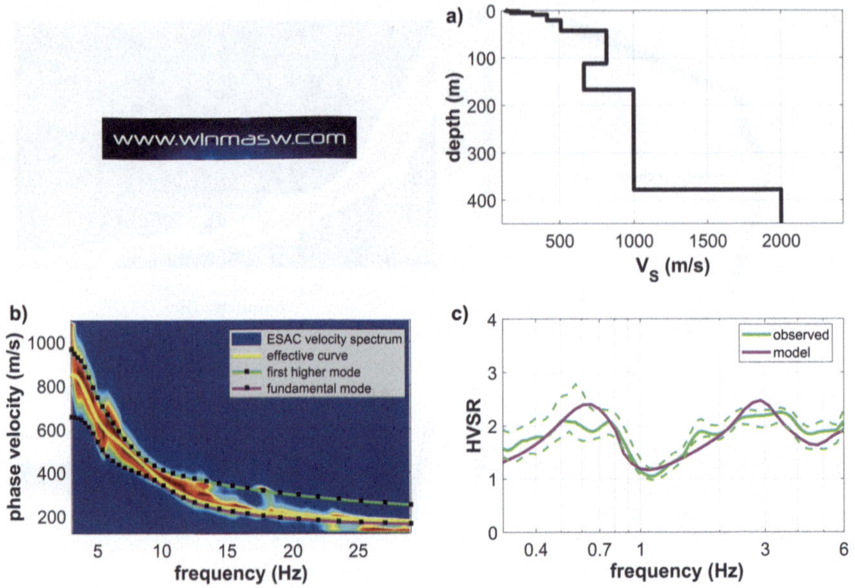

Fig. 3.32 Result of the joint analysis of the *effective* dispersion curve obtained via ESAC and average HVSR curve (data presented in Fig. 3.31). The *effective* dispersion curve (thick yellow line) was modelled according to Tokimatsu et al. (1992) and the HVSR according to Arai and Tokimatsu (2004). Also shown the *modal* dispersion curves for the first two modes (green and magenta dotted lines) of the identified model. See text for further comments

the retrieved V_S profile increases with the depth. When we invoke the λ/3 (or, in case we are extremely optimistic, λ/2) approximation, we do not say anything about the actual V_S uncertainty at that depth.

Some simple rule of thumb is anyway useful or necessary. In general terms we can say that, through the ESAC, we are collecting information useful to characterize the subsurface conditions down to a depth approximately equal to half the maximum distance between the channels. But it is the heavy non-uniqueness of the solution that defines the actual limitations and we should always consider that, in order to really constrain the subsurface model, we always need more than one observable (Fig. 1.8 and related text). In fact, having information about something does not mean that those information are sufficient to fully clarify the model (see for instance the joint ESAC + HVSR example presented in Figs. 3.31 and 3.32).

Robustness of the ESAC methodology for the determination of the dispersion curve

ESAC is certainly an extremely robust technique. The presence of dead traces or noise does not significantly affect the accuracy of the obtained dispersion curve and, therefore, it can be used with good confidence in complex and noisy zones (e.g. urban areas).

Furthermore, during the data processing, the most important parameter that needs to be set is just the window length and its value does not affect too much the solution (for array of 70–200 m, standard values usually range between 5 and 20 s).

References

AA.VV.—SESAME (2005) Guidelines for the implementation of the H/V spectral ratio technique on ambient vibrations measurements, processing and interpretation, 62 pp. Open file: ftp://ftp.geo.uib.no/pub/seismo/SOFTWARE/SESAME/USER-GUIDELINES/SESAME-HV-User-Guidelines.pdf. Accessed May 2020

Albarello D, Lunedei E (2013) Combining horizontal ambient vibration components for H/V spectral ratio estimates. Geophys J Int 194:936–951

Arai H, Tokimatsu K (2004) S-wave velocity profiling by inversion of microtremor H/V spectrum. Bull Seismol Soc Am 94:53–63

Arai H, Tokimatsu K (2005) S-wave velocity profiling by joint inversion of microtremor dispersion curve and horizontal-to-vertical (H/V) spectrum. Bull Seismol Soc Am 95:1766–1778

Asten MW (1978) Geological control on the three-component spectra of Rayleigh-wave microseisms. Bull Seismol Soc Am 68:1623–1636

Asten MW (2006) On bias and noise in passive seismic data from finite circular array data processed using SPAC methods. Geophysics 71:153–162

Asten MW, Hayashi K (2018) Application of the spatial auto-correlation method for shear-wave velocity studies using ambient noise. Surv Geophys 39:633–659

Asten MW, Henstridge JD (1984) Array estimators and the use of microseisms for reconnaissance of sedimentary basins. Geophysics 49:1828–1837

Asten MW, Dhu T, Lam N (2004) Optimised array design for microtremor array studies applied to site classification; comparison of results with SCPT logs. In: Proceedings of the 13th world conference on earthquake engineering, Vancouver. Paper no 2903

Asten MW, Aysegul A, Ezgi EE, Nurten SF, Beliz U (2014) Site characterisation in north-western Turkey based on SPAC and HVSR analysis of microtremor noise. Explor Geophys 45:74–85

Borcherdt RD (1970) Effects of local geology on ground motion near San Francisco Bay. Bull Seismol Soc Am 60:29–61

Bowden DC, Tsai VC (2017) Earthquake ground motion amplification for surface waves. Geophys Res Let 44(1):121–127

Dal Moro G (2010) Insights on surface-wave dispersion curves and HVSR: joint analysis via Pareto optimality. J Appl Geophys 72:29–140

Dal Moro G (2015) Joint analysis of Rayleigh-wave dispersion and HVSR of Lunar seismic data from the Apollo 14 and 16 sites. Icarus 254:338–349

Dal Moro G (2020) On the identification of industrial components in the Horizontal-to-Vertical Spectral Ratio (HVSR) from microtremors. Pure Appl Geophys. https://doi.org/10.1007/s00024-020-02424-0

Dal Moro G, Coviello V, Del Carlo G (2014) Shear-wave velocity reconstruction via unconventional joint analysis of seismic data: a case study in the light of some theoretical aspects. In: Engineering geology for society and territory, vol 5. Springer, pp 1177–1182

Diagourtas D, Tzanis A, Makropoulos K (2001) Comparative study of microtremors analysis methods. Pure Appl Geophys 158:2463–2479

Dimitriu P, Kalogeras I, Theodulidis N (1999) Evidence of nonlinear site response in horizontal-to-vertical spectral ratio from near-field earthquakes. Soil Dyn Earthq Eng 18:423–435

Fasan M, Magrin A, Amadio C, Romanelli F, Vaccari F, Panza GF (2016) A seismological and engineering perspective on the 2016 Central Italy earthquakes. Int J Earthq Impact Eng 1:395–420

Ikeda T, Matsuoka T, Tsuji T, Hayashi K (2012) Multimode inversion with amplitude response of surface waves in the spatial autocorrelation method. Geophys J Int 190:541–552
Lanchet C, Bard PY (1994) Numerical and theoretical investigations on the possibilities and limitations of Nakamura's technique. J Phys Earth 42:377–397
Louie JN (2001) Faster, better: shear-wave velocity to 100 meters depth from refraction microtremor arrays. Bull Seismol Soc Am 91:347–364
Mark N, Sutton GH (1975) Lunar shear velocity structure at Apollo sites 12, 14, and 15. J Geophys Res 80:4932–4938
Mittal H, Kamal KA, Singh SK (2013) Estimation of site effects in Delhi using standard spectral ratio. Soil Dyn Earthq Eng 50:53–61
Mohammadioun B (1997) Nonlinear response of soils to horizontal and vertical bedrock ground motion. J Earthq Eng 1:93–119
Nakamura Y (1989) A method for dynamic characteristics estimation of subsurface using microtremor on the ground surface. Q Rep Railw Tech Res Inst (RTRI) 30:25–33
Nakamura Y (1996) Realtime information systems for seismic hazard mitigation. Q Rep Railw Tech Res Inst (RTRI) 37:112–127
Nakamura Y (2000) Clear identification of fundamental idea of Nakamura's technique and its applications. In: Proceedings of the XII world conference on earthquake engineering, New Zealand. Paper no 2656
Nakamura Y (2008) On the H/V spectrum. In: The 14th world conference on earthquake engineering, 12–17 Oct 2008, Beijing, China
Nakamura Y (2019) What is the Nakamura method? Seismol Res Lett 90:1437–1443
Ohori M, Nobata A, Wakamatsu K (2002) A comparison of ESAC and FK methods of estimating phase velocity using arbitrarily shaped microtremor arrays. Bull Seismol Soc Am 92:2323–2332
Okada H (2003) The microseismic survey method. In: Geophysical monograph, series no 12. Society of Exploration Geophysicists of Japan, Tulsa
Okada H (2006) Theory of efficient array observations of microtremors with special reference to the SPAC method. Explor Geophys 37:73–85
Olsen KB (2000) Site amplification in the Los Angeles basin from three-dimensional modeling of ground motion. Bull Seismol Soc Am 90:S77–S94
Panza GF, La Mura C, Peresan A, Romanelli F, Vaccari F (2012) Seismic hazard scenarios as preventive tools for a disaster resilient society. Adv Geophys 53:93–165
Panza GF, Kossobovok V, Peresan A, Nekrasova A (2014) Why are the standard probabilistic methods of estimating seismic hazard and risks too often wrong. In: Wyss M (ed) Earthquake hazard, risk, and disasters. Elsevier, London, UK, pp 309–357. https://doi.org/10.1016/b978-0-12-394848-9.00012-2. ISBN: 978-0-12-394848-9
Perron V, Gélis C, Froment B, Hollender F, Bard P-Y, Cultrera G, Cushing EC (2018) Can broad-band earthquake site responses be predicted by the ambient noise spectral ratio? Insight from observations at two sedimentary basins. Geophys J Int 215:1442–1454. https://doi.org/10.1093/gji/ggy355
Poggi V, Fäh D (2010) Estimating Rayleigh wave particle motion from three-component array analysis of ambient vibrations. Geophys J Int 180:251–267
Schnabel PB, Lysmer J, Seed HB (1972) SHAKE—a computer program for earthquake analysis of horizontally layered sites. Report No. EERC 72-12, Earthquake Engineering Research Center, University of California, Berkeley
Tada T, Cho I, Shinozaki Y (2007) Beyond the SPAC method: exploiting the wealth of circular-array methods for microtremor exploration. Bull Seismol Soc Am 97:2080–2095
Tokimatsu K, Tamura S, Kojima H (1992) Effects of multiple modes on Rayleigh wave dispersion characteristics. J Geotech Eng ASCE 118:1529–1543
Zuccolo E, Vaccari F, Peresan A, Panza GF (2011) Neo-deterministic and probabilistic seismic hazard assessments: a comparison over the Italian territory. Pure Appl Geophys 168:69–83

4 New Trends: HS, MAAM and Beyond

A circle, a line: they look good, they are abstract, they are common knowledge.
They belong to everyone and equally to the past, the present, the future.
Richard Long

Less is More
Ludwig Mies van der Rohe

G. Dal Moro, *Efficient Joint Analysis of Surface Waves and Introduction to Vibration Analysis: Beyond the Clichés*,
https://doi.org/10.1007/978-3-030-46303-8_4

Abstract

This chapter focuses on two state-of-the-art techniques for the analysis of surface wave dispersion: HS (*Holistic analysis of Surface waves*) and MAAM (*Miniature Array Analysis of Microtremors*). HS is based on the analysis of active data recorded by a single 3-component (3C) geophone deployed at a fixed distance (offset) from the source. MAAM is a passive methodology aimed at the determination of the Rayleigh-wave dispersion curve from the data collected by very few geophones (4 or 6) deployed symmetrically around a circle with a radius of just few (2–3) meters. From a practical point of view, MAAM and HS have in common the fact that the field equipment is extremely simple and the acquisition procedures extremely straightforward. HS and MAAM data can be easily integrated with the HVSR and provide a well-constrained (i.e. robust) subsurface model. Since the HS methodology also allows to analyze the actual Rayleigh-wave Particle Motion (RPM), a series of facts and case studies are presented in order to make the reader familiar with this observable (to include in the data joint inversion both in case of single- and multi-offset data).

4.1 Introducing HS and MAAM

Two introductory notes about the two techniques presented in this chapter:

1. the first methodology (HS—*Holistic analysis of Surface waves*) is based on the analysis of active data and represents the improvement of a very classic seismological technique (MFA [Multiple Filter Analysis]/FTAN [Frequency Time ANalysis]—e.g. Dziewonski et al. 1969; Levshin et al. 1972; Bhattacharya 1983). It requires a single 3-component (3C) geophone and, consequently, greatly simplifies the field procedures;
2. the second (MAAM—*Miniature Array Analysis of Microtremors*) is a passive methodology aimed at the determination of the Rayleigh-wave (vertical-component) dispersion curve. It requires very few geophones (4 or 6) and, for investigating the frequency range typical of near surface applications (about 3–30 Hz), the dimension of the array is of just few (2–4) meters.

From a practical point of view, it is then clear that MAAM and HS have in common the fact that the field equipment is extremely simple and the acquisition procedures particularly simple and straightforward.

From the aesthetic point of view (aesthetics is the discipline that analyzes the relationship between *form* and *content*), HS is a line that connects two points (the source and the receiver), while the MAAM is a circle.

Although this chapter is dedicated to HS and MAAM, at the end of it we will also see how to expand the HS approach (which requires only one 3C geophone and is therefore about the analysis of multi-component and single-offset data—see later) for the acquisition and analysis of multi-offset and multi-component data.

Let us start with the HS case (i.e. multi-channel, multi-component and single-offset case).

The HS acronym stands for *Holistic analysis of Surface waves.*

The term *holistic* clearly refers to a joint approach in which different components and "objects" (or, to be more rigorous, *observables*) are considered in such a way that we are able to *fully* consider the propagation of surface waves (dispersion is only one of the aspects involved and we can consider both Rayleigh and Love waves and several aspects that can be used to describe the way waves propagate).

HS is a recent improvement of a very classical seismological method (the analysis of group velocities via MFA/FTAN) and was described in a series of papers providing the overall theoretical framework as well as a series of case studies (Dal Moro et al. 2015a, b, 2016, 2017b, 2019; Dal Moro and Puzzilli 2017; Dal Moro 2019b).

On the other side, MAAM stands for *Miniature Array Analysis of Microtremors* (Cho et al. 2006a, b, 2013; Tada et al. 2007) and, although its mathematics is quite different, from the data-acquisition point of view, is a sort of mini-SPAC (SPatial Autocorrelation), with the great advantage of being able to investigate a relatively wide frequency range in spite of the limited dimensions of the array. Roughly speaking, in case we wish to define the Rayleigh-wave dispersive properties between 3 and 40 Hz, SPAC (which is practically a mere circular ESAC) would require an array with a radius of several tens of meters (about 60–80 m, depending on the site characteristics). On the opposite, in order to investigate the same frequency range through the MAAM approach, a triangle with a radius of about 3 m is usually enough.

What do we need to perform an HS acquisition?

The acquisition of HS data is quite simple. Fundamentally, we need a source that generates the waves that are then recorded by a single 3C geophone deployed at a certain distance (*offset*) from the source (Fig. 4.1). The "zero time" (or *origin time* in the seismological lingo) is the time of the shot. This means that, like for any standard refraction/reflection survey, we need a trigger geophone that defines the start of the recording. Record length and sampling rate follow the same simple rules and criteria as for any surface-wave survey (general information and discussed examples are provided in the following pages and in Dal Moro 2014).

In case we mean to analyze the propagation of Rayleigh waves, we produce them with a very classical vertical impact source (e.g. a sledgehammer over a plate, as for any classical survey aimed at analyzing Rayleigh or P waves). The Z and R traces will then contain the data of the vertical and radial components of Rayleigh waves (that can be processed so to obtain a wealthy amount of information—see later on). In case we wish to analyze Love waves, we produce them through a shear source (see Chap. 1) and consider the data recorded by the T (transversal) component.

Choosing a meaningful file name is a key fact too often little understood and considered. The files shown in Fig. 4.2 were named so to give all the necessary information (there is no need to explain *what* is *what*): *VF_off50* tells everything we need to know (the source-receiver distance is 50 m, the source is a Vertical Force

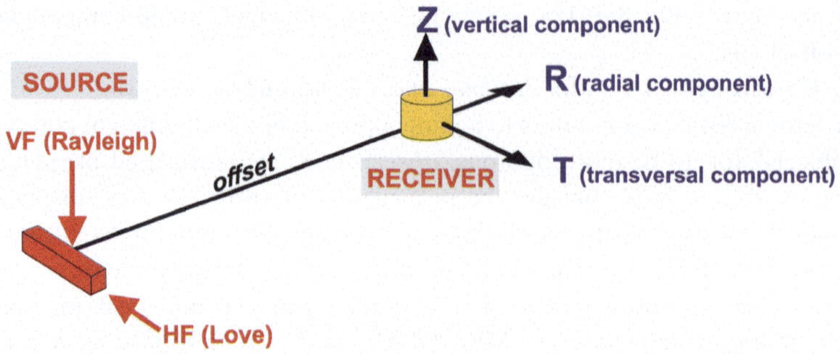

Fig. 4.1 HS acquisition: the scheme highlights the names of the components and the source (VF and/or HF) used to produce the Rayleigh waves associated to the Z and R traces, and/or the Love waves (the T component). Since the HS method is active, it requires the stack of several shots (see Fig. 4.2). Of course, in this case, the receiver is a 3-component geophone

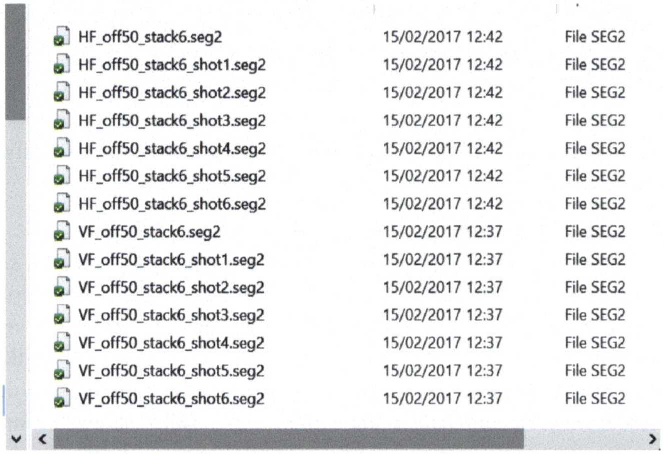

Fig. 4.2 Folder with the HS data named in order to give a physical meaning to the file names. The vertical stack is six (both the single shots and the stacked one are saved). Recorded both Rayleigh (VF) and Love (HF) waves. The distance between the source and the receiver (*offset*) is 50 m. All this information is unambiguously clear already from the file name, so there is no need for verbose accompanying notes: the file names say everything (see also guidelines in Appendix A)

and we must therefore consider only the data recorded by the Z and R components —i.e. the Rayleigh waves); on the other side, *HF_off50* is about Love waves and we should then consider just the T trace.

Further practical details and suggestions on how to effectively handling HS data are reported in Appendix A.

Multi-channel, multi- and single-offset, multi-component: a bit of clarity

The correct and precise understanding of terms and expressions is necessary and allows an enormous leap in the quality of our data acquisitions and analyses.

It is well known that, in the terminology associated with the old "MASW" approach, the M stands for *multi-channel*.

Is it possible to associate the term *multi-channel* to the HS methodology? Since HS deals with the active data recorded by a single 3-C geophone, we can surely define the HS as a multi-channel methodology since we deal with three channels/traces.

So what differentiates a single-component MASW and the HS approach?

While the classical single-component MASW provides a multi-channel dataset that refers to different offsets but to a single component (multi-channel, multi-offset, single-component), the HS acquisition provides a multi-channel, multi-component and single-offset dataset.

Consequence: with a classical single-component MASW we deal with a single observable (usually the phase velocity spectrum of the Z component) while through the HS data we can define (and jointly analyze) up to five observables (this chapter illustrates how to do that).

This means that in the conceptual scheme summarized in Fig. 1.8 we deal with just one polygon (*object/observable*) in the case of classical MASW while with up to five polygons (*objects/observables*) in case of HS data.

What is the main point that characterizes the HS approach?

A key point is that the HS methodology is based on the analysis of group velocities of the three (Z, R and T) components (Fig. 4.1). In fact, HS represents an improvement (or evolution) of the classic MFA (Multiple Filter Analysis—Dziewonski et al. 1969) or FTAN (Frequency-Time ANalaysis—Levshin et al. 1972; Ritzwoller and Levshin 2002; Natale et al. 2004) techniques which represent a very classical way to analyze seismological data (e.g. Ritzwoller and Levshin 1998; Ritzwoller et al. 2002; Zhou et al. 2004; Nguyen et al. 2009; Fang et al. 2010).

Compared to these very classical methodologies, the HS approach benefits from the FVS analysis of the group-velocity spectra and two more observables are considered (see next sections).

Incidentally, the whole book also attempts to build a bridge between seismologists and applied geophysicists. Group velocities (which are often ignored/unknown by the "applied geophysicists") have in fact a very long history in seismological/crustal studies and the fact that they are not used in near-surface applications is quite bizarre indeed.

What are the group velocities and why should we use them?
While the mathematical definition can be found in several textbooks (group velocities are a sort of derivative of the phase velocities), in extremely simple and operative terms, the group velocity is a kind of ratio between *space* and *time*. We must in fact define, frequency by frequency, how much time it takes for the signal produced by our hammering to reach the geophone deployed at a certain distance (the *offset*).

Phase velocities (determined from multi-offset data) are the slope of the signal while the group velocities are (just to be extremely practical) a measure of how much it takes a certain frequency to travel from the source to the receiver.

This means that a so-called pre-trigger time is irrelevant for the computation of the phase velocities (the slope is a relative measurement) but not allowed in case we intend to define the group velocities (which require the measure of the "absolute" time—in other terms the first sample must correspond to the origin time, i.e. the hammering).

It is therefore important to underline some basic and crucial aspects that are often not properly and fully considered:

1. phase velocities (determined for example by MASW, ReMi, ESAC/SPAC and MAAM—all methods that require arrays with several geophones) are just one way to look at the surface wave dispersion;
2. as a matter of fact we can (also) work with the group velocities (as seismologists are used to do—e.g. Fang et al. 2010);
3. in order to compute the group velocities, a single trace (i.e. geophone) is sufficient (so our 3C geophone allows to determine the group-velocity spectra for three different components);
4. especially when dealing with small offsets, group-velocity spectra are often clearer compared to phase-velocity spectra (compare the phase and group velocity spectra presented in Figs. 4.3 and 4.4, respectively).

In the introductory chapter, we showed that, for a standard MASW survey, the number of channels does not significantly affect the quality of the phase velocity spectra.

On the other side, is much important to get a profound understanding of the scheme reported in Fig. 1.8. What are those merging polygons? A polygon represents an *observable*, i.e. "something" that can be analyzed in order to get information about the subsurface conditions (in our case the V_S values). An large number of observables necessarily reduces the ambiguities of the solution.

Since these facts are often not sufficiently clear, let us clarify this point a bit further. When we consider the phase-velocity spectrum obtained from a standard MASW survey (e.g. 24 vertical geophones) and the HVSR curve, we deal with two polygons (i.e. observables to jointly analyze).

In case we consider the HVSR and 12 vertical geophones (instead of 24) we still deal with the same two polygons/observables (increasing the number of geophones does not help in increasing the robustness of the retrieved V_S model).

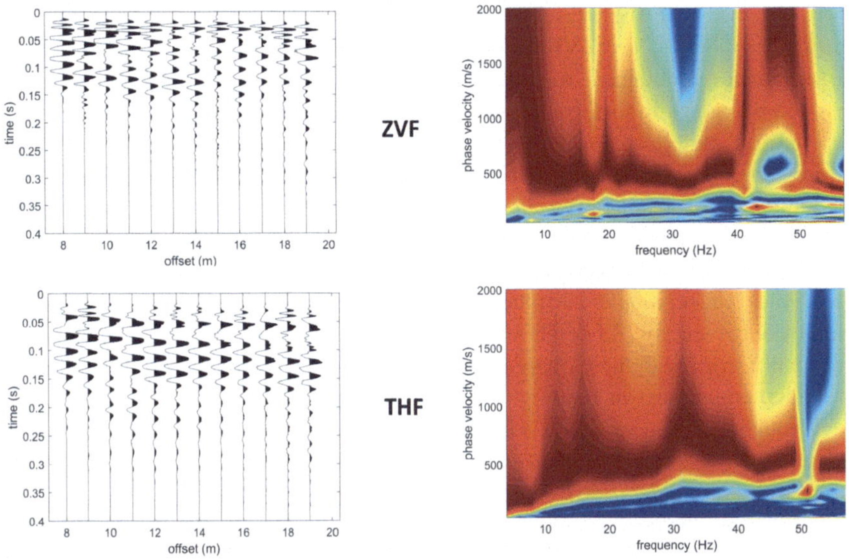

Fig. 4.3 Standard phase velocity spectra (MASW) computed for the ZVF and THF components in case of limited offsets (ranging from 8 to 19 m). The interpretation of the obtained phase velocity spectra in terms of modal dispersion curves is practically impossible. This is a very common issue especially while working with data with limited offsets. On the other side (Fig. 4.4), once we compute the *group* velocity spectra for the very last traces (offset 19 m), the obtained spectra are apparently "simpler" and can be proficiently inverted

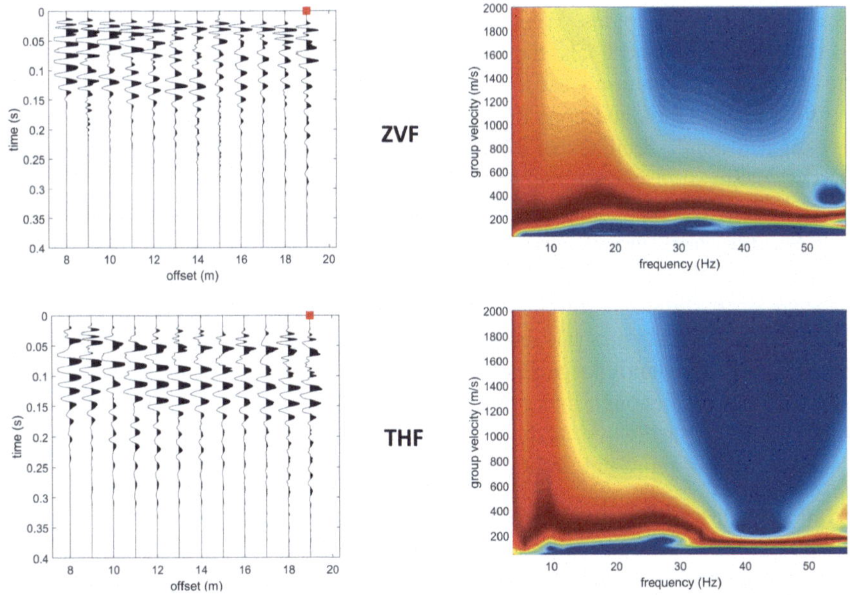

Fig. 4.4 ZVF and THF components: traces (on the left) and group-velocity spectra (on the right) computed considering just the last traces (offset 19 m). Compared to the phase-velocity spectra reported in Fig. 4.3, it is clear that group-velocity spectra are significantly clearer and can be soundly analyzed. Further comments in the text

If, together with the phase-velocity spectrum from the standard MASW (with vertical geophones, i.e. ZVF component) and the HVSR, we also consider the dispersion curve obtained from the ESAC, fundamentally we are still playing with just two polygons. In fact, both ESAC and ZVF-MASW provide information about the vertical component of Rayleigh waves: ESAC is potentially able to retrieve information about lower frequencies (but only if the array is sufficiently large) but the ambiguity of the effective curve retrieved from ESAC is somehow larger when compared with the dispersion obtained from active data (see Chap. 2).

In case, instead of using a set of vertical geophones, we use 12 horizontal geophones and compute the phase velocity spectra for Love waves and for the radial component of Rayleigh waves (see Sect. 1.2 and Appendix A), we can eventually work with three observables: the HVSR and the dispersive properties of the radial component of Rayleigh waves and Love waves.

Through the HS approach, we can define up to five observables: the group velocity spectra of the three components (Z, R and T) and the RPM and RVSR curves (i.e. the *Rayleigh-wave Particle Motion* and the *Radial-to-Vertical Spectral Ratio* curves—see next sections). In the conceptual scheme reported in Fig. 1.8 we have thus five polygons which, by adding the HVSR curve (from the passive data recorded by the same 3C geophone), would become six.

We might summarize these observations by saying that *more* does not necessarily mean *better*. Multi-component analysis based on the smart explotation of the data gathered by a single 3C geophone can in fact provide much more information so that one of the best known aphorisms of Mies van der Rohe (one of the fathers of modernist architecture) seems to apply perfectly: *less is more*.

While planning a seismic survey, it is important to choose carefully the most appropriate methodologies. We should be ready to understand whether, for those specific conditions and goals, the ESAC + HVSR approach is better or worse than the HS + HVSR solution (just to mention two possible approaches).

What are the pros and cons of these two possible solutions with respect to the investigated depth, accuracy of the retrieved model and logistical problems during the data acquisition?

As usual, universal answers do not exist since everything is necessarily *site specific*. For very noisy environments (e.g. industrial areas), ESAC is for instance often (but not necessarily) "better" because active data can suffer from the presence of the large microtremors induced by the industrial activities.

The group velocity analysis requires high-quality data and this means that a good source and a sufficiently-large stack is important. A 5-kg sledgehammer is usually unable to produce enough energy in the low-frequency range (which is extremely important in surface-wave analysis) and the 10-kg version is recommended (the 8-kg version can represent an honorable compromise).

What do we need for the MAAM?

Being a passive methodology, we just need four (in case we intend to follow a triangular geometry—see Fig. 4.5) or six (in case we want to deal with a pentagonal geometry) vertical geophones with excellent general characteristics and

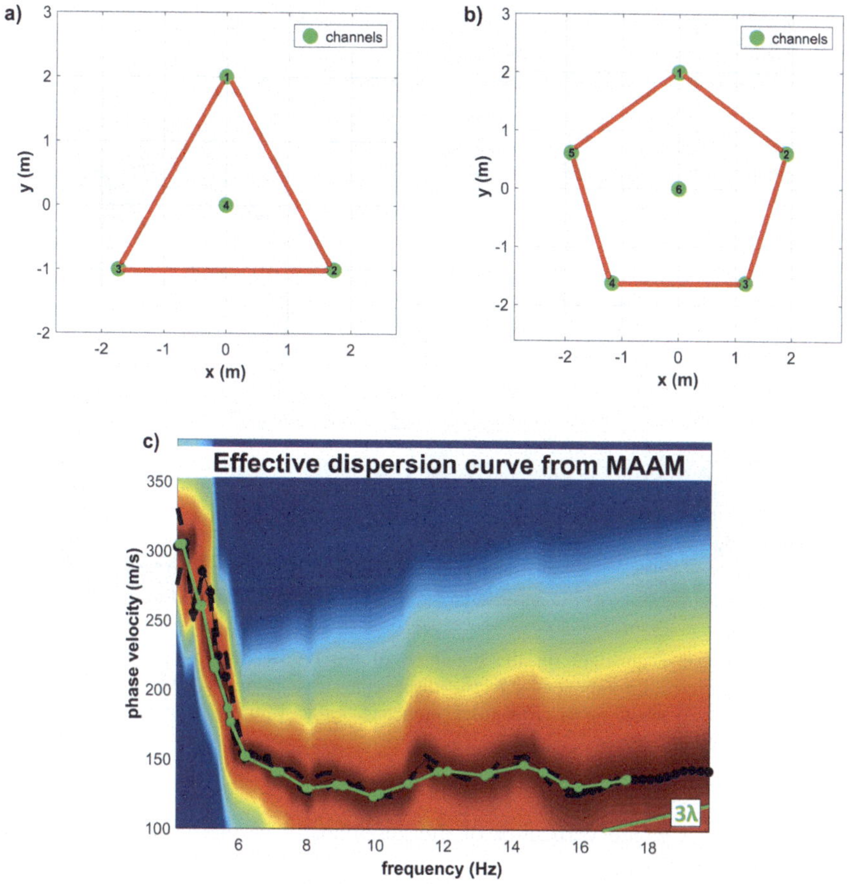

Fig. 4.5 MAAM acquisition geometries (triangle or pentagon) for the determination of the Rayleigh-wave effective dispersion curve of the vertical component. The influence of the noise is compensated through the procedure described in Cho et al. (2006a, b, 2013). Compared to ESAC/SPAC, the array dimension (i.e. the radius) is much smaller but the data processing definitely trickier

identical response curves. In addition, of course, we need a seismograph capable of recording data with an excellent signal-to-noise ratio.

Needless to say that, in case we have a suitable *acquisition system*, we can jointly record all the data necessary for the MAAM and the HVSR (see example in Fig. 4.6).

This means that this kind of surveys can be carried out by means of an extremely simple equipment (easily transportable even in remote and complex areas where, due to logistical reasons, it would be extremely complicated to apply more complex multi-channel techniques). Since the processing is strongly influenced by the data quality, it must be underlined that all the elements (geophones, seismic cables and A/D unit) must be of high quality and the field operations must be accomplished

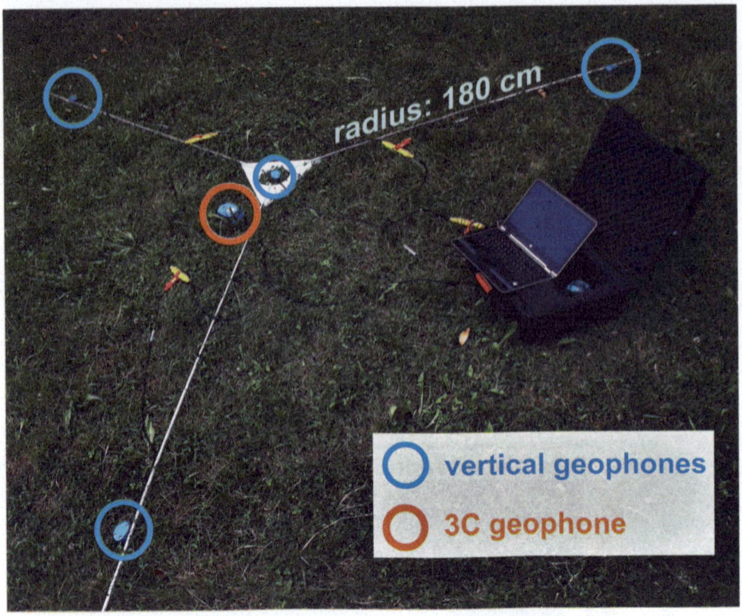

Fig. 4.6 Joint data acquisition for the definition of the HVSR (the 3C geophone) and the Rayleigh-wave (vertical component) *effective* dispersion curve from MAAM (the four vertical geophones deployed according to a triangular geometry—radius 1.80 m). Note that the equipment is reduced to the bone

very carefully (e.g. the verticality of the geophones and their coupling must be meticulously verified).

4.2 HS and MAAM: How They Work from an Early Case Study

The case study presented in this section is one of the first works where HS was used for a large-scale project aimed at characterizing a wide perilagoon area in NE Italy.

Since this work dates back to 2012 (and was presented in Dal Moro et al. 2015b), the RPM curve was not used. The RPM curve (that defines the actual prograde or retrograde Rayleigh-wave motion, frequency by frequency) was in fact defined in a work published in 2017 (Dal Moro et al. 2017a) and will be jointly inverted with the other observables in the next case studies.

Joint inversion is performed through a MOEA (Multi-Objective Evolutionary Algorithm) approach based on the Pareto criterion (Van Veldhuizen and Lamont 2000; Pardalos et al. 2008; Dal Moro 2014) and considering, depending on the goals, different possible multi- or single-offset multi-component data (e.g. Dal Moro

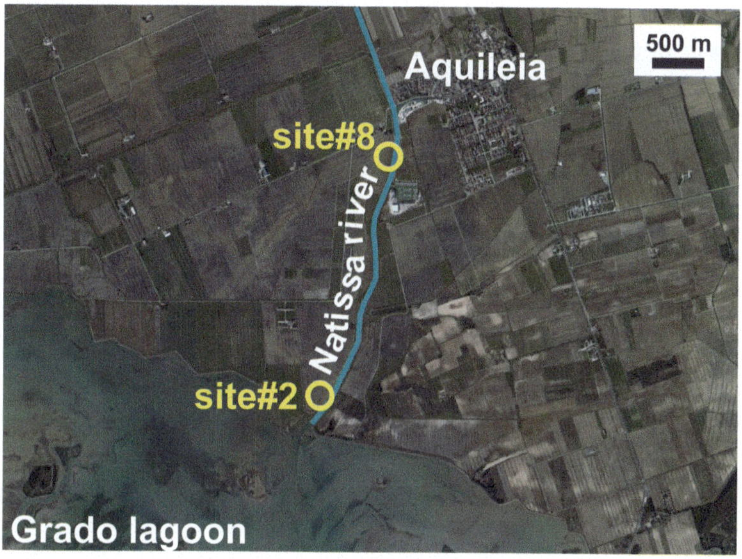

Fig. 4.7 The Grado-Marano perilagoon area (NE Italy): highlighted the two sites considered in this early case study

et al. 2015c, 2016, 2017a, b, 2019 and the case studies presented in this chapter as well in the Appendices).

The work was part of a geotechnical study aimed at the evaluation of the stability of a river bank along the Natissa river (Grado lagoon, NE Italy—Fig. 4.7). The main goal was to define the local stratigraphy and, in particular, the depth of the sands that lie below the superficial silty sequence.

For the HS active data the source was a standard 8 kg sledgehammer.

The aim of the work was to test some unconventional approaches (HS and MAAM) that, compared to others, require a lighter equipment and, consequently, simple and straightforward field operations (because of the wide area to explore, this was particularly important).

In the following, we present the data and analyses for just two sites (Fig. 4.7) and consider different observables so to get gradually familiar with the ideas (and practice) possible by means of the HS approach.

In fact, although this approach allows the definition and joint inversion of up to six observables (in addition to the five observables from the active HS data we can also include the HVSR), depending on the goals and site characteristics, we can decide to define and analyze just some of them (since it is not always necessary to deal with all of them).

Data and analyses

Site #2 was investigated by considering two possible approaches:

(a) joint analysis of the ZVF group velocity spectrum and HVSR curve;
(b) analysis of the ZVF group velocity spectrum only.

For the exploration of site #8 we considered two different approaches:

(a) MAAM + HVSR (i.e., a purely-passive approach);
(b) active HS only (joint analysis of the Z and R group-velocity spectra together with the RVSR curve—i.e. the *Radial-to-Vertical Spectral Ratio* that somehow defines the ellipticity of the Rayleigh waves).

It is quite important to underline once again that, in spite of the very simple equipment (one 3C geophone and four vertical geophones), the collected data can be processed and (jointly) inverted according to several possible procedures since the different observables can be combined in various different ways, depending on the specific needs and data characteristics.

Site #2—first approach: joint inversion of the Z-component group-velocity spectrum + HVSR

The Rayleigh-wave group velocity spectra were computed via MFA while considering the (active) data obtained with an 8-kg sledgehammer (offset 40 m—see seismic traces in Fig. 4.8b) while the H/V spectral ratio was computed by considering a 15-min passive dataset recorded by the same 3C geophone used to record the active HS data.

For the active data (aimed at analyzing the Rayleigh-wave dispersion), the *offset* was fixed considering that the goal of the work was the characterization of the first 10–20 m and keeping in mind that, as a rule of thumb, the investigated depth can be estimated in about half or two thirds of the *offset*. A rigorous method for estimating the actual investigated depth is described in Dal Moro et al. (2019) and briefly summarized later or (see Sect. 4.5).

Clearly, when the active data are analyzed jointly with the H/V spectral ratio, the investigated depth increases considerably and allows the reconstruction of the Vs profile down to several tens of meters (depending on the lowest considerd frequency of the HVSR curve).

The results of the joint inversion of the ZVF group-velocity spectrum (processed according to the *Full Velocity Spectrum* approach) and HVSR curve are summarized in Fig. 4.8c and d. The background colors represent the field data velocity spectrum while the overlaying black contour lines relate to the synthetic velocity spectrum for the identified model (shown in Fig. 4.9). The overall agreement is apparently extremely good both with respect to the HVSR (Fig. 4.8d) and the group velocities (Fig. 4.8c).

Also shown the P-wave refraction (green rectangle in Fig. 4.8b) due to the water-table that the CPTU (*piezocone penetration test*) data indicates at a depth of about 2 m and which appears extremely clear along the vertical (Z) component but with a smaller amplitude along the radial (R) component.

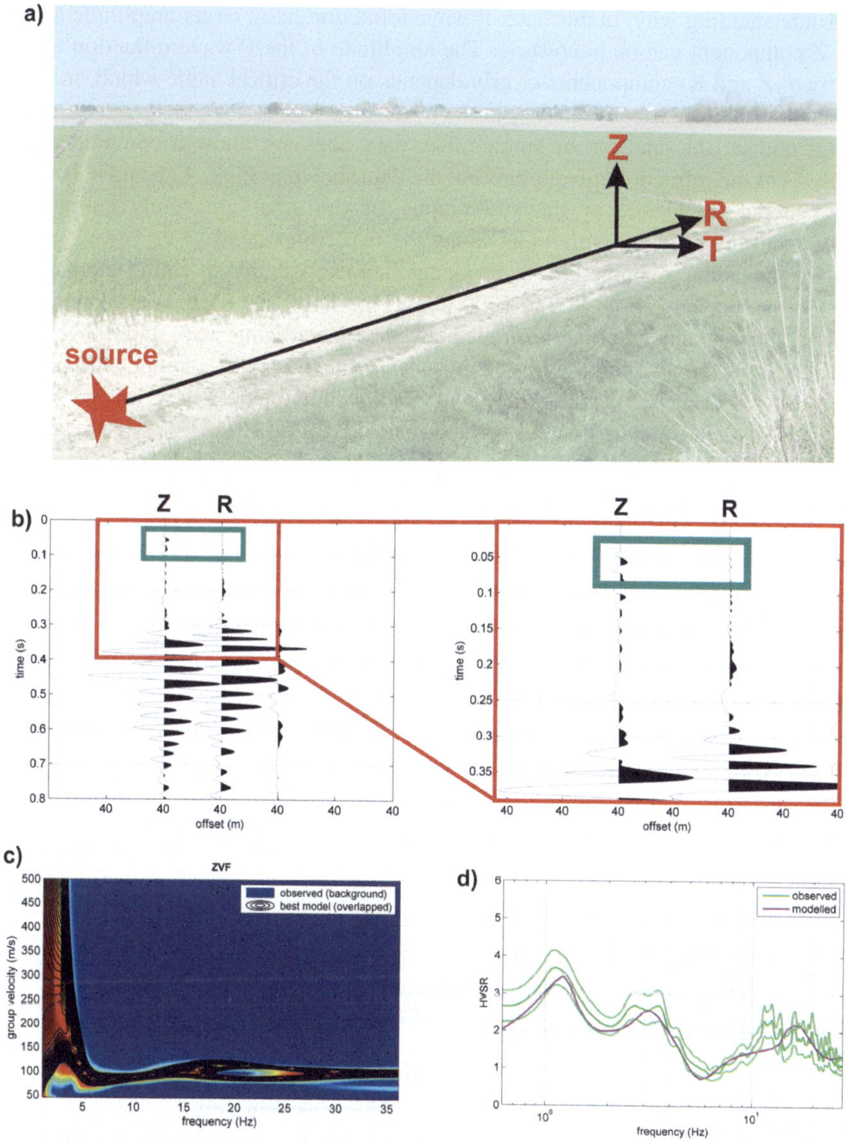

Fig. 4.8 Site #2—joint inversion of the ZVF group-velocity spectrum (FVS approach) and the HVSR curve: **a** HS acquisition (one source, one 3C geophone); **b** Z and R seismic traces (the rectangle highlights the P-wave refraction due to the groundwater at a depth of about 2 m—more details in Dal Moro et al. 2015b); **c** group-velocity spectra for the field data (background colors) and for the identified model (overlaying black contour lines); **d** HVSR curves (in green the experimental curve, in magenta the synthetic one). The identified model is shown in Fig. 4.9. From Dal Moro et al. (2015b)

Understanding why, in this case, P-wave refraction has a larger amplitude along the Z component can be interesting. The amplitude of the P-wave refraction along the two (Z and R) components clearly depends on the critical angle which, in turn, depends on the ratio between V_{P1} and V_{P2} (Snell's law).

Of course, the analysis of single-offset data does not allow a comprehensive analysis of the refraction travel times but the data shown in Figs. 4.8b and 4.10 (and the general knowledge of the stratigraphy of the area) are surely enough to understand the different amplitudes along the two components.

If we consider a simple two-layer model characterized by an abrupt increase of the P-wave velocity (from 300 to 1700 m/s) (Fig. 4.10b and d), the consequence is simple: the critical angle (i.e. the angle between the emerging wave and the vertical axis) is quite small (about 10°). This means that the radial component of the emerging ray (which is the amplitude of the radial component of the refracted wave) is much smaller compared to the vertical one.

We must remember that Rayleigh-wave analysis clearly indicates very low shear-wave velocities typical of unconsolidated sediments (see model in Fig. 4.9). If we consider the different amplitude of the P wave along the Z and R axes, it is clear that the refraction is due to the shallow water table that saturates soft unconsolidated sediments. In fact, the lower the *critical angle* (which depends on the ratio between V_{P1} and V_{P2}), the smaller the amplitude along the radial component.

Site #2—second (purely active) approach: FVS inversion of the group-velocity spectrum of the ZVF component only

Since the analysis previously presented visibly indicates that the sandy layers are not too deep (about 10 m), was also performed the simplest possible analysis of the active data: the ZVF group-velocity spectrum only. The goal was to compare the results so to assess the potentiality of the active HS data for this kind of conditions and surveys.

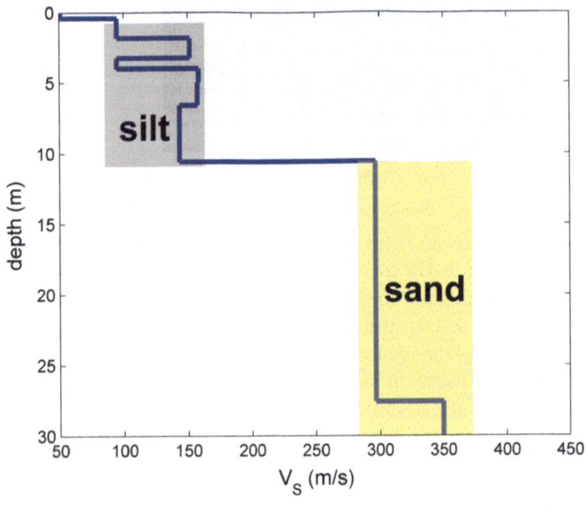

Fig. 4.9 Site #2: the V_S model interpreted in the light of the available geotechnical data and identified through the joint analysis of the group velocities (Z component) and the HVSR curve (see also Fig. 4.8)

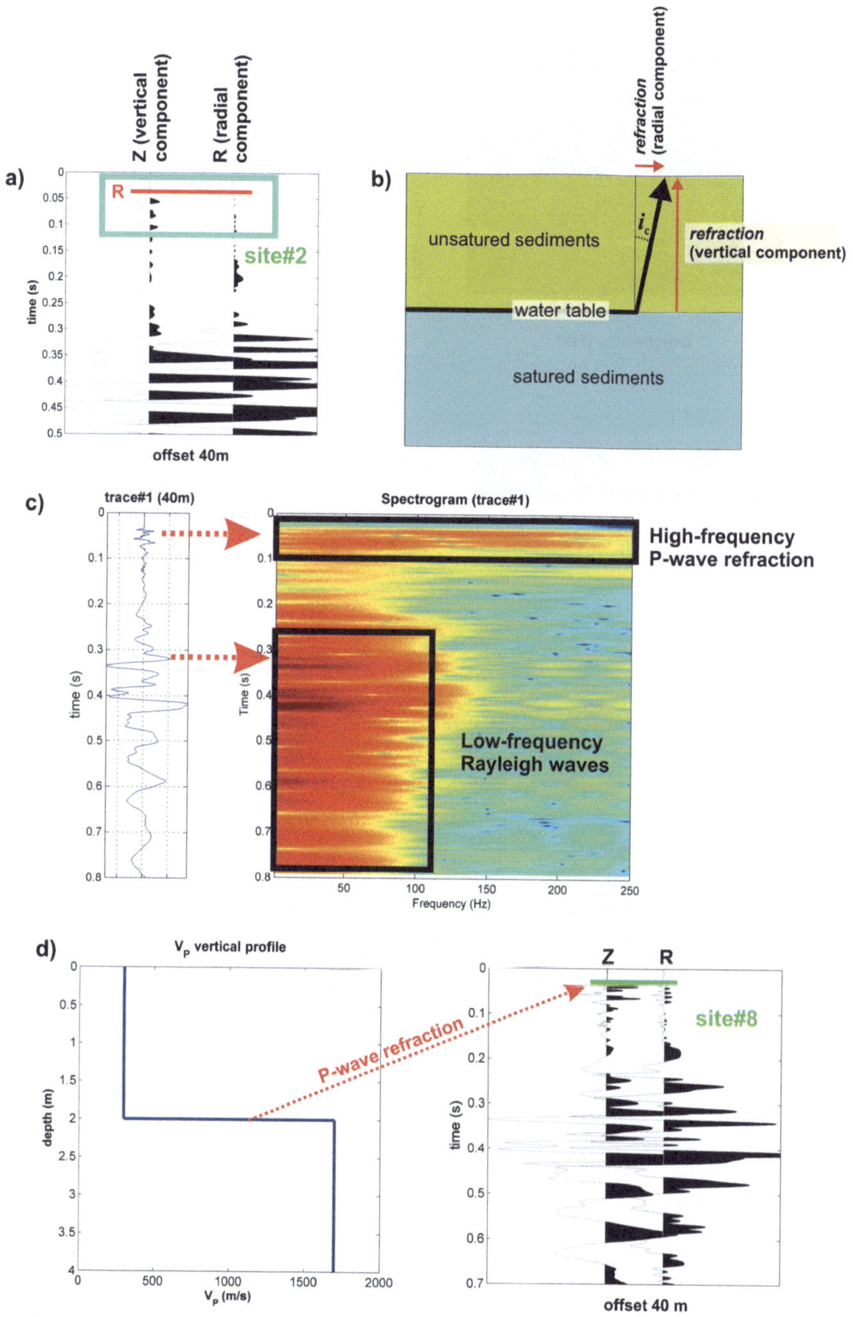

Fig. 4.10 Site #2—P-wave refraction along the vertical (Z) and radial (R) components: **a** close up of the field traces; **b** subsurface model aimed at highlighting the small critical angle responsible for the different amplitudes along the two (Z and R) components (very small for the radial component, very large for the vertical one); **c** spectrogram of the vertical component showing the P-wave refracted wave (high frequencies) and the Rayleigh waves (slower and dominated by lower frequencies); **d** V_P model and vertical and radial traces with highlighted the computed arrival time. Further comments in the text and in Dal Moro et al. (2015b)

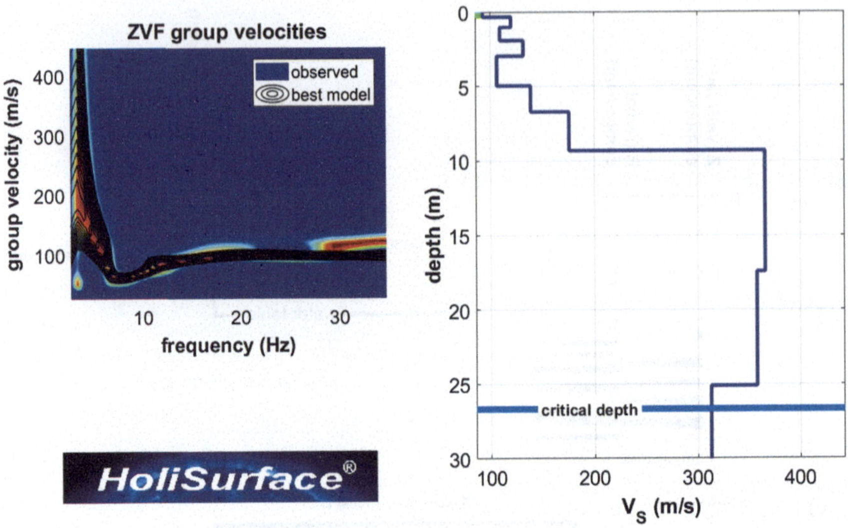

Fig. 4.11 Site #2—V_S profile obtained through the FVS inversion of the group-velocity spectrum of the ZVF component only. Compare with the results presented in Figs. 4.8 and 4.9, when HVSR was also included

The FVS inversion of the ZVF group-velocity spectrum (the same previously inverted together with the HVSR curve) was then performed with the results presented in Fig. 4.11. Thanks to the limited depth of the silt-sand contact, the obtained subsurface model appears quite similar to the one previously obtained by considering also the HVSR.

We should remember that this is an early case study and the current state-of-the-art of the HS methodology significantly improved (see next case studies and Dal Moro 2019b; Dal Moro et al. 2019).

The CPTUs (*piezocone penetration test*) carried out down to a depth of 10 m showed values that match quite well with the obtained V_S profiles: in the upper layers, *qc* (cone resistance—average value around 1 MPa), *fs* (*sleeve friction*—average value around 0.02 MPa) and F_R (*friction ratio*, average value of about 2.5%) are typical of poor silty materials, in agreement with the low V_S values identified through the analysis of the seismic data.

The considerable increase of the V_S values at about 9–10 m can be easily interpreted as a stratigraphic transition from the shallow soft (silty) sediments (saturated at a depth of about 2 m—see P-wave refraction analysis) to a sandy sequence.

Site #8

About 2.5 km North from site #2 (Fig. 4.7), in an area where the CPTU indicates that the thickness of the silty cover is approximately the same, a series of seismic data were acquired so to perform HS, MAAM and HVSR analyses.

Site #8—first approach: purely active seismics (HS approach)

For the previous site, the active data were analyzed considering the group velocities only, but the active 3-component data that can be obtained by means of a 3C geophone can go well beyond this.

As presented for instance in Dal Moro (2019b), by using a single 3C geophone it is possible to define up to five *observables* (to jointly analyze so to obtain a robust subsurface model free from major ambiguities): three group-velocity spectra (related to the three component—Z and R about Rayleigh wave and T about Love waves), the *Radial-to-Vertical Spectral Ratio* (RVSR) and the recently-introduced RPM (*Rayleigh-wave Particle Motion*) curve (Dal Moro et al. 2017b, 2019).

Since for this old pilot-work we only considered a vertical-impact source (therefore only Rayleigh waves), in the following we present the result of the joint inversion of the Z and R group-velocity spectra (FVS approach) together with the RVSR curve, i.e. the amplitude of the radial component with respect to the vertical one (at the time of this study the RPM curve was not yet defined/available).

Figure 4.12 reports all the five possible observables that can be defined through the processing a 3-component active data (details in the figure caption and in the

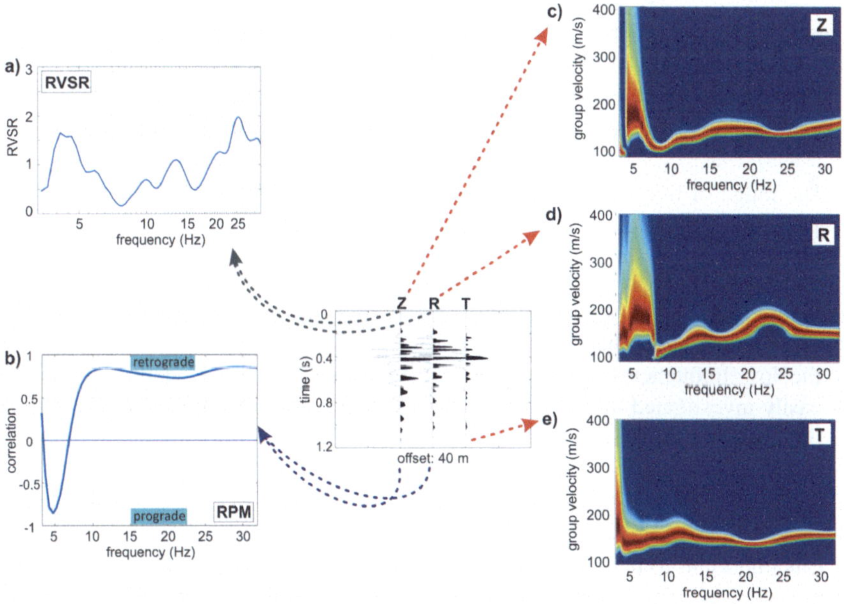

Fig. 4.12 Scheme with all the elements that can be considered during the holistic analysis of the surface waves recorded by means of a single 3-component geophone (HS methodology). The Z and R traces can be used to compute up to four *observables*: the group velocity spectrum of the Z component; the group velocity spectrum of the R component; the *radial-to-vertical spectral ratio* (RVSR); the RPM (Rayleigh-wave Particle Motion) curve. In addition to this, we can clearly also compute the Love-wave group-velocity spectrum (from the THF component). Eventually, the acquisition of microtremor data allows to define the HVSR curve, a sixth observable that can be considered in our joint inversion procedure. From Dal Moro (2019a)

mentioned literature). To these five observables, by further exploting the same 3C geophone, we can add the HVSR curve obtained from the analysis of microtremor (i.e. passive) data.

Figure 4.13 presents the results of the joint inversion of the three *objects/observables* here considered (the Z and R group-velocity spectra and the the RVSR curve obtained from an active HS dataset—offset 40 m): the general agreement between the experimental and synthetic data is clear both for the velocity spectra and the RVSR curve.

Incidentally, the HS data of this case study are available for downloading (the link is provided in the preface of the book). In case you decide to analyze them, we would highly recommend to perform the Z + R + RPM joint analysis (here not shown for the sake of brevity) and compare the results with the solution shown in Fig. 4.13.

Site 8—second approach: purely-passive seismics (MAAM + HVSR)
A purely-passive acquisition was carried out with the aim of determining the V_S vertical profile via joint analysis of the HVSR curve and the Rayleigh-wave effective dispersion curve defined via MAAM.

In this case, the radius of the triangular array for the MAAM was fixed to 1.5 m (recording time 14 min—see data and grometry in Fig. 4.14).

The 16-bit seismograph used for this old work required the use of high-sensitivity 2 Hz geophones but the quality of the up-to-date 24-bit seismographs is definitely higher. Nowadays it is then possible to obtain the same results by using high-quality (high-sensitivity) 4.5 Hz geophones that, usually, are useful to define the dispersion curve down to about 2 Hz (depending on the quality of the seismograph and on the characteristics of the background microtremor field).

The range of frequencies that can be analyzed via MAAM is a function of the adopted radius, which should be fixed considering that:

1. the greater the radius, the lower the minimum frequency that can be identified (of course, due to spatial aliasing, higher frequencies are lost and we should always find a reasonable compromise);
2. the low frequencies (which allow the investigation of the deeper layers) are easily investigated by the HVSR (a series of practical recommendations will be provided in the following).

The result of the joint inversion of the HVSR and the effective dispersion curve obtained via MAAM is presented in Fig. 4.15.

It is important to underline that the dispersion curve retrieved via MAAM must be modelled as *effective* dispersion curve, i.e. as the result of the contribution of all the exited modes (Tokimatsu et al. 1992; Ikeda et al. 2012) (we put some emphasis on this point also in Chap. 3 while speaking about the dispersion curves obtained via ESAC/SPAC/ReMi).

The modeling of the HVSR curve was performed by taking into account the contribution of both Rayleigh and Love waves (about the role of Love waves on the HVSR curve see Chap. 3).

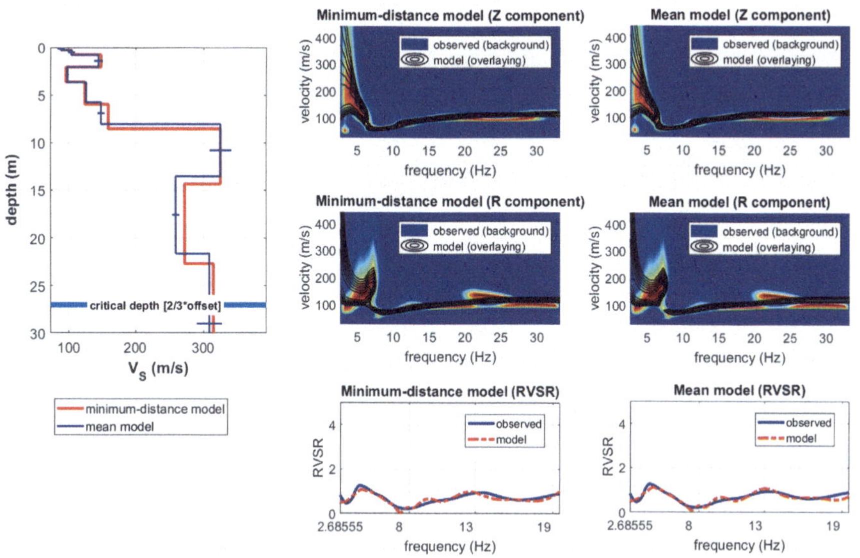

Fig. 4.13 Site #8 (HS active data—offset 40 m): results of the holistic inversion of the group velocity spectra of the Z and R components (according to the FVS approach) together with the RVSR curve. Reported the data about the two most important V_S models (the minimum-distance and the mean model—Dal Moro et al. 2019)

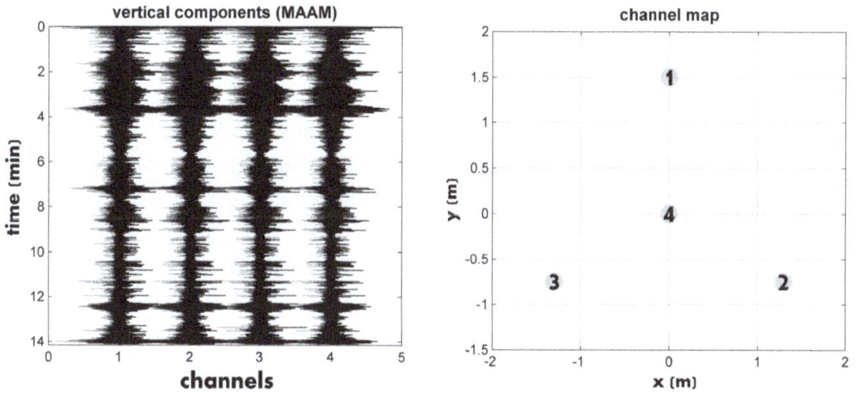

Fig. 4.14 Site #8: MAAM data and geometry. Traces are apparently quite similar to each other (since the four geophones are so close to each other, this is normal and necessary and in case one trace would show an average amplitude significantly different from the other three traces, that would be the evidence of some acquisition problem—see we will see later on)

The agreement with the profile obtained considering the purely active HS approach (Fig. 4.13) is apparent and demonstrates the overall consistency of both the adopted procedures. The reader should consider the limited offset of the HS data

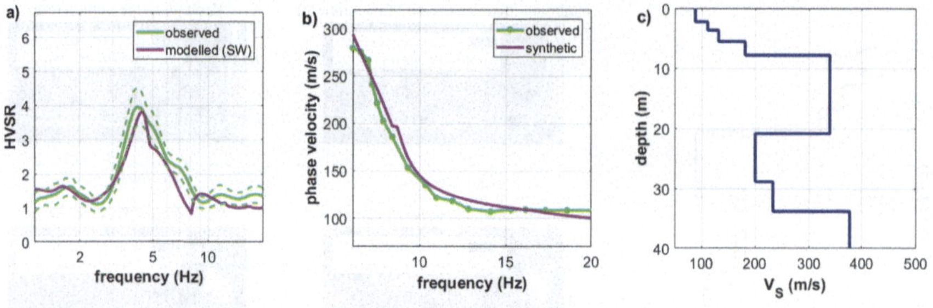

Fig. 4.15 Site #8: results of the joint analysis of the effective dispersion curve obtained via MAAM and the HVSR curve: **a** observed and synthetic HVSR curves; **b** field and synthetic *effective* dispersion curves; **c** obtained V_S subsurface model

(40 m) and, conseqeuntly, the limited investigated depth of the data presented in Fig. 4.13 (the V_S profiles reported in Figs. 4.15c and 4.13 should be compared only down to about 20 m).

There is a very important note to highlight in order to catch a crucial point that should be deeply understood and digested. The reason why the author of this manuscript is usually not inclined to use the term “method” (or “methodology”) is that, as should be clear from all the examples presented throughout the book, there is no “method”: there are several possible “objects” (*observables*) that can be computed and combined in several different ways depending on the site, on the data and on the goals (the idea that surface waves are useful just for defining the well-known Vs30 parameter is equivalent to the idea that a knife is useful just to cut the cheese).

Perhaps the most appropriate term is therefore *approach.* In fact, such a term can be used to indicate a multitude of possibilities: MAAM + HVSR approach, HS (Z + R + RVSR) approach, HS (Z + R + RPM) approach, HS (Z + R + RPM) + HVSR approach, multi-component multi-offset MASW (Z + R + RPM), multi-component multi-offset MASW (T + R) + HVSR and so on and so on (the list can be extremely long).

4.3 From the RPM Curve to the Advanced HS Analysis: Some Remarks and Two More Case Studies

In the first chapter, we mentioned the problem—often misunderstood—of the North-South direction (what is “North” and what is “South”?).

Let us now briefly consider the crucial importance of this issue in case we need to analyze the actual particle motion induced by the propagation of the Rayleigh waves generated by an earthquake or by our hammering over a plate.

An effective way to analyze the particle motion induced by the Rayleigh-wave propagation was introduced in Dal Moro et al. (2017a) through the computation, frequency by frequency, of the correlation coefficients between the radial component and the Hilbert transform of the vertical component (of Rayleigh waves).

There are at least two reasons to compute the RPM (Rayleigh-wave Particle Motion) curve:

(a) it is useful to further constrain the holistic inversion of Rayleigh waves (thus obtaining a robust, i.e. precise, V_S profile—Dal Moro et al. 2017, 2019; Dal Moro and Puzzilli 2017; Dal Moro 2019b);
(b) it provides information regarding possible Rayleigh-wave prograde motion, identified as a potential critical factor for the structural stability (Trifunac 2009).

The RPM curve provides in fact detailed information about the actual Rayleigh-wave motion: the correlation value equals to +1 in case of perfectly retrograde motion and −1 in the case of prograde motion. Of course, in case of single-offset data we can compute a single RPM frequency curve (for that specific *offset*) while in case of multi-offset data we can define the RPM frequency-offset surface (i.e. we compute the RPM curve for each offset).

In order to correctly compute the RPM curve, it is clearly necessary to know the polarity of the *acquisition system* and properly orientate the geophone(s).

In the first chapter, we introduced the problem of the data polarity. Let us now imagine having a 3C geophone to define the RPM curve. How should we orientate the sensor? A series of conventions need to be defined, first about the sequence of the recorded traces, then about the polarities. Usually, the first trace is about the vertical component and the second about the radial one (it goes without saying it that the third trace will be about the T component).

The next problem is to establish how the geophone should be rotated (see Fig. 1.27 and related text).

In both cases shown in Fig. 1.27 the *direction* is the same but the *verse* is obviously different (a vector has both a *direction* and a *verse* and when we rotate the geophone by 180°, we keep the same *direction* but modify the *verse*).

In order to correctly compute the RPM curve, we must define the correct orientation of the geophone (which should be tested with respect to the whole *acquisition system*—see Chap. 1).

It might sound complicated and verbose but the point is simple (Fig. 4.16): if we orientate the geophone correctly, we get the correct RPM curve, if we rotate it by 180° we get a specular (wrong) RPM curve.

As repeatedly emphasized throughout the book, we must underline once more that:

(1) the polarity of the data does not depend only on the geophone but on the seismograph-cable-geophone combination, i.e. on the whole *acquisition system* (see Sect. 1.6);
(2) the North arrow often present on 3C geophones has nothing to do with the *verse* to be used in case of active acquisitions: the standard HVSR does not distinguish between North and South (in case we rotate the geophone by 180° we obtain exactly the same HVSR) while for an active acquisition the correct *verse* is crucial (we cannot rotate the geophone by 180°—see Fig. 4.16).

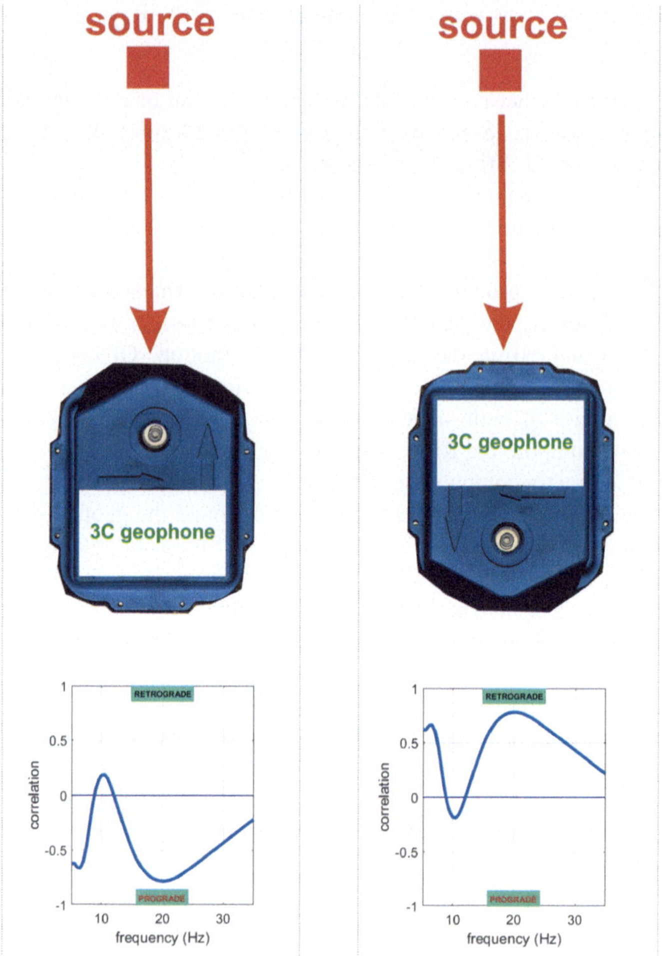

Fig. 4.16 The effect of an erroneous geophone orientation for the computation of the RPM frequency curve: by rotating the 3C geophone by 180° (see also Figs. 1.17, 1.18 and 1.27 and related text) two specular RPM curves are obtained. Of course, only one is correct and, in order to know the correct orientation for our geophone, the whole *acquisition system* must be tested specifically for the polarity (and, even more specifically, for the correct determination of the RPM curve)

If we want to use a 3C geophone to record an active dataset to use for the computation of the RPM curve we must ask for a field equipment (i.e. an *acquisition system*) explicitly designed so to fully consider all the issues involved in the definition of the correct polarity.

Although there is no universally-accepted rule, in seismology it is often assumed that:

1. positive values of the vertical-component trace correspond to the ground moving up;
2. positive values of the radial-component trace correspond to ground moving away from the source (in the direction of increasing radial distance);
3. positive values of the transverse-component trace correspond to ground moving in a clockwise direction when looking down to the surface.

On the other side, things can be seen and assessed also from different points of view (e.g. Brown et al. 2002).

As a matter of facts, there is nothing complicated: it is just about knowing how the acquisition system works and what is the actual meaning of the recorded traces.

Of course, if we want to analyze the RPM frequency-offset surface (see next pages), this recommendation also applies for the single-component geophones used to record the Z (vertical) and R (radial) components during a multi-offset and multi-component survey (see also case studies in Appendix B and G).

Case study #2: the RPM into play

In cooperation with *Lorenz Keller* (roXplore gmbh—Switzerland).

In the previous case study (relatively outdated), Rayleigh-wave propagation (from active HS data) has been analyzed considering three observables: the velocity spectra of the two Z and R components and the RVSR curve.

As part of a series of surveys for the characterization of some of the monitoring stations of the Swiss Seismological Service (*Swiss Seismological Service—Schweizerischer ErdbebenDienst—SED*), we also tested the HS approach.

Compared to the previous example, for this case study we will get deeper in the data also considering the RPM frequency curve and the group-velocity spectrum of the THF component (Love waves) (also jointly with the HVSR curve obtained from microtremor data).

We will therefore consider the joint analysis of six observables (collected by using a single 3C geophone—see scheme in Fig. 4.12).

The active HS data were recorded according to the acquisition parameters reported in Table 4.1 (the HVSR was computed thanks to a 1 h passive dataset) and the site (not far from Zürich—Switzerland) is characterized by silts, marls and fine-grained sandstones typical of that area of the Swiss basin (Dal Moro et al. 2015a).

Table 4.1 Case study #2—HS acquisition parameters

Record length	2 s (then reduced to 0.75 s—Fig. 4.17a)
Sampling rate	1000 Hz (1 ms)
Offset	70 m
Stack	5
Source	8-kg sledgehammer used for VF and HF data (i.e. for Rayleigh and Love waves)

Recorded seismic traces are shown in Fig. 4.17a: the Z and R traces are those associated to the VF acquisition while the T trace is the one obtained by means of a HF source (see Sect. 1.14 and the data combination procedure illustrated in Appendix A—Fig. A.2)

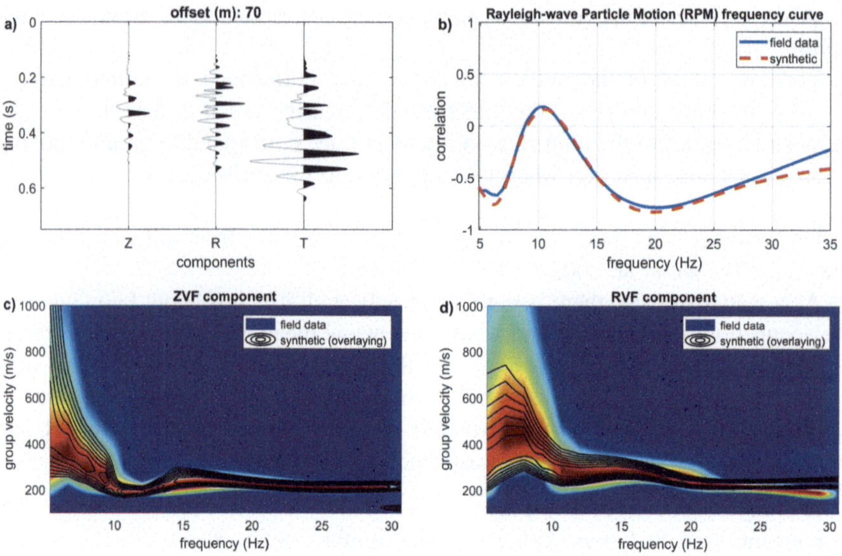

Fig. 4.17 Case study #2—joint inversion of the six considered observables (see also Fig. 4.18): **a** active HS seismic traces (Z, R and T components—see also Table 4.1); **b** field and synthetic RPM curves (Rayleigh waves appear strongly prograde along the whole frequency range); in **c** and **d** are shown the group velocity spectra for the Z and R components (the background colors represent the field data while the overlying black contour lines the synthetic spectrum of the minimum-distance model reported in Fig. 4.18d)

The results of the accomplished joint inversion are summarized in Figs. 4.17 and 4.18 (the "minimum distance" model represents the model with the minimum geometric distance from the utopian point—for details on these aspects see Dal Moro, 2017b, 2019a; Dal Moro et al. 2019).

The obtained V_S profile (Fig. 4.18d) is in excellent agreement both with the results obtained through the analysis of standard multi-offset and multi-component data and with the model retrieved via Vertical Seismic Profiling (VSP) (see V_S profiles in Fig. 4.18d and Dal Moro et al. 2015a).

Few remarks:

- Compared to the RVSR curve, the RPM frequency curve reveals less sensitive to small stratigraphic details but more robust (less sensitive to small data pre-processing and/or cleaning) (see also the synthetic modelling presented in Dal Moro et al. 2019).
- A joint inversion is necessarily a sort of compromise between the misfits of all the considered observables and, consequently, it is not always possible to obtain a *perfect* misfit for each single *object* (see also Dal Moro and Puzzilli 2017). From this point of view, this case study is a quite exceptional example in terms of overall agreement for all the single observable.

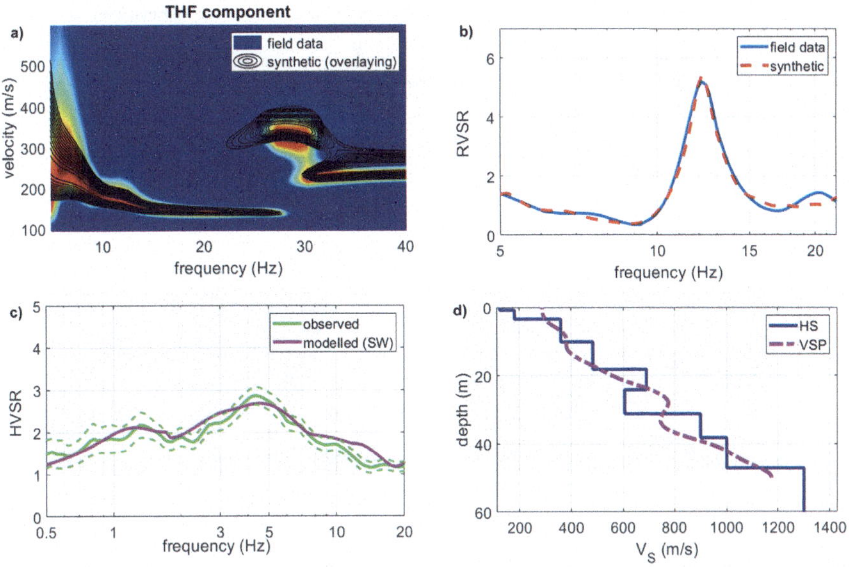

Fig. 4.18 Case study #2—joint inversion of the six considered observables (see also Fig. 4.17): **a** Love wave group-velocity spectra (THF component); **b** field and synthetic RVSR curves; **c** field and synthetic HVSR curves; **d** V_S profile obtained through the joint inversion of the six considered observables and compared with the V_S profile obtained from the Vertical Seismic Profiling (VSP)

- Despite the common belief that Rayleigh waves propagate according to a retrograde motion, prograde motion occurs very frequently even for very simple subsurface models far from the conditions classically invoked to explain it (e.g. abrupt changes in the V_S values and high Poisson's ratio values—see Tanimoto and Rivera 2005; Malischewsky et al. 2008 and compare also with Dal Moro et al. 2017a, b).

Case study #3: a vintage asphalt road dataset

In the first chapter (Sect. 1.13), we showed the multi-offset and multi-component (RVF + THF) traces collected in an urban area (an asphalt road—see Figs. 1.34, 1.35 and 1.36 and related text). During that survey (that took place in 2012), we also performed an experimental HS acquisition (the methodology was still "in progress"). Figure 4.19 presents the recorded data. Since the equipment was not optimized, the data quality is not excellent but the obtained group-velocity spectra (Fig. 4.19d and e) appear quite clear and suitable for the HS analysis.

Data (the two group velocity spectra and the RPM frequency curve) were then jointly inverted with the results presented in Fig. 4.20. The obtained V_S profile is apparently in excellent agreement with the subsurface model obtained while considering the multi-offset and multi-component (RVF + THF) MASW data presented in Chap. 2 (Fig. 2.14) while introducing the FVS approach to dispersion analysis.

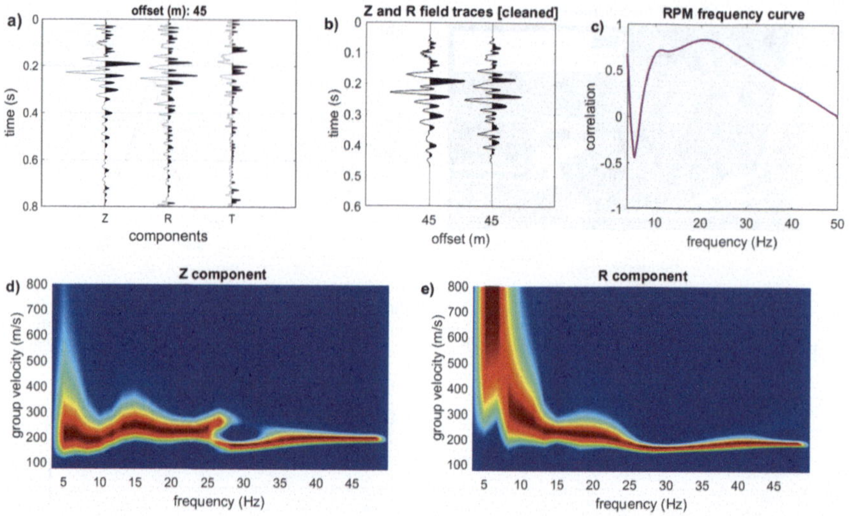

Fig. 4.19 A vintage HS dataset recorded in 2012 over an asphalt street (see Fig. 1.34—the multi-offset and multi-component MASW data (RVF + THF) were presented in Sect. 1.13). At that time, the equipment was somehow rudimentary and unable to provide high-quality data: **a** raw HS traces; **b** Z and R traces after some cleaning; **c** RPM frequency curve; **d** group-velocity spectrum for the Z component; **e** group-velocity spectrum for the R component. In spite of the relatively-low quality of the seismic traces, after some simple cleaning the quality of the Z and R group velocity spectra and RPM curve appear quite good. The analysis is presented in Fig. 4.20

4.4 Back to Purgessimo (NE Italy): Group Velocity Spectra (Z and R Components) + RPM Frequency Curve + HVSR

In one of the test sites we often consider in order to test and verify some ideas and methodologies (e.g. Dal Moro 2010, 2014—case study #2), we recorded an HS + HVSR dataset. The site (about 18 km East from Udine—Italy) is a small basin characterized by a soft-sediment cover of about 25–30 m laying over a calcarenitic bedrock (the basin is a reclamation area recovered from marshy conditions during the fascist regime).

Figure 4.21 presents the considered data and summarizes the results of the joint HS (Z + R + RPM) + HVSR analysis accomplished via MOEA (e.g. Dal Moro and Pipan 2007; Dal Moro et al. 2019). Since the site is relatively simple, the data are well characterized and it is possible to highlight a very educational fact/phenomenon: the HVSR peak frequency (about 2 Hz) corresponds to the frequency where the RPM changes from retrograde (at higher frequencies) to prograde. This happens due to the abrupt variation of the V_S value (the contact between the soft shallow sediments and the calcarenite *bedrock* at about 30 m).

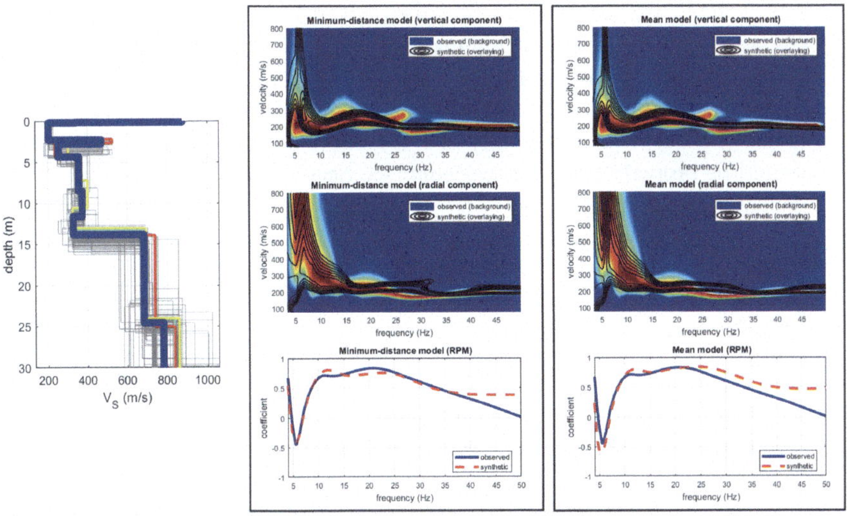

Fig. 4.20 Results of the joint analysis of the HS data shown in Fig. 4.19. On the left the V_S profiles of the models belonging to the Pareto front and, along the two columns, the field data (background colors) and the synthetic data for the two most important models (the *minimum-distance model* and the *mean model* determined as the average of all the models of the Pareto front)

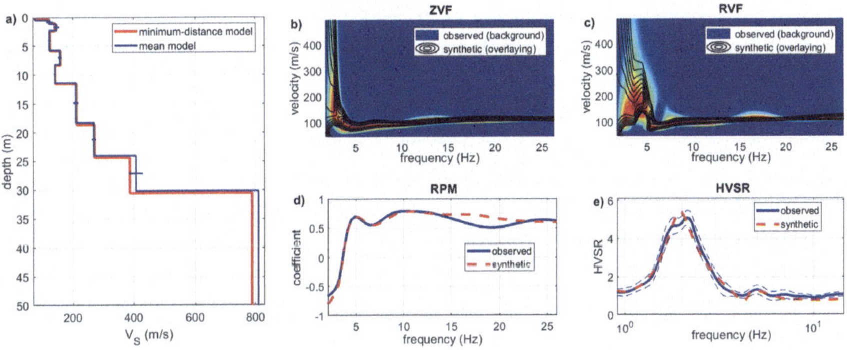

Fig. 4.21 Results of the joint analysis of the Purgessimo (NE Italy) HS (Z + R + RPM) + HVSR data. Compare with the analysis of multi-component (multi-offset) MASW data presented in Dal Moro (2014—case study #2)

Few additional comments to the data/analysis:

(1) the excellent consistency between field and synthetic data (for the identified subsurface model) is apparent;

(2) the V_S models identified through the HS approach (which is multi-component *per se*) and via multi-component multi-offset MASW (which clearly requires a much heavier field effort) are fundamentally identical [compare the V_S model

in Fig. 4.21a and the solutions presented in Dal Moro 2010, 2014 (case study #2)—a perfect identity is clearly unconceivable, also because the two sites are about thirty meters one from the other].

(3) in some cases, the modeling of the HVSR cannot be performed just according to the Rayleigh-wave ellipticity and the effect of Love and/or body waves should be properly taken into account (see Sect. 3.1.1, Bonnefoy et al. 2008; Albarello and Lunedei 2010). This is probably the case of this site, which is discussed in some detail in Dal Moro (2010).

The HS (VF) and HVSR data of this site are available for the download (the link is provided in the preface of the book).

4.5 Group Velocities and Penetration Depth

While working with group velocities, the definition of the maximum investigated depth is not straightforward. While considering the phase velocities (determined for instance via MASW, ReMi, MAAM, SPAC/ESAC) the maximum penetration depth is usually estimated from the maximum wavelength λ identified from the minimum considered frequency. The maximum wavelength λ is defined as the ratio between the phase velocity and the corresponding (minimum) frequency ($\lambda = v/f$).

The obtained value is then divided by a constant value which is usually fixed to 2 or, more conservatively, 3 (e.g. Yang et al. 2007). Actually, such a constant is not universal since it significantly depends on the Poisson's ratio of the materials (Karray and Lefebvre 2008) as well as on how we define the sensitivity (e.g. Urban et al. 1993; Cercato 2018; Pan et al. 2019). In any case, the obtained value must be considered as a very rough approximation and not as a precise number, carved in stone.

Since from the group velocities it is not possible to define the wavelengths, seismologists (who are routinely working with group velocities) usually consider the analysis of the eigenfunctions as a function of the depth (and for different frequencies—e.g. Urban et al. 1993; Fang et al. 2010). Since such approach cannot be applied in a straightforward way, it is necessary and possible to adopt a sort of hybrid (*a posteriori*) approach.

Once the final model is obtained (from the analysis of the group velocities, possibly jointly with the RVSR/RPM/HVSR curves), it is possible to compute the theoretical phase-velocity dispersion curve and, from the minimum considered frequency, the related maximum wavelength which can be then divided by three so to obtain the maximum investigated depth according to the classical λ/3 approximation (for details see Dal Moro et al. 2019).

The investigated depth depends on the characteristics of the site and on the lowest frequency that we (arbitrarily) decide to consider during the analysis. Clearly, such a frequency depends on the data quality, which, in turn, depends on the noise and on the characteristics of the source and acquisition system (i.e. a series of facts that cannot be generalized).

On the other side, we often need some simple (quick-and-dirty) rule of thumb usefull to plan the field acquisition parameters and, from a number of comparative HS tests, we can conservatively estimate the maximum investigated penetration in about half or two thirds of the offset.

4.6 MAAM: Data Quality, Radius and Weather Conditions

As repeatedly underlined, the *Miniature Array Analysis of Microtremors* (MAAM) is very sensitive to data quality. In order to process the data according to such a methodology, it is necessary to carefully consider:

1. the overall quality of the *acquisition system* (geophones, cables and A/D unit);
2. the accuracy of the acquisition procedures.

For instance, an inaccurate soil-geophone coupling or a flawed verticality of the sensors can be critical factors.

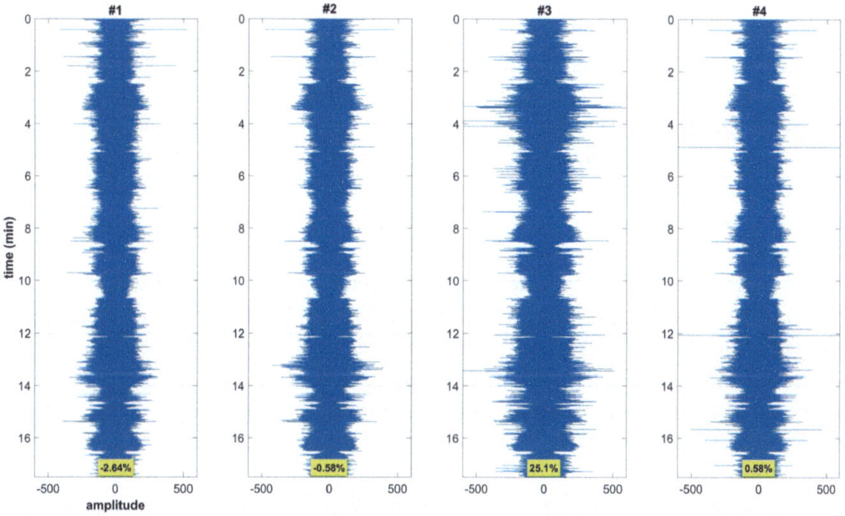

Fig. 4.22 MAAM data quality check. Computation of the mean amplitudes (standard deviations) for the four traces recorded considering a triangular geometry (Fig. 4.5a). If the standard deviation of one trace is too different from the median value computed considering all the four traces, there is probably some acquisition problem. In this example, the third trace has a significantly-higher amplitude compared to the other traces (its standard deviation is 25.1% larger than the median value) and the dataset can be unsuitable for the MAAM. This kind of quality check should be carried out already on the field on a short dataset (2 or 3 minutes are sufficient). In case we notice this kind of problems, we should immediately verify if the problematic sensor (in this case the third one) is correctly coupled and connected (we assume that the curves of the four geophones are identical) and once we solve the problem we can proceed with the full (longer) data acquisition. After Dal Moro (2019b)

From the practical point of view, in order to avoid the acquisition of low-quality data that would not be suitable for the MAAM, it would be useful to run a preliminary quick quality check on a small test dataset (a couple of minutes are usually sufficient).

Figure 4.22 presents an example of possible procedure (details in Dal Moro 2019b).

Fundamentally, it is about the computation and comparison of the average amplitudes for all the traces. These can be estimated through the standard deviations of each trace: if one trace deviates too much (say more than 10–15%) from the median value, there is some problem.

We should in fact consider that the geophones are very close to each other (usually the radius is just 2–3 m) and in such a limited area the amplitude of the microtremors must be homogenous. In case one trace has a significantly-different amplitude (see third trace in Fig. 4.22) is probably because the geophone is not properly set up (of course we should check the response curves of all the geophones and be sure they are identical).

In case we identify this kind of problems on the test data, we must check the problematic channel/geophone and verify that everything is correctly set up. When everything is fine (the standard deviations of the traces are homogenous), we can finally record the definitive dataset (the recording time clearly depends on the lowest frequency we are interested in, but for our ordinary near-surface applications 15–30 min are usually sufficient).

In very general terms, if we are interested in determining the dispersion curve between say 4 and 30 Hz and we are working on unconsolidated sediments (silt, clay, sand, gravel), the radius can be a couple of meters, while if we need to analyze lower frequencies, the radius should be significantly increased following a sort of exponential law (see Cho et al. 2006a). However, one must consider that the deeper layers (i.e. frequencies) can be investigated through the H/V spectral ratio (easily determinable even for very low frequencies).

We might therefore consider the dispersion curve from MAAM in the relatively-high frequency range (4–30 Hz) and the HVSR for lower frequencies (e.g. between 0.5 and 10 Hz).

What is the best moment to collect passive data?

The amplitude of the microtremors can influence the quality of the analysis to accomplish via MAAM. A weak microtremor field produces low-amplitude signals that can mirror in problematic (confused) analyses. Cloudy (but not windy) days are often the best moment to obtain high-quality passive data (in particular for the MAAM technique, which is very sensitive to the characteristics and quality of the data) (Fig. 4.23).

4.7 RPM Curves: More Insights and the Multi-offset Case

The RPM frequency curve (defined through the processing of the Z and R traces—details in Dal Moro et al. 2017a) describes, frequency by frequency, the actual particle motion due to the Rayleigh-wave propagation.

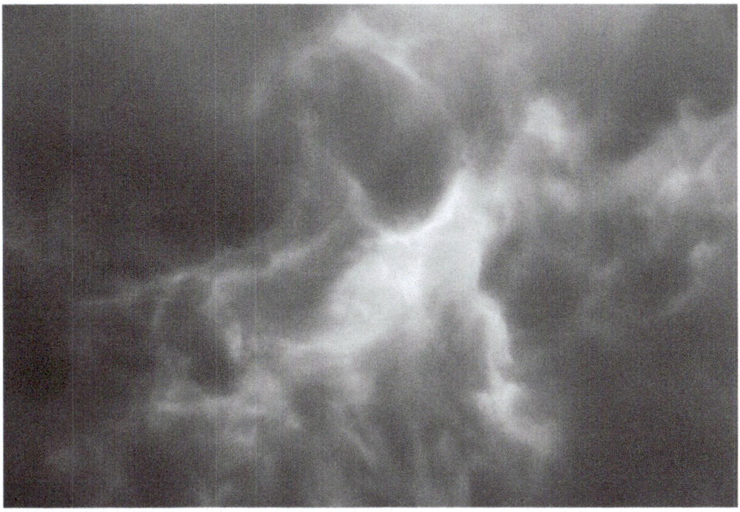

Fig. 4.23 Cloudy sky and no wind: an excellent day for passive seismology

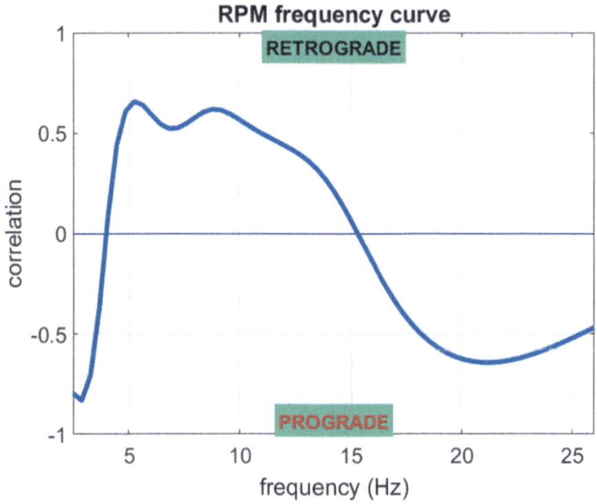

Fig. 4.24 Example of RPM frequency curve for a given offset: the particle motion is prograde for frequencies lower than 3 Hz and higher than 15 Hz and predominantly retrograde in-between. Figure 4.25 shows the particle motion at 3 Hz

It is important to list few crucial facts about the actual Rayleigh-wave particle motion:

(1) in spite of the common belief, Rayleigh waves are very rarely retrograde. Motion is usually a complex mix of *prograde* and *retrograde* motion and, as we saw in some of the presented case studies, prograde motion often dominates;
(2) in general, the actual motion is a function of both the frequency and offset;

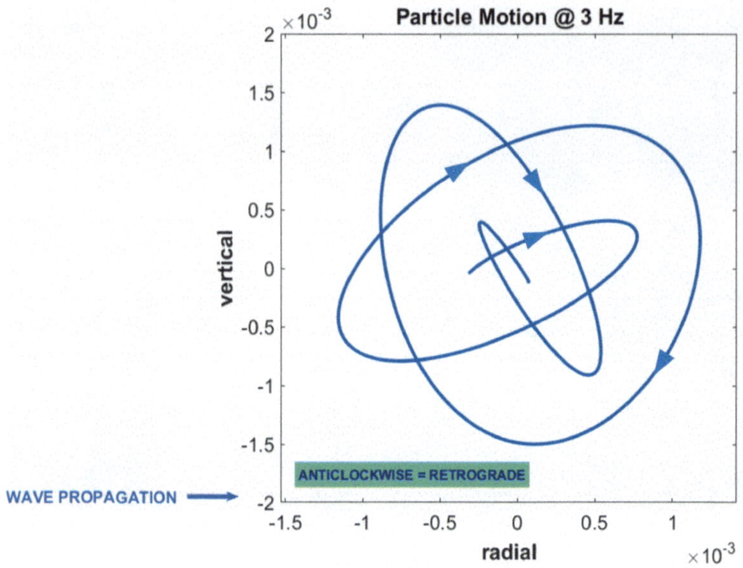

Fig. 4.25 Particle motion induced at 3 Hz because of the Rayleigh-wave propagation along the Z-R plane (see RPM curve shown in Fig. 4.24). Rayleigh wave propagates from left to right: the particle motion is not retrograde as too often simplistically believed, but prograde (clockwise). Arrows indicate the progression of the particle motion

(3) in case of single-offset data (HS approach), the particle motion is described by the RPM frequency curve (the correlation coefficient is just a function of the frequency—see for instance the curve reported in Fig. 4.24 and the related Fig. 4.25);
(4) in case of Z + R multi-offset data (multi-component MASW approach) the particle motion is described by the RPM frequency-offset surface (the correlation coefficient depends on both the offset and frequency) (see Appendices B and G as well as the data reported in the following of this section);
(5) to properly compute the RPM curve, it is necessary to work with an *acquisition system* (Sect. 1.6) explicitly designed so to correctly define the polarity of the traces (Figs. 1.17, 1.18, 1.27 and 4.16 and related text).

Figure 4.26 shows a synthetic example of the RPM frequency-offset surface that can be defined in case of multi-offset (Z+R) data. In this case, the particle motion significantly changes both with respect to the offset and frequency.

A comprehensive real-world multi-component dataset is shown in Fig. 4.27 and the result of the performed joint inversion is summarized in Fig. 4.28 (data are available for the download from the link provided in the preface of the book). The joint inversion was performed (as usual) via MOEA by considering the three phase-velocity spectra (for the three components) and the RPM frequency-offset

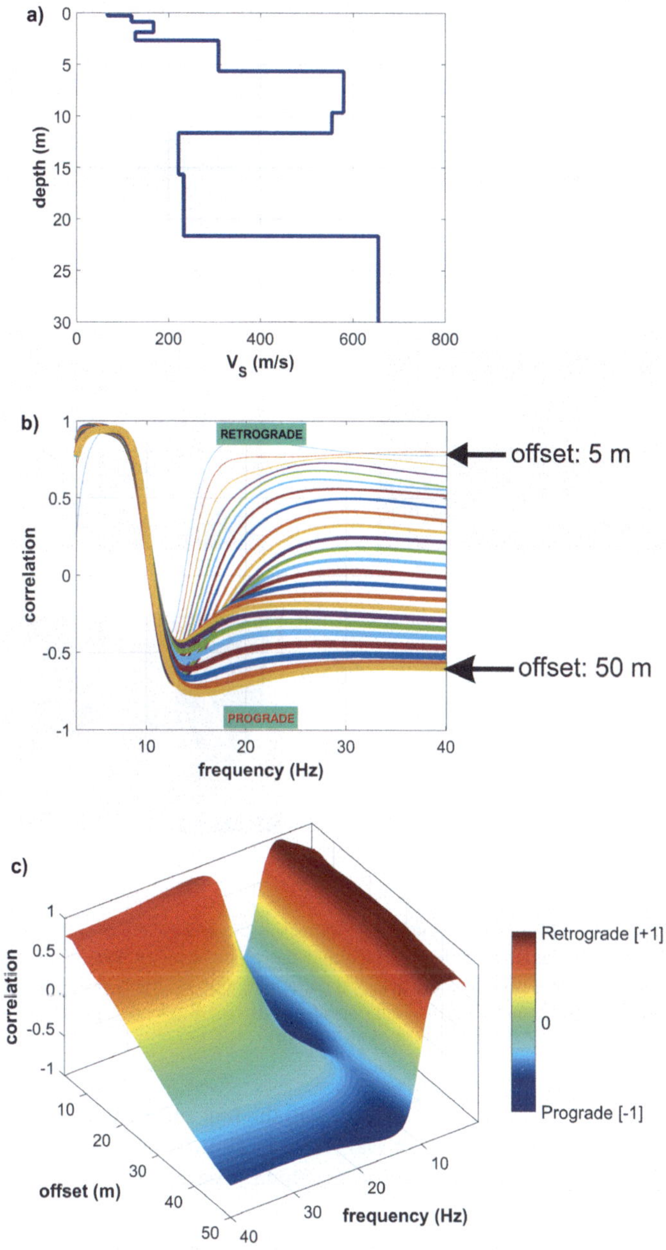

Fig. 4.26 Example of RPM frequency-offset surface for a synthetic data: **a** V_S model; **b** RPM frequency curves for different offsets; **c** 3D RPM representation as a function of both frequency and offset. Appendices B and G present two case studies where, in order to effectively characterize two urban areas, the phase velocity spectra of the Z and R components are analyzed together with the RPM frequency-offset surface

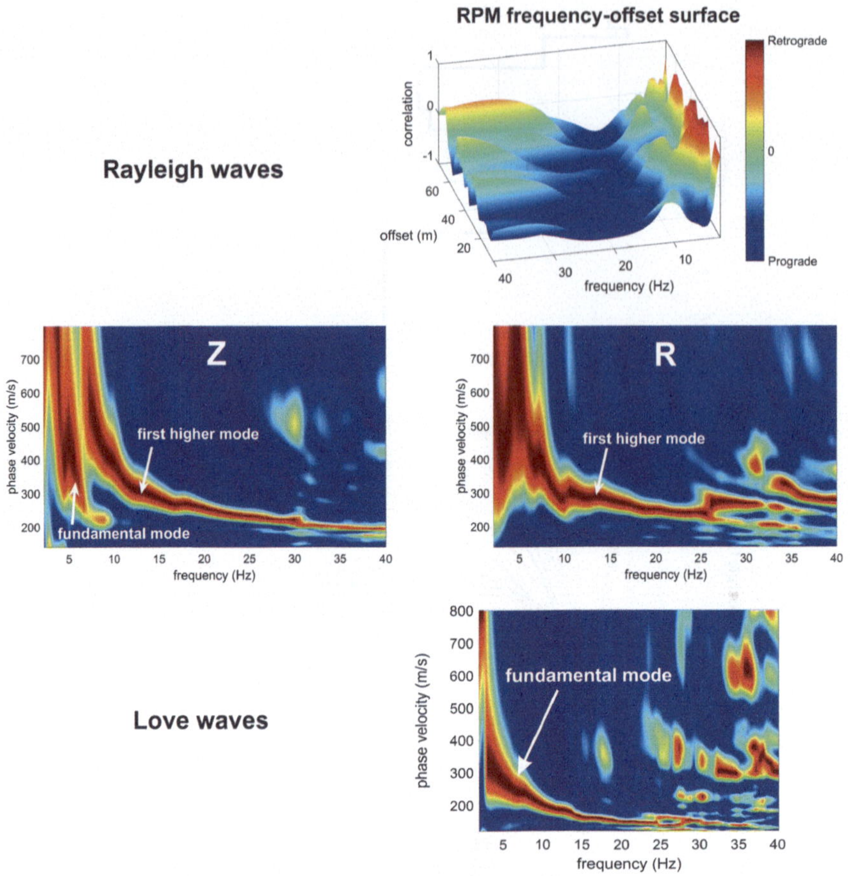

Fig. 4.27 Rayleigh and Love wave data for an industrial site in Tuscany (IT): in the upper block are reported the RPM frequency-offset surface and the phase-velocity spectra for the Z and R components (higher modes dominate the data and particle motion is largely prograde). In the lower panel is shown the Love-wave phase-velocity spectrum (unlike Rayleigh waves, the fundamental mode largely dominates). After Dal Moro et al. (2017a)

surface (the obtained model is consistent for all the four observables and is therefore extremely robust—further details in Dal Moro et al. 2017a).

These data/analyses can be useful to highlight a series of facts:

- Rayleigh waves (both the components) are dominated by the first higher mode even if there is no significant LVL (Low Velocity Layer) [a urban legend states that higher mode are generated by LVLs but this idea is completely unfounded as also shown in Dal Moro et al. (2015c)];
- Love waves are (as usual) very simple: unlike Rayleigh waves, the fundamental mode largely dominates;

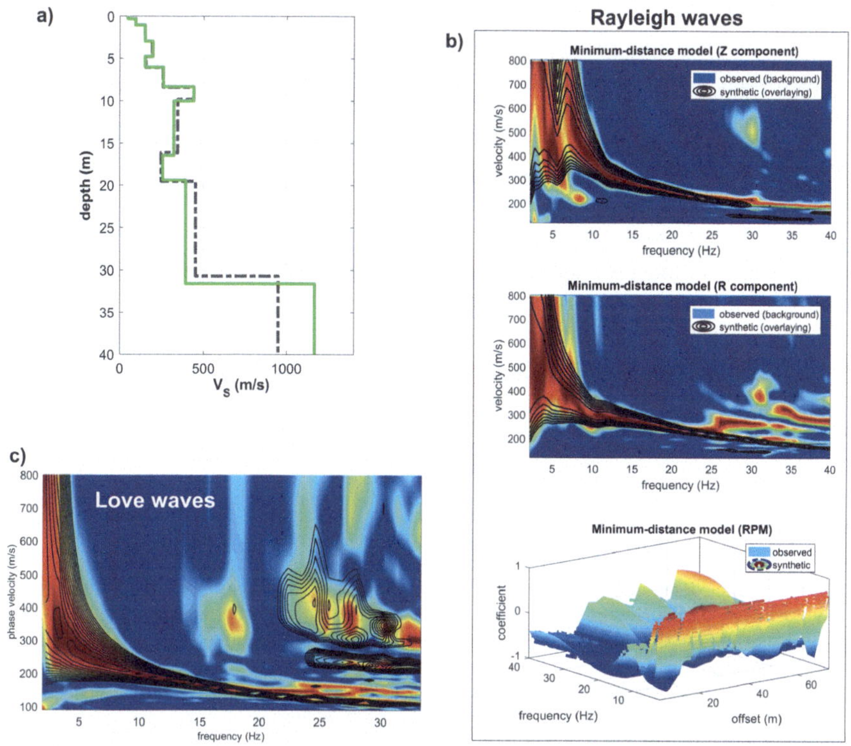

Fig. 4.28 Result of the joint inversion of the data presented in Fig. 4.27: **a** the two obtained subsurface V_S models (dotted curve refers to the mean model, continuous line to the minimum-distance model—in case the joint inversion was properly set up, they must be pretty similar); **b** Rayleigh waves (Z and R phase-velocity spectra and RPM frequency-offset surfaces); **c** Love waves. According to the FVS representation adopted throughout the book, background colors refer to the field data while the overlying black contour lines to the model (in this case we considered the minimum-distance model but the mean one is practically identical)

- Rayleigh wave motion is prograde along almost the entire frequency range and for all the offsets (it becomes retrograde only for the lowest frequencies);
- this kind of joint analysis is the extension to the multi-offset case of the single-offset HS approach.

The dataset is the umpteenth example of Rayleigh waves largely dominated by higher modes and would risk to be misinterpreted and lead to overestimated V_S values. In fact, both in the literature and in applied works, it is possible to find several examples of misinterpreted Rayleigh-wave velocity spectra where, because the fundamental mode is hardly visible, higher modes are interpreted (and picked) as fundamental mode and the retrieved V_S values are consequently significantly overestimated.

In case of multi-offset data, the acquisition of the Z and R components can be accomplished according to different field strategies.

For instance, in case we have a 12-channel acquisition system and want to work with 12 offsets, we must first record the Z component and then the R component.

On the other side, in case we have a 24-channel seismograph with two distinct 12-channel cables, we could stretch the two cables parallel to each other and connect the vertical geophones to the first cable and the horizontal ones (for the R component) to the second cable (see Fig. 1.11).

Of course, we must pay attention to the polarity and be sure about the orientation (the verse) of the horizontal geophones (see Figs. 1.17 and 1.18).

Appendices B and G present two examples for this kind of multi-offset and multi-component (Z + R) surveys. Both the case studies are about the exploration of urban areas (Appendix B for the determination of a single 1D vertical profile and Appendix G for the reconstruction of a 2D V_S section).

References

Albarello D, Lunedei E (2010) Alternative interpretations of horizontal to vertical spectral ratios of ambient vibrations: new insights from theoretical modeling. Bull Earthq Eng 8:519–534

Bhattacharya SN (1983) Higher order accuracy in multiple filter technique. Bull Seismol Soc Am 73:1395–1406

Bonnefoy-Claudet S, Köhler A, Cornou C, Wathelet M, Bard P-Y (2008) Effects of Love waves on microtremor H/V ratio. Bull Seismol Soc Am 98:288–300

Brown RJ, Stewart RR, Lawton DC (2002) A proposed polarity standard for multicomponent seismic data. Geophysics 67:1028–1037

Cercato M (2018) Sensitivity of Rayleigh wave ellipticity and implications for surface wave inversion. Geophys J Int 213:489–510

Cho I, Tada T, Shinozaki Y (2006a) New methods of microtremor exploration: the centerless circular array method and two-radius method. In: Proceedings of the third international symposium on the effects of surface geology on seismic motion, Grenoble (France), 30 Aug–1 Sept, pp 335–344

Cho I, Tada T, Shinozaki Y (2006b) Centerless circular array method: inferring phase velocities of Rayleigh waves in broad wavelength ranges using microtremor records. J Gophys Res 111: B09315. https://doi.org/10.1029/2005JB004235

Cho I, Senna S, Fujiwara H (2013) Miniature array analysis of microtremors. Geophysics 78: KS13–KS23

Dal Moro G (2010) Insights on surface-wave dispersion curves and HVSR: joint analysis via Pareto optimality. J Appl Geophys 72:29–140

Dal Moro G (2014) Surface wave analysis for near surface applications. Elsevier, Amsterdam, The Netherlands, 252 pp. ISBN 978-0-12-800770-9

Dal Moro G (2019a) Surface wave analysis: improving the accuracy of the shear-wave velocity profile through the efficient joint acquisition and Full Velocity Spectrum (FVS) analysis of Rayleigh and Love waves. Explor Geophys. https://doi.org/10.1080/08123985.2019.1606202

Dal Moro G (2019b) Effective active and passive seismics for the characterization of urban and remote areas: four channels for seven objective functions. Pure appl Geophys 176:1445–1465

Dal Moro G, Ferigo F (2011) Joint inversion of Rayleigh and Love wave dispersion curves for near-surface studies: criteria and improvements. J Appl Geophys 75:573–589

Dal Moro G, Pipan M (2007) Joint inversion of surface wave dispersion curves and reflection travel times via multi-objective evolutionary algorithms. J Appl Geophys 61:56–81

Dal Moro G, Puzzilli LM (2017) Single- and multi-component inversion of Rayleigh waves acquired by a single 3-component geophone: an illustrative case study. Acta Geodyn Geomater 14 (4, 188):431–444. https://doi.org/10.13168/agg.2017.0024

Dal Moro G, Keller L, Poggi V (2015a) A comprehensive seismic characterization via multi-component analysis of active and passive data. First Break 33:45–53

Dal Moro G, Ponta R, Mauro R (2015b) Unconventional optimized surface wave acquisition and analysis: comparative tests in a perilagoon area. J Appl Geophys 114:158–167

Dal Moro G, Moura RM, Moustafa SR (2015c) Multi-component joint analysis of surface waves. J Appl Geophys 119:128–138

Dal Moro G, Moustafa SR, Keller L, Al-Arifi N, Moustafa SR (2016) Shear-wave velocity profiling according to three alternative approaches: a comparative case study. J Appl Geophys 134:112–124

Dal Moro G, Al-Arifi N, Moustafa SR (2017a) Improved holistic analysis of Rayleigh waves for single- and multi-offset data: joint inversion of Rayleigh-wave particle motion and vertical- and radial-component velocity spectra. Pure Appl Geophys 175:67–88. https://doi.org/10.1007/s00024-017-1694-8 (open access)

Dal Moro G, Al-Arifi N, Moustafa SR (2017b) Analysis of Rayleigh-wave particle motion from active seismics. Bull Seismol Soc Am 107:51–62

Dal Moro G, Weber T, Keller L (2018) Gaussian-filtered Horizontal Motion (GHM) plots of non-synchronous ambient microtremors for the identification of flexural and torsional modes of a building. Soil Dyn Earthq Eng 112:243–245

Dal Moro G, Al-Arifi N, Moustafa SR (2019) On the efficient acquisition and holistic analysis of Rayleigh waves: technical aspects and two comparative case studies. Soil Dyn Earthq Eng 125. https://www.sciencedirect.com/science/article/pii/S0267726118310613

Dziewonski A, Bloch S, Landisman N (1969) A technique for the analysis of transient seismic signals. Bull Seismol Soc Am 59:427–444

Fang L, Wu J, Ding Z, Panza GF (2010) High resolution Rayleigh wave group velocity tomography in North China from ambient seismic noise. Geophys J Int 181:1171–1182

Ikeda T, Matsuoka T, Tsuji T, Hayashi K (2012) Multimode inversion with amplitude response of surface waves in the spatial autocorrelation method. Geophys J Int 190:541–552

Karray M, Lefebvre G (2008) Significance and evaluation of Poisson's ratio in Rayleigh wave testing. Can Geotech J 45:624–635

Levshin AL, Pisarenko VF, Pogrebinsky GA (1972) On a frequency-time analysis of oscillations. Ann Geophys 128:211–218

Malischewsky PG, Scherbaum F, Lomnitz C, Tuan TT, Wuttke F, Shamir G (2008) The domain of existence of prograde Rayleigh wave particle motion for simple models. Wave Motion 45:556–564. https://doi.org/10.1016/j.wavemoti.2007.11.004

Natale M, Nunziata C, Panza GF (2004) FTAN method for the detailed definition of Vs in urban areas. In: 13th world conference on earthquake engineering, Vancouver, BC, Canada, p 2694

Nguyen XN, Dahm T, Grevemeyer I (2009) Inversion of Scholte wave dispersion and waveform modeling for shallow structure of the Ninetyeast ridge. J Seismol 13:543–559

Pan L, Chen X, Wang J, Yang Z, Zhang D (2019) Sensitivity analysis of dispersion curves of Rayleigh waves with fundamental and higher modes. Geophys J Int 216:1276–1303

Pardalos PM, Migdalas A, Pitsoulis L (eds) (2008) Pareto optimality, game theory and equilibria. Springer, New York. ISBN 978-0-387-77247-9

Ritzwoller MH, Levshin AL (1998) Eurasian surface wave tomography: group velocities. J Geophys Res 103:4839–4878

Ritzwoller MH, Levshin AL (2002) Estimating shallow shear velocities with marine multi-component seismic data. Geophysics 67:1991–2004

Ritzwoller MH, Shapiro NM, Barmin MP, Levshin AL (2002) Global surface wave diffraction tomography. J Geophys Res 107:B12

Tada T, Cho I, Shinozaki Y (2007) Beyond the SPAC method: exploiting the wealth of circular-array methods for microtremor exploration. Bull Seismol Soc Am 97:2080–2095

Tanimoto T, Rivera L (2005) Prograde Rayleigh wave motion. Geophys J Int 162:399–405. https://doi.org/10.1111/j.1365-246X.2005.02481.x

Tokimatsu K, Tamura S, Kojima H (1992) Effects of multiple modes on Rayleigh wave dispersion characteristics. J Geotech Eng ASCE 118:1529–1543

Trifunac MD (2009) The role of strong motion rotations in the response of structures near earthquake faults. Soil Dyn Earthq Eng 29:382–393

Urban L, Cichowicz A, Vaccari F (1993) Computation of analytical partial derivatives of phase and group velocities for Rayleigh waves with respect to structural parameters. Stud Geophys Geod 37:14–36

Van Veldhuizen DA, Lamont GB (2000) Multiobjective evolutionary algorithms: analyzing the state-of-the-art. Evolut Comput 8:125–147

Yang Y, Ritzwoller MH, Levshin AL, Shapiro NM (2007) Ambient noise Rayleigh wave tomography across Europe. Geophys J Int 168:259–274

Zhou Y, Dahlen FA, Nolet G (2004) 3-D sensitivity kernels for surface-wave observables. Geophys J Int 158:142–168

Introduction to Vibration Monitoring and Building Characterization via GHM

5

A mistake is always forgivable, rarely excusable and always unacceptable.

Robert Fripp

G. Dal Moro, *Efficient Joint Analysis of Surface Waves and Introduction to Vibration Analysis: Beyond the Clichés*,
https://doi.org/10.1007/978-3-030-46303-8_5

Abstract

Since vibration data can be recorded with the same field equipment used to collect seismic data and several national and international building codes require the identification of the eigenmodes of the existing structures, we decided to include a chapter aimed at introducing the Earth scientist/professional to vibration analysis. The chapter cannot be a comprehensive guide to vibration analysis but intends to shown that the vibration monitoring at a construction site and the characterization of the eigenmodes of a building or structure can be relatively simple. A new technique (GHM—Gaussian-filtered Horizontal Motion) that relies on the data collected by a single 3C geophone and can discriminate flexural and torsional modes is presented through three case studies.

We decided to include the present chapter in a book otherwise focused on surface-wave analysis for two reasons:

(1) vibration data can be recorded with the same field equipment used to collect seismic data;
(2) several national and international building codes (the same that require the determination of the V_S profile) put a special emphasis on the characterization of the vibration modes of a building.

From several points of view, vibration acquisition and analysis is simpler compared to surface waves. The theoretical aspects are not particularly complex and can be explored in a reasonable amount of time and with a relatively-small effort.

Of course this chapter does not intend to be a comprehensive guide to vibration analysis but just to introduce some basic facts and show that working in this sector can be relatively simple.

In any case, it is important to underline that we should always try to go beyond all those countless simplifications that are often put forward and try to get a genuine and profound understanding of the phenomena and analyses.

We must also consider that the standards and regulations about the monitoring and assessment of the vibrations induced by human activities (e.g. Kouroussis et al. 2014) are often modified. Therefore, since acquisition procedures and threshold values depend on the locally-adopted regulations, it is recommended to obtain updated information about the standards in use in our country (the values considered in the following pages cannot be taken as universally and everlastingly valid).

From a practical point of view, for recording and analyzing vibration data, we need to deal with data recorded in *physical units* (usually mm/s) and not in *counts* (arbitrary units that cannot be always or easily converted into physical units). In other terms, our *acquisition system* (see Chap. 1) should provide data in mm/s (or any other physical unit convenient for our analyses).

This is mandatory in order to verify whether the vibrations induced at a construction site exceed or not the thresholds defined by the national (or international) regulations but it is useful (not compulsory but recommended) also for the vibration data recorded in order to characterize the behavior of a building.

5.1 Vibrations Induced by Construction Activities, Quarry Blasts and Vehicles

Construction activities (for instance, driving or removing sheet piles), quarry blasts and the passage of large vehicles (trucks, trains, etc.) produce vibrations that can jeopardize the integrity of the structures or cause nuisance and discomfort to people (de Silva 2005).

In order to assess the actual danger of this kind of vibrations, we can refer to a plethora of national and international regulations and documents that define a series of threshold limits that should not be exceeded.

These limits (thresholds) are usually expressed in mm/s (velocity) for the evaluation of the impact on buildings and structures and in mm/s^2 (accelerations) for the assessment of the possible effects on the human body.

Accelerations can be easily obtained from velocities by means of a simple derivative operation while obtaining velocity values from acceleration data is more problematic. This is why it is preferable to record velocities by means of an ordinary geophone (i.e. a velocimeters) and then compute the accelerations rather than recording the accelerations and try to obtain the velocities through data integration.

A general note: depending on the country, regulations can be more or less rigid (sometimes they are considered in a very stringent way while in other cases are a sort of "suggested values").

Anyway, as underlined in the forward of this chapter, it is necessary to work with an *acquisition system* that provides data expressed in physical units (i.e. in mm/s in the case of geophones or mm/s^2 in the case of accelerometers).

The correct positioning (and orientation) of the 3C sensor (geophone or accelerometer) is of course decisive. Since everything depends on the specific problem to solve and on the local logistic conditions, it is pretty hard to state where the sensor should be deployed. The only recommendation is not to place it over, for instance, a piece of furniture or on any other point not relevant for the problem we are facing.

Similarly to the seismic case, while monitoring the vibration at a construction site three components are defined: the vertical (Z), radial (R) and transverse (T) (in some cases ambiguously indicated as z, x and y).

The 3C geophone (throughout this chapter we consider vibration data recorded by means of a geophone) must be therefore properly oriented so to be able to clearly define and identify the three components. In other words, the geophone must be oriented so that its radial direction corresponds to the line that connects the sensor with the vibration source (see Chap. 1).

In order to get familiar with vibration monitoring, in the following pages are reported the analyses performed in order to address two simple cases: (1) the analysis of the vibrations produced by railway convoys; (2) the monitoring of the vibrations created by sheet pile driving at a construction site.

For both the case studies, we refer to the limits recommended by the DIN4150 (about the buildings/structures) and UNI9614 (about the human body) standards.

DIN standards (*Deutsches Institut für Normung*—German Institute for Standardization) are very well known and considered almost universally as reference while UNI9614 (Italian National Unification body) are here considered about the effects on human body because the illustrated data refer to Italian case studies.

Fundamentally, UNI9614 follows a very simple approach: vibrations are considered as potentially disturbing if the accelerations exceed certain threshold values (slightly different for the vertical and horizontal axes). This might be considered a simplistic approach and a series of further methodologies (e.g. ISO2631—ISO stands for *International Organization for Standardization*) attempt to consider not only the amplitude of the accelerations but the duration of the phenomenon as well (the bottom-line idea is clearly that the damages to the human body depend also on how long we are exposed to the vibrations).

In the assessment of transient (i.e. non-continuous) vibrations, DIN4150 identifies three classes of buildings:

(1) class 1: industrial buildings
(2) class 2: residential buildings
(3) class 3: buildings that do not belong to the first two categories and are worthy of being protected (e.g. buildings/structures of historical/archaeological value).

The threshold limits for the three classes clearly change (see the curves in Fig. 5.1—lower panel) and, in case of transient vibrations/events, vary as a function of the (peak) frequency. For class 1 buildings, for instance, the threshold value goes from a minimum of 20 mm/s (at low frequencies) up to 50 mm/s (for higher frequencies).

5.1.1 Example #1: Vibrations Induced by Trains (Railway)

In this first and simple illustrative example, the sensor (a 3C geophone) is deployed about 10 m away from a railway while trains passes by. The data presented in Figs. 5.1 and 5.2 highlight a rather common fact for this kind of sources/situations.

Figure 5.1 shows the amplitude spectra (upper plot) and the velocity peaks (lower plot) for all the three components (shown the three DIN4150 curves), while Fig. 5.2 present the accelerations (derived from the velocities) assessed with respect to the UNI9614 limits.

The comparison of the data reported in the two figures clearly show that the vibrations produced by the train convoy do not exceed the DIN4150 limits (which are based on the velocity) while the accelerations (easily obtained from the

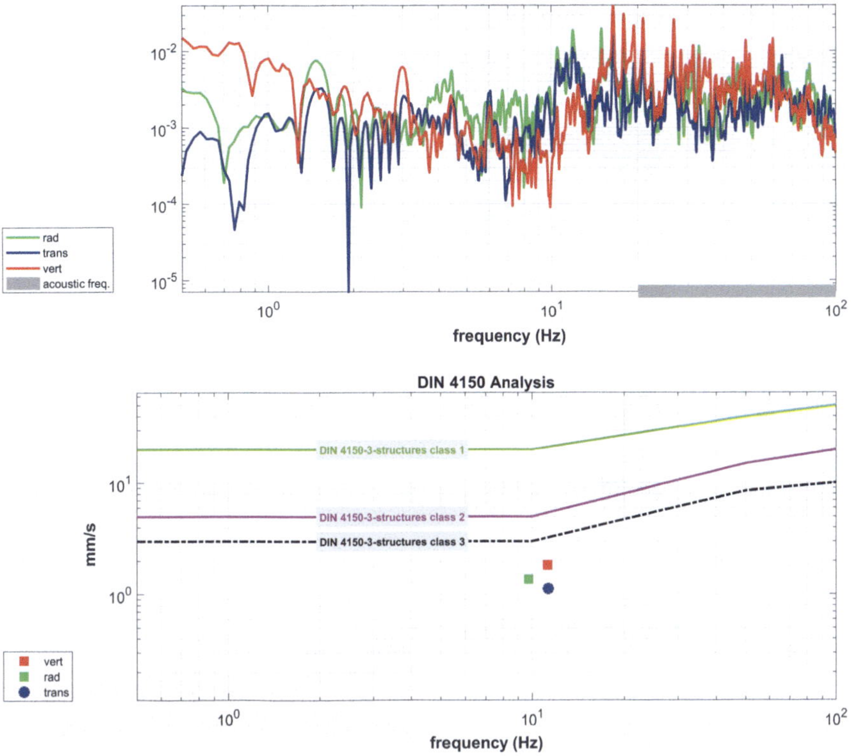

Fig. 5.1 Vibrations induced by a train convoy at a point about 10 m away from the railway track. Upper plot: amplitude spectra for the three components (vertical, radial and transverse); lower plot: velocity peaks of the same components as a function of the frequency. Shown the three DIN4150 reference curves

velocities) appear well beyond the UNI9614 threshold values (30 mm/s^2 for the vertical component and 21.6 mm/s^2 for the two horizontal axes).

From a practical point of view this means that, at about 10 m from the railway, the vibrations produced by the trains are not dangerous for the stability of the buildings but are anyway quite disturbing from the point of view of human perception.

5.1.2 Example #2: Vibrations at a Construction Site (Sheet Pile Driving)

Several industrial or construction activities can produce large-amplitude vibrations that can represent a problem for the integrity of the nearby buildings and structures or cause disturbances to people.

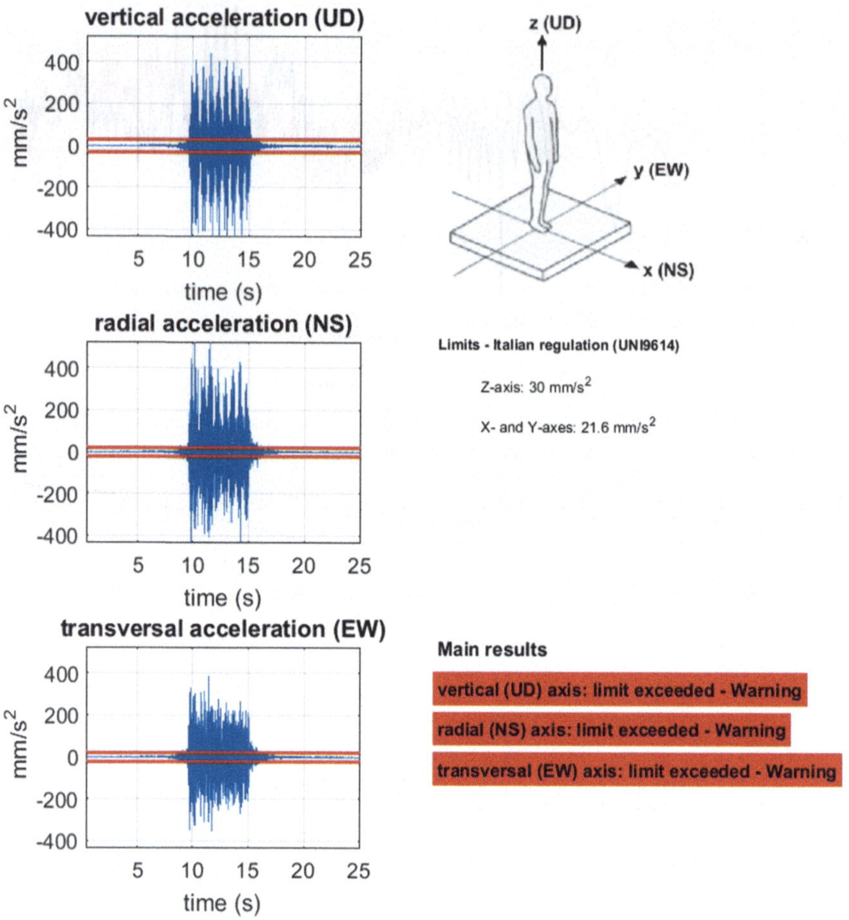

Fig. 5.2 Vibrations induced by a train convoy about 10 m away from the railway track. The three graphs on the left show the accelerations along the three axes (the passage of the small and fast train is apparent between about 10 and 15 s) and the red lines represent the UNI9614 limit values (which are largely exceeded)

In this case study, we consider the analysis of the vibrations induced by sheet pile driving (a very common activity that takes place daily in several construction sites all over the world—e.g. Jongmans 1996). In our case, the machine used was similar to the one shown in the scheme of Fig. 5.3 and the goal of the accomplished monitoring was to verify that the amplitude of the produced vibrations was small enough not to damage the nearby buildings.

The amplitude, of course, decreases with the distance from the source of vibrations (Figs. 5.4 and 5.5) and the two main phenomena responsible for the signal attenuation (e.g. Sato and Fehler 2009) are:

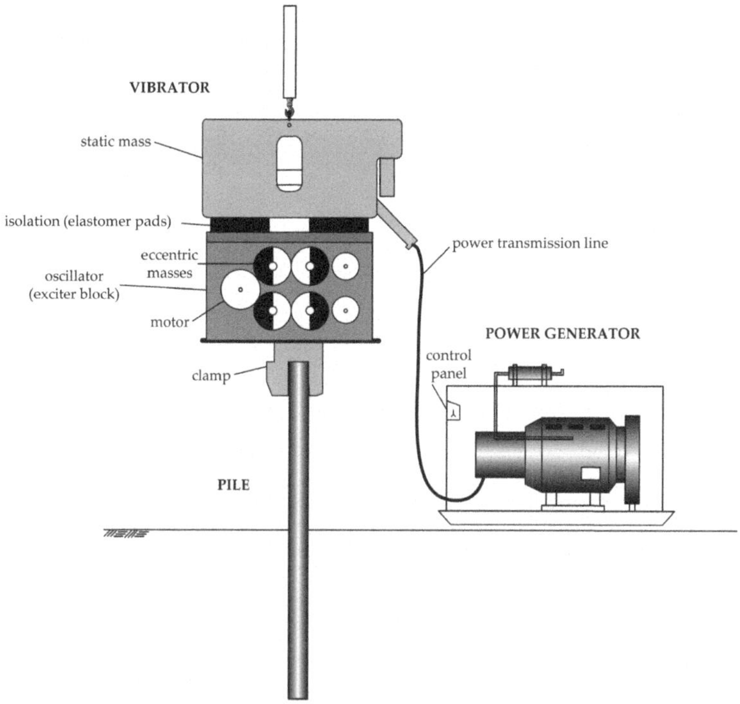

Fig. 5.3 Example of vibratory equipment for pile driving (from Deckner 2013): piles are driven into the soil (or removed from it) thanks to combination of a vertical force and a vibration applied in order to favor the penetration/extraction of the pile into/from the ground

(1) geometrical factors (since the emitted energy expands over a larger and larger volume/front, the amplitude decreases);
(2) intrinsic attenuation due to the fact that the energy is partly converted into plastic deformations and thermal phenomena.

Figures 5.6 and 5.7 show the raw data recorded about 5 m from the pile driving machine. The first figure reports the three traces (amplitude vs. time) while in the second one are shown the particle motions along the three planes (Z-T, Z-R, R-T). Figure 5.8 presents the amplitude spectra and the peak velocities for the three components (as a function of the frequency).

These graphs provide the evidence of some important facts:

(1) The amplitude spectra clearly show the presence of both the fundamental (between about 30 and 40 Hz) and some upper harmonics (see also Figs. 5.9 and 5.10);
(2) The peak value for the vertical component exceeds the DIN4150 threshold for class 3 (historical/archaeological structures) but not for classes 1 and 2 (Fig. 5.8, lower panel);

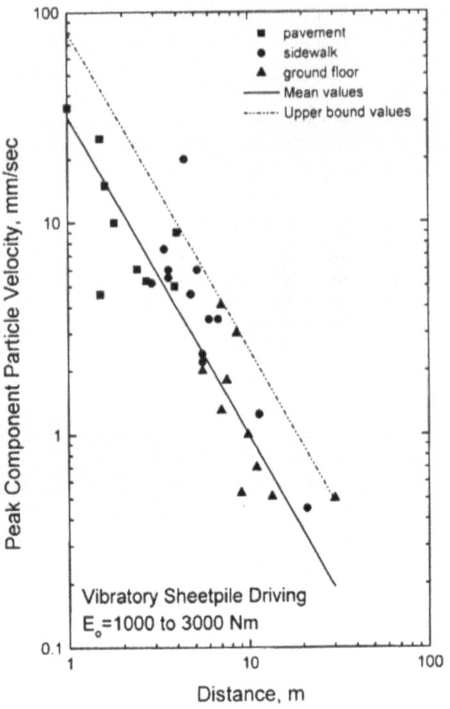

Fig. 5.4 Amplitude decrease of the maximum velocity (peak velocity) as a function of the distance from the vibratory source. From Athanasopoulos and Pelekis (2000)

(3) The fact that only the vertical component exceeds that limit is however reassuring since this component is of relatively little interest (the most dangerous vibrations are those along the horizontal plane);

(4) We should consider that these reference curves refer to transient events whereas for continuous phenomena the limits considered are smaller and independent on the frequency (for common residential buildings, DIN 4150-3 recommend 5 mm/s for the horizontal components and 10 mm/s for the vertical one). Are defined as continuous, those vibrations that last longer than five times the time constant τ_0:

$$\tau_0 = \frac{1}{2\pi\zeta_0 f_0} \tag{5.1}$$

being f_0 the fundamental frequency of the building and ζ_0 the damping value.

(5) In this case the geophone was rather close to the vibration source (just a few meters) but quite far (several tens of meters) from the residential buildings: the attenuation (e.g. Figs. 5.4 and 5.5) is such that after few more meters the amplitude rapidly and significantly decreases (see also attenuation data predented later on).

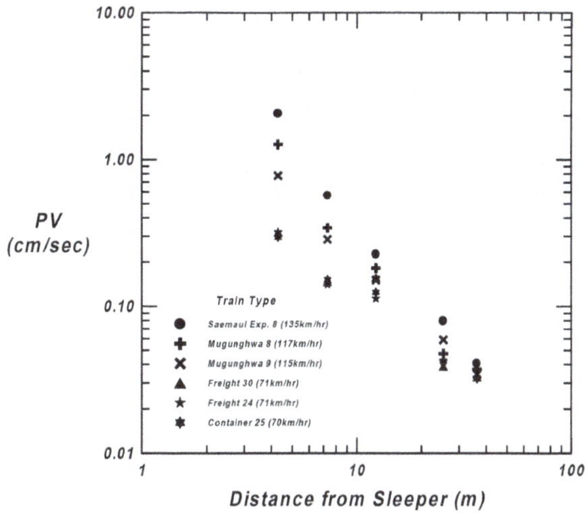

Fig. 5.5 Further example of amplitude decrease of the maximum velocity (peak velocity—PV) as a function of the distance for a pile driving machine similar to the one used in the case study presented in these pages (from Kim and Lee 2000). The overall trend is clearly equivalent to the data shown in Fig. 5.4. Pay attention to the different units: here cm/s while mm/s in case of the data shown in Fig. 5.4

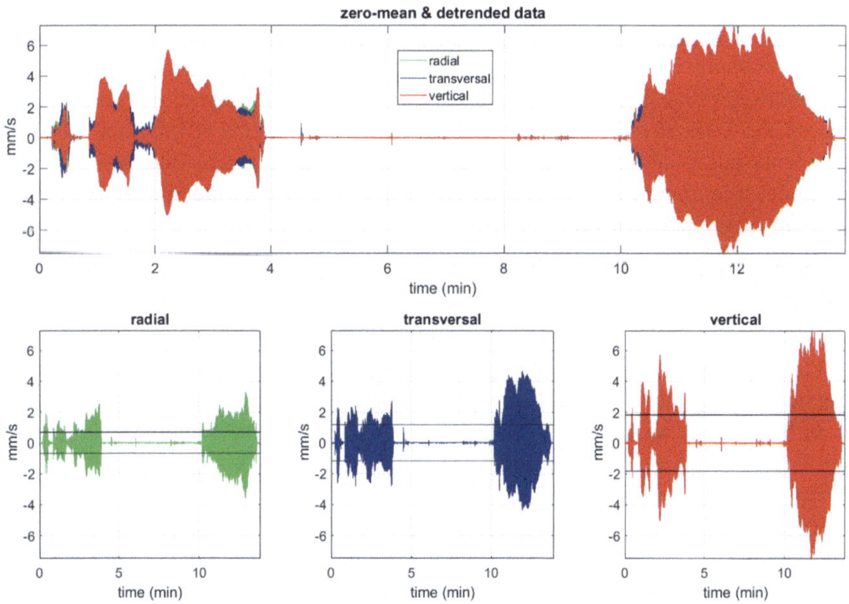

Fig. 5.6 Vibration data (fourteen-minute raw data) at about 6 m from the pile driving machine. The vertical component has a peak value of about 7 mm/s

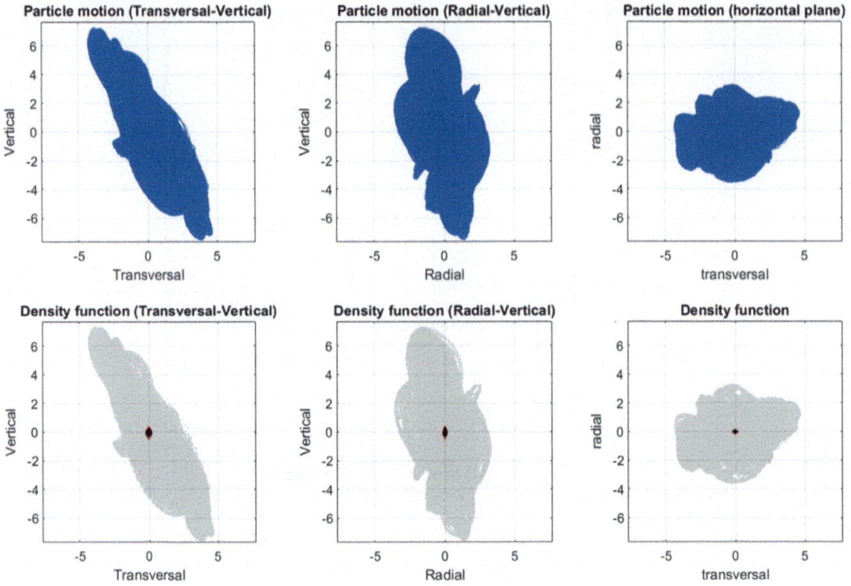

Fig. 5.7 Particle motion along the three planes for the data shown in Fig. 5.6 (vertical-transverse, vertical-radial and transverse-radial—upper plot the particle motion, lower plot the density functions). Data in mm/s

In the next section we will see how is possible to quantitatively estimate the vibration amplitude at any arbitrary distance from the source (Fig. 5.11 presents the acceleration values that, since we are very close to the source, not surprisingly exceed the UNI 9614 limits). We should in fact consider that this kind of measurements and analyzes are often carried out in case of legal disputes and it is therefore important to be scientifically rigorous and provide a comprehensive scenario of the phenomenon.

Analysis of the attenuation of the vibration data (in short)

In some cases, impediments due to logistical (or even legal) conditions prevent from the possibility to record data at any arbitrary point.

Furthermore, in some cases, we might be interested in preventively estimate the amplitude of the vibrations induced by a certain machine at various distances from it.

In both cases, we can use the equations that describe the attenuation (e.g. Sato and Fehler 2009) so to obtain the value of the amplitude at any distance from the (known) source of vibrations.

The equation commonly used to express the amplitude as a function of the offset (i.e. the distance between the source and the measurement point) is reported for instance in Massarsch and Fellenius (2008):

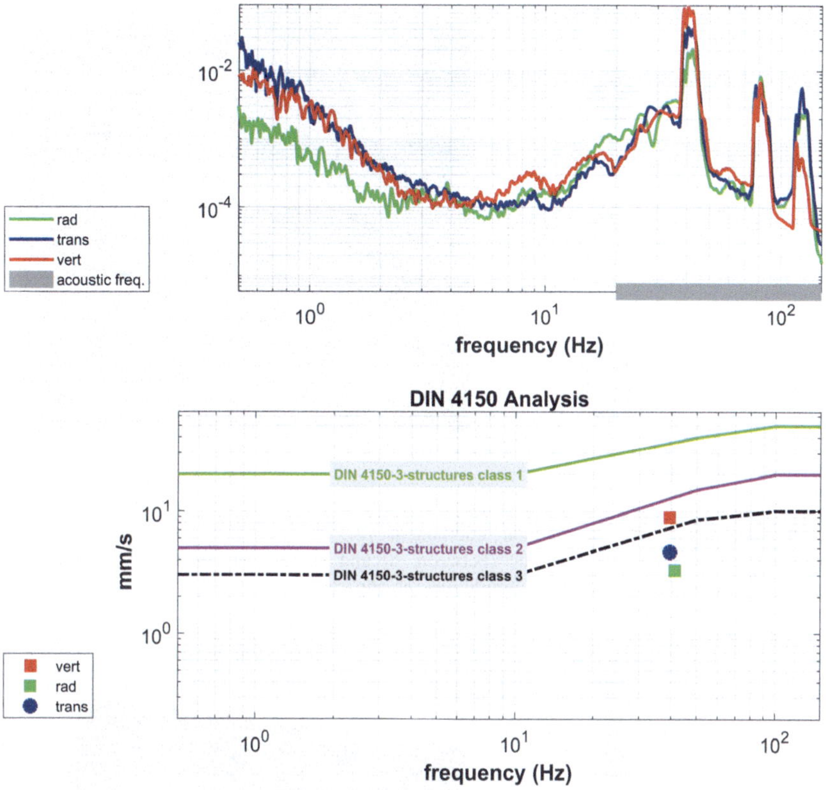

Fig. 5.8 Spectral analysis of the data shown in Fig. 5.6. Upper panel: amplitude spectra for the three components (the primary signal produced by the machine is around 40 Hz but are also apparent two upper harmonics at about 80 and 120 Hz); lower panel: peak velocities and DIN4150 curves for the three building classes. Further comments in the text

$$A_2 = A_1 \left(\frac{R_2}{R_1}\right)^{-n} e^{-\alpha(R_2 - R_1)} \tag{5.2}$$

where

A_1 = amplitude of the vibration at the distance R_1 (from the source)
A_2 = amplitude of the vibration at the distance R_2 (from the source)
α = absorption coefficient.

For surface waves (which, compared to body waves, have significantly larger amplitude), the coefficient n is fixed to 0.5.

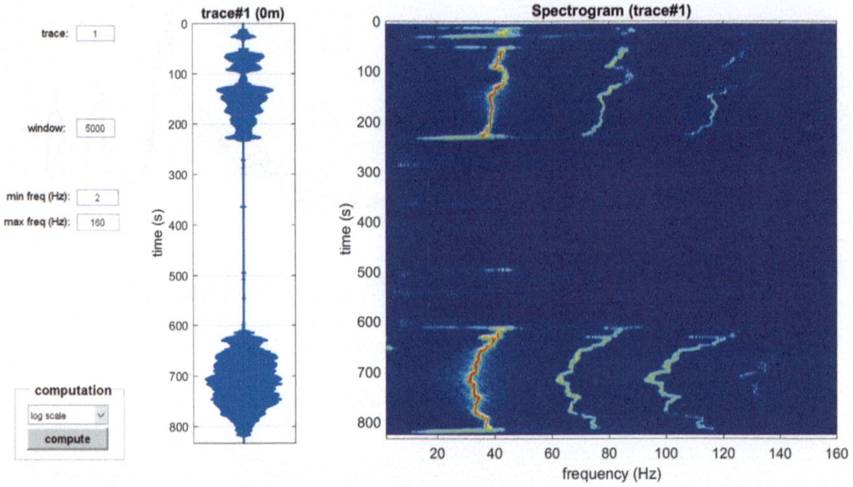

Fig. 5.9 Spectrogram of the vertical component (2D perspective): the presence of the fundamental large-amplitude harmonic (between 30 and 40 Hz) and of the two/three upper harmonics (with gradually decreasing amplitudes) is apparent. See also Figs. 5.8 and 5.10

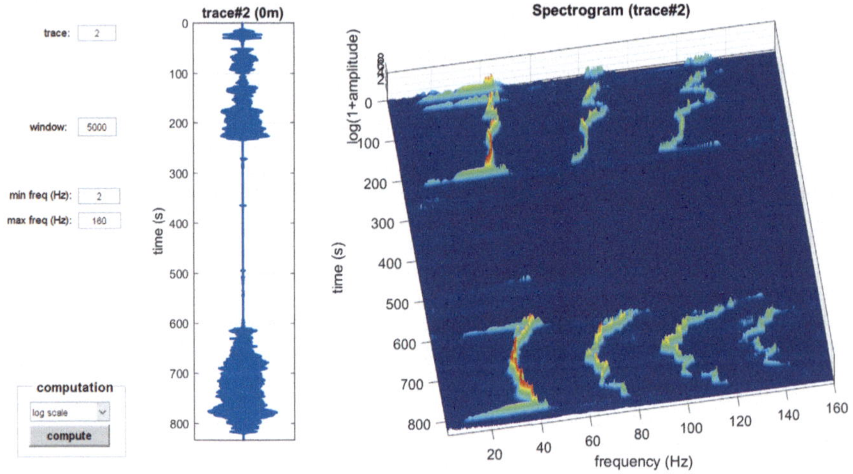

Fig. 5.10 Spectrogram of the radial component (3D perspective): fundamental and upper harmonics are extremely clear (compare with Fig. 5.9)

The α coefficient can be easily computed from the following relationship:

$$\alpha = \frac{2\pi f D_{\mathrm{M}}}{c_{\mathrm{R}}} \tag{5.3}$$

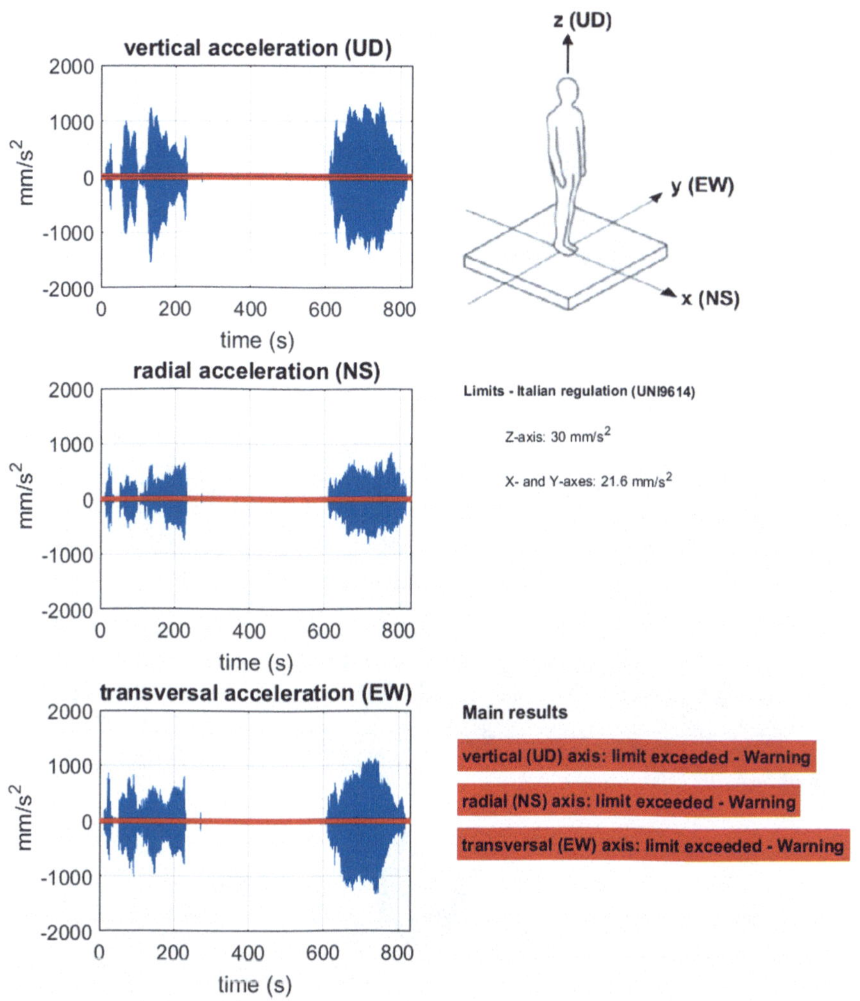

Fig. 5.11 UNI 9614: recorded acceleration values compared to the UNI threshold limits. It is clear that, few meters from the source, the acceleration values are inevitably above the regulation limits

where

D_M = damping
f = frequency of the considered vibration
c_R = surface-wave velocity.

The damping can be experimentally obtained from the analysis of multi-offset data or, if these are not available, assumed between 3 and 5% depending on the soil characteristics (Massarsch and Fellenius 2008). Of course, surface-wave velocity

(c_R) can be obtained from the analysis of surface-wave propagation (most of this book deal with it) or estimated from the general characteristics of the local soils.

From the practical point of view, once we know the quality factors Q (or, correspondingly, the damping) of the local sediments and the vibration amplitude at a given point, by means of the reported equations it is possible to estimate the amplitude at any arbitrary distance.

Figure 5.12 shows the acquisition setting used to record the experimantal data necessary to analyze the signal attenuation (just behind the fence there is a pile driving machine at work).

Analyzing the amplitude decrease as a function of the offset (Fig. 5.13) and considering Eqs. 5.2 and 5.3, it is therefore possible to estimate the vibration

Fig. 5.12 Multi-offset (vertical component) acquisition for the analysis of the attenuation of the vibrations produced by a pile driving machine (hardly visible just beyond the fence): six vertical geophones are deployed at six different offsets (4, 5, 6, 7, 8 and 9 m) from the source. It is also visible a 3C geophone (nearby the second vertical geophone). Maximum recorded amplitudes are shown in Fig. 5.13

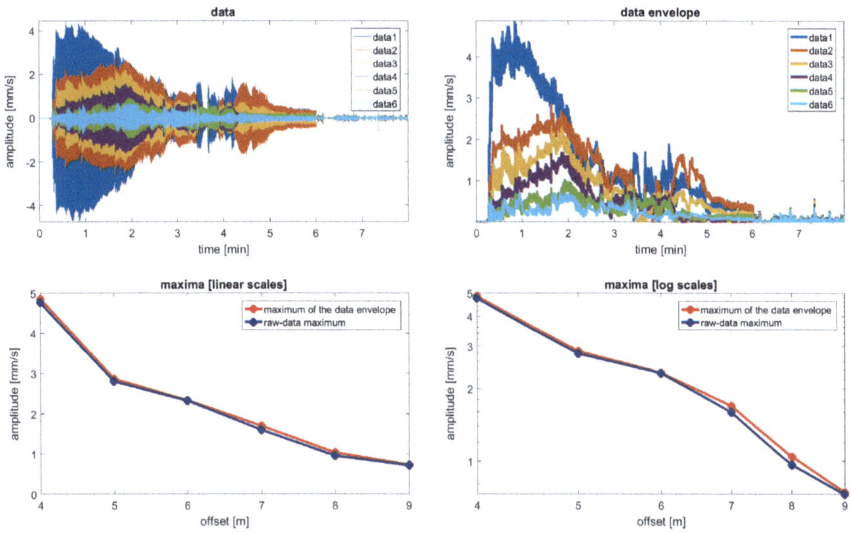

Fig. 5.13 Attenuation analysis (vertical component). Upper panel: seismic traces for the six considered offsets (see array shown in Fig. 5.12) (shown the traces and, on the right, the envelopes); lower panel: the peak values as a function of the distance from the source (offset). Peak velocities are shown with a linear scale (on the left) and with a log scale (on the right). At a distance of 5 m the peak velocity is 3 mm/s and decreases to 1 mm/s at 8 m (the offset is the distance from the source)

amplitude at any distance from the source. For doing that, it is necessary to fix the reference frequency (in our case between 30 and 40 Hz—see data shown in Figs. 5.8, 5.9 and 5.10) and know the velocity (c_R) of the local (shallow) sediments. Figure 5.14 reports the peak velocity curve reconstructed (as a function of the distance from the source) while considering the characteristics of this site and the vibrations generated by this specific pile driving machine (amplitude at 1 m from the source fixed to 20 mm/s, average c_R 250 m/s and reference frequency 35 Hz).

In general terms, we can highlight that the attenuation is larger when we are dealing with soft sediments (for instance peat or clay—smaller quality factors and, consequently, larger damping) while smaller while working on compact sediments (e.g. compacted sands, gravels or conglomerates—larger quality factors, i.e. smaller damping).

5.2 Characterization of the Behavior of a Building: The GHM Approach

In this section, we intend to present the fundamentals for the characterization of a building in terms of vibration modes: determination of the *eigen* (or natural) frequencies (and the respective damping) and identification of the type of vibration

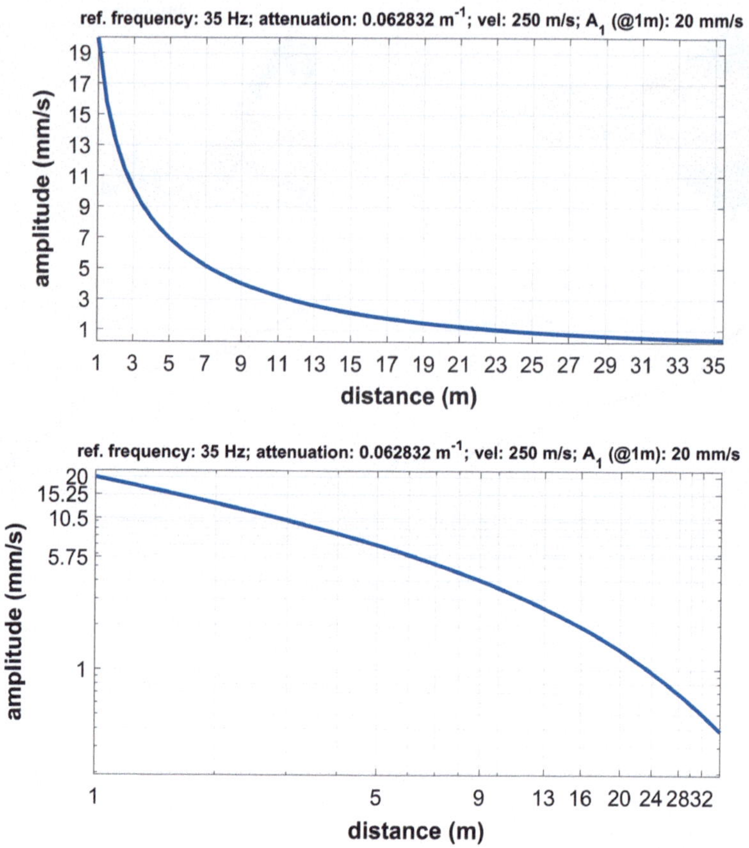

Fig. 5.14 Reconstruction of the maximum velocity amplitude as a function of the distance from the source by means of Eqs. 5.2 and 5.3 (reference frequency 35 Hz, c_R 250 m/s and 20 mm/s at 1 m from the source). Once the absorption coefficient is computed and the vibration amplitude at one (arbitrary) point is known (in this case a 20 mm/s peak value was recorded at 1 m from the source), it is possible to estimate the maximum amplitude at any offset. The two graphs present the same curve using a linear-linear (upper plot) and a log-log (lower plot) scale. About 7 m away from the source, the peak value reaches the threshold limit for the continuous vibrations for common residential buildings suich a value is 5 mm/s)

mode (flexural, torsional or mixed). Of course, it is not possible to provide a comprehensive and detailed guide to building vibration analysis (there are already several books that focus on it) and the main goal is to show how the seismic equipment can be used for the determination of the behavior of a building.

In order to describe the dynamic behavior of a structure, we can rely on both numerical simulations and empirical methods based on the analysis of vibration data (e.g. Gupta 1990; Chopra 1995; Naito and Ishibashi 1996; Bonev et al. 2010; Hong and Hwang 2000; Brownjohn 2003; Ghosh 2003; Takada et al. 2004; Su et al. 2005; Trifunac 2009; Zembaty 2009; Charney 2010; Michel et al. 2010, 2011; Garevski 2012).

In case we need to verify the behavior of an existing structure, *in situ* vibration measurement and analysis is a common practice that can be integrated with numerical models so to verify the overall congruity between the numerical model and the experimental evidence.

Although there are some empirical relationships that link the height of a building and the frequency of the fundamental flexural mode, in order to verify the actual behavior of a building there are several possible experimental approaches (the "fundamental mode" is a pure abstraction since, most of the times, there are at least two flexural and one torsional modes with, clearly, different frequencies).

To avoid verbose and complicated descriptions, in order to understand what are the flexural and torsional modes, one of the simplest things to do is a quick search over *youtube* by using keywords such as *building vibration torsional mode* and/or *building flexural vibration modes* and so on. The several available videos will immediately clarify what kind of motions we are speaking about: in practice, a torsional mode is a sort of rotation of the whole building around a vertical axis while a flexural mode is the vibration mode that most people imagine when they see a skyscraper.

In case we want to determine whether a certain mode is flexural or torsional, the classic methods require the acquisition of synchronous data at two or more points and the computation of the cross-spectra (which enable us to determine the vibration frequencies) and phase functions (which are used to understand whether the mode is flexural or torsional).

The synchronization of the sensors is possible via GPS (which is sometimes a bit painful but allows the sensors to be positioned very freely) or by connecting the sensors to the same cable (which transmits the signals to the "data logger"—i.e. the seismograph).

From a practical point of view, it is clear that, in both cases, the acquisition procedures can be relatively cumbersome.

In the following pages, together with the standard analysis (which requires the acquisition of synchronous data at two or more points), a simpler approach based on the analysis of non-synchronous data recorded at two (or more) points is also presented.

This latter method is based on the analysis of non-synchronous data acquired from a single 3C geophone positioned, sequentially, at two corners of the building (on the same floor). A series of narrow (Gaussian) band-pass filters are applied and the obtained GHM (*Gaussian-Filtered Horizontal Motion*) plots are assessed so to identify whether a certain mode is flexural or torsional. Needless to underline that the same techniques can be efficiently applied also to synchronous data.

The methodology is illustrated through three case studies regarding the analysis of ambient microtremor data. The first one (a small 3-storey family house in the NE Italy seismic area) is used to compare the traditional method with the results obtained via GHM.

The second case study is about the analysis of a 25-storey building in Zürich (Switzerland) and represents a further application of the GHM methodology.

In this case, the experimental results obtained by means of the GHM technique are compared with the outcome of a numerical simulation performed via Finite Element (FE) modelling in order to verify whether the model parameters were adequate to simulate the actual building behavior. In fact, it is clear that the experimental results must match sufficiently well with the numerical simulations (see Takada et al. 2004; Dal Moro et al. 2018).

The third case study is about a 13-storey reinforced concrete frame building realized in the 60s in Italy. Once again, the experimental modes defined via GHM are compared with the data from a FE modelling so to understand whether the several parameters involved in the numerical model are properly set.

A note about the positioning and orientation of the geophone

Of course (and as always), in order to obtain meaningful data, the 3C geophone must be positioned and oriented in a sensible manner. Few recommendations:

(1) The geophone should be positioned close to the perimeter walls (points in the middle of the rooms or nearby stairwells, elevator shafts etc. should be avoided);

(2) The NS direction of the 3C geophone/accelerometer must be fixed by considering the axes of the investigated structure and not the geographical North.
Once the two measurement points (corners) are chosen, the best orientation to adopt in order to obtain clear GHM plots (see later on), is to set the EW axis as the direction connecting the two points. In other words, the NS direction of the geophone(s) should be the axis perpendicular to the imaginary line that connects the two measurement points.

Needless to say that the polarity of the data (which depends on the combinations of all the elements of our *acquisition system*—see Chap. 1) must be correct, since, otherwise, we risk to obtain misleading GHM plots.

5.2.1 Case Study #1: The GHM Technique for a 3-Storey Building

The first case (for details see Dal Moro et al. 2018) is about a 3-storey building (Fig. 5.15) in the Friuli (NE Italy) seismic area. The masonry building was built in 1974 and reinforced after the 1976 Friuli earthquake (Cagnetti and Pasquale 1979).

In order to identify the vibration modes according to the classic approach, we placed two 3C geophones on the top floor (points P1 and P2 in Fig. 5.15). Both the geophones were connected to the seismograph by means of an ordinary seismic cable. Sampling frequency and recording times were fixed to 2 ms and 15 min, respectively (during the processing, data were then resampled to 64 Hz).

During the recording, weather conditions were quite good (sunny day with a mild breeze).

The standard approach (analysis of synchronous data)

According to the standard approach, the synchronous data are processed so to determine the cross-spectra and the phase functions. Since from the engineering point of view the movements along the vertical axis are usually not very relevant, we will just focus on the two horizontal components.

Their analysis, incidentally, allow determining whether a mode is flexural or torsional: for a given eigenfrequency, in case the motions of the two points are in phase (i.e. the two points move together along the same direction) the mode is flexural, while in case the data are out of phase by 180° is torsional.

Figure 5.16 shows the cross-spectra and phase functions for both the horizontal components (NS and EW).

First of all we need to highlight two main facts about the cross-spectra (let us neglect minor features):

(1) the NS component has two distinct peaks at 9.40 and 12.40 Hz (Fig. 5.16a);
(2) the EW component has two distinct peaks at 8.37 and 12.40 Hz (Fig. 5.16b).

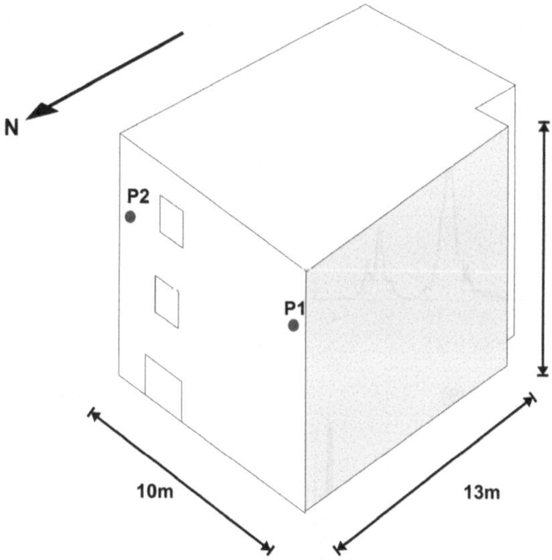

Fig. 5.15 First GHM case study: schematic representation of the 3-storey house with the two measurement points (P1 and P2) at the top floor (where the amplitude of the vibrations are usually larger and therefore easier to characterize). The North direction is fixed perpendicular to the imaginary line that connects the two measurement points (this way the GHM plots can be understood in a straightforward manner)

This means that the building has three vibration modes and the phase functions (Fig. 5.16c and d) will help identifying their nature (flexural or torsional).

The EW components are in fact clearly in phase for all the three identified frequencies (i.e. modes) while for the 12.40 Hz mode, the NS components are out of phase by 180°.

These facts clarify that we have two flexural (8.37 and 9.4 Hz) and one torsional (12.4 Hz) mode (see Table 5.1). The 8.37 Hz flexural mode is about a vibration along the EW direction (see cross-spectra in Fig. 5.16) while the 9.40 Hz flexural mode is mainly along the NS direction but with a smaller component also along the EW axis (see cross-spectra).

The 12.40 Hz torsional mode is such that, while the motion is the same along the EW direction (the EW data are in phase—see Fig. 5.16d), is reversed (out of phase by 180°) along the NS direction (Fig. 5.16c). In other words, while point P2 (see Fig. 5.15) is moving (for instance) towards North-West, point P1 moves South-West; then, when P2 moves back towards South-East, P1 moves towards North-East.

So far, we presented the analyzes performed according to the traditional approach (based on the assessment of the cross spectra and phase functions of synchronous data), but in case it is not possible to record synchronous data, it is not possible to compute the phase functions that are used to check whether a certain mode is flexural or torsional.

The GHM technique presented in the next section can be used to verify whether a mode is flexural or torsional even by considering non-synchronous data collected at two different points (on the same floor) of the structure (e.g. points P1 and P2 in Fig. 5.15).

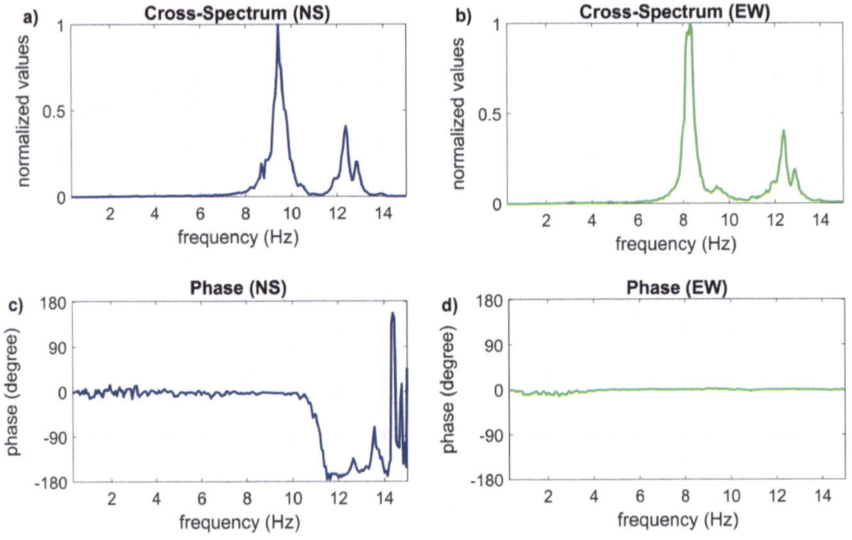

Fig. 5.16 First GHM case study: normalized cross spectra (upper plots) and phase functions (lower plots) for the synchronous data of the NS and EW components. See text for comments. After Dal Moro et al. (2018)

Table 5.1 First GHM case study: identified modes (compare with cross spectra and phase functions in Fig. 5.16)

Frequency (Hz)	Type of mode
8.37	EW flexural mode
9.40	Flexural mode (N20W axis)
12.40	Torsional

The GHM approach

The GHM technique (Dal Moro et al. 2018) is accomplished through the application of a series of narrow Gaussian filters centered on the vibration frequencies identified through the computation of the amplitude spectra.

The process is accomplished in two simple steps and can be applied both on synchronous and non-synchronous data recorded at two (or more) points.

First of all, the amplitude spectra of the NS and EW components are used to identify the eigenfrequencies of the building.

After some general cleaning (to remove possible large-amplitude transient and spurious events), data are divided into a series of windows and the amplitude spectra computed for each segment so to obtain an average amplitude spectrum for both the components (Fig. 5.17).

Of course, the length of the adopted windows depends on the lowest frequencies of the building and we might for instance adopt the same 10-cycle rule used for the computation of the HVSR (see Chap. 3). As a simple and rough rule of thumb we can consider that the lowest eigen frequency (in Hz) of a building can be estimated by dividing 55 by the height (in meters) of the building (of course this is a very approximate value to adopt only in order to fix the length of the window to consider for the data analysis).

Figure 5.17 reports the amplitude spectra for the two horizontal components (for both the points): we can easily recognize the same three frequencies previously identified through the analysis of the cross-spectra of the synchronous data (Fig. 5.16a and b).

If we would plot the particle motion (on the horizontal plane) while considering the raw data we would not obtain any evidence about the building vibration modes because, in general, the particle motion is the complex combination of all the modes (see data reported in Dal Moro et al. 2018).

On the other hand, once we apply a series of Gaussian filters centered on the three vibration frequencies previously identified (see again the amplitude spectra in Fig. 5.17), some evidences clearly emerge.

Figures 5.18, 5.19 and 5.20 present the velocity plots of the particle motion for the points P1 and P2 after applying a series of Gaussian filters centered at 8.37, 9.40 and 12.40 Hz, respectively.

For brevity, such plots are called *GHM plots* (Gaussian-filtered Horizontal Motion plots) and provide the evidence of the type of mode (flexural or torsional).

At 8.37 Hz (Fig. 5.18), the particle motion is exactly the same for both the points and this fact provides the evidence that this is a EW flexural mode (of course, the noise inevitably present in the data can produce some small discrepancies in the two graphs and we must learn to understand whether possible observed discrepancies are

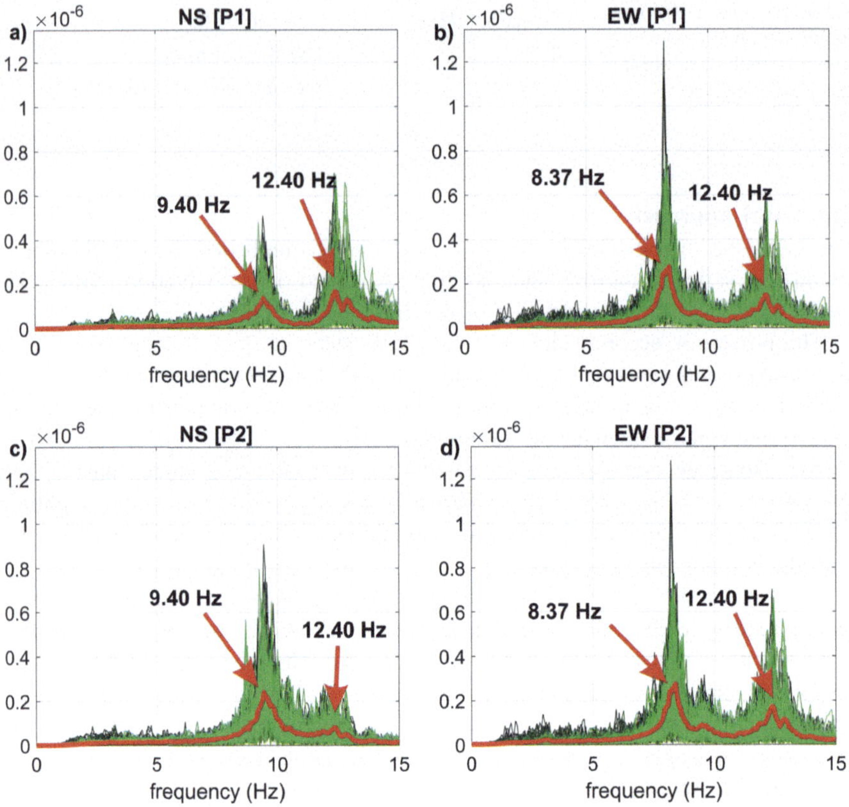

Fig. 5.17 Non-synchronous data recorded on P1 and P2 (Fig. 5.15): amplitude spectra for the NS and EW components. The thin green curves in the background are the amplitude spectra for each single 10 s window while the thick red line is the average spectrum. The three vibration modes at 8.37, 9.40 and 12.40 Hz are evident. Their identification as flexural or torsional is performed through the assessment of the respective GHM plots (see text and next figures). After Dal Moro et al. (2018)

completely irrelevant or not—incidentally this is the reason why data should be firstly cleaned so to remove possible large-amplitude transient events).

The GHM plot clearly demonstrate that the 9.40 Hz mode (Fig. 5.19) is flexural (N20W-S20E axis).

Finally, Fig. 5.20 presents the 12.40 Hz GHM plot. The particle motion along the horizontal plane is the one typical of a torsional mode. In fact, while point P1 moves along the SW-NE direction, point P2 moves approximately along a NW-SE direction.

If we consider the analysis of the synchronous data, at this frequency the NS data are out of phase by 180° (Fig. 5.16c) while EW data are perfectly in phase.

As obvious from the angles and amplitudes, the rotation (torsion) axis is not exactly the geometric center of the building. We should remember that the sensors were positioned symmetrically (Fig. 5.15) and consequently, in case the rotation

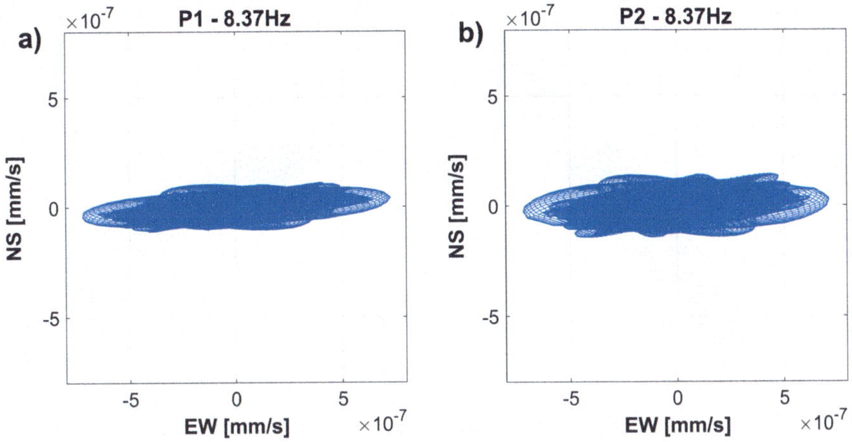

Fig. 5.18 GHM plots (points P1 and P2) while considering a narrow 8.35 Hz Gaussian filter: velocity data for the points P1 (**a**) and P2 (**b**). The 8.37 Hz mode is clearly flexural along the EW axis. This type of evidence is obtained both in the case of synchronous and non-synchronous data. After Dal Moro et al. (2018)

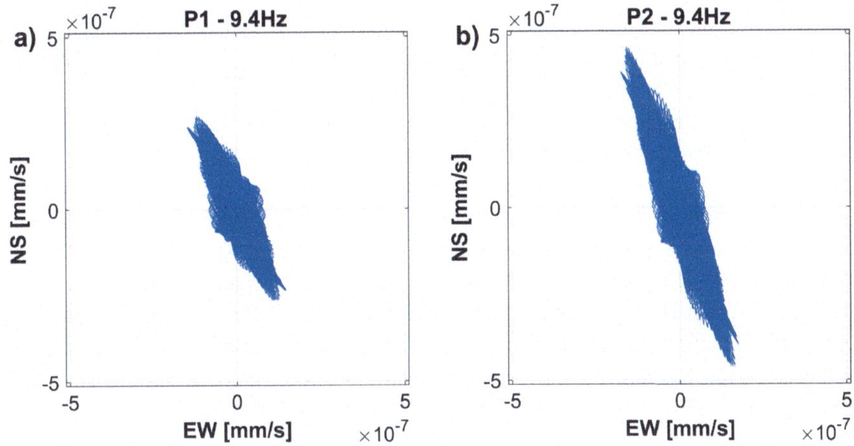

Fig. 5.19 GHM plots (points P1 and P2) while considering a narrow 9.40 Hz Gaussian filter: velocity data for the points P1 (**a**) and P2 (**b**). The 9.40 Hz mode is clearly flexural along a N20W axis

axis would be at the center of the building the amplitudes would be identical and the two angles would be exactly specular. In other words, different amplitudes and angles indicate a different distance of the measurement points from the rotation axis.

This is not so uncommon since the actual behavior depends on several construction details such as for instance the presence and location of stairwells, the mass distribution in the building etc. (see also the different amplitudes of the particle motion at P1 and P2 for the 9.40 Hz flexural mode).

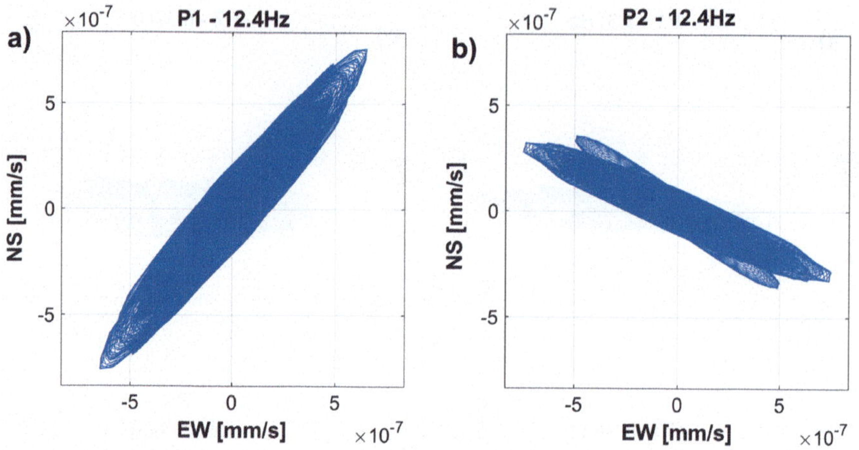

Fig. 5.20 GHM plots (points P1 and P2) while considering a narrow 12.40 Hz Gaussian filter: velocity data for the points P1 (**a**) and P2 (**b**). In this case the mode is clearly torsional

As widely underlined while describing the correct computation of the RPM curve (Chaps. 1 and 4), the data polarity is critical also for the building vibration data. In order to provide a concrete evidence of this, the reader should for instance consider the GHM plot in Fig. 5.20 and imagine that the polarity of the NS trace is reversed. What kind of plots would we obtain? P1 would move along a NW-SW direction and P2 along a NE-SW direction which, from the physical point of view, would be clearly meaningless.

It can be interesting to assess the time series of the synchronous data once the data are filtered with a 12.40 Hz Gaussian filter. Figure 5.21 reports the filtered data for the two points and the two components. The facts highlighted while assessing the phase functions (Fig. 5.16) and the GHM plot (Fig. 5.20) are now even more apparent: while along the EW direction the motion is perfectly synchronous (the two time series perfectly overlap), along the NS direction they are out of phase by 180°. This, once again, means that the 12.40 Hz mode is torsional and that can be identified both through the classic approach (see the joint assessment of the cross-spectra and phase functions) both through the GHM technique.

It is important to emphasize once more that in order to compute the phase functions (considered in the classical approach) and compare the time series as in Fig. 5.21, it is necessary to work with synchronous data. On the other side, the GHM plots shown in Figs. 5.18, 5.19 and 5.20 can be computed considering either synchronous or non-synchronous data (in both cases they provide the same "evidence" about the torsional or flexural nature of the identified mode).

GHM technique can therefore be applied to the data recorded by a single 3C geophone positioned (in sequence) at two (or more) points of the building (on the same floor).

In addition to the frequencies of the eigenmodes, we can also compute the respective damping values according to the classic $1/\sqrt{2}$ method (e.g. Arakawa and

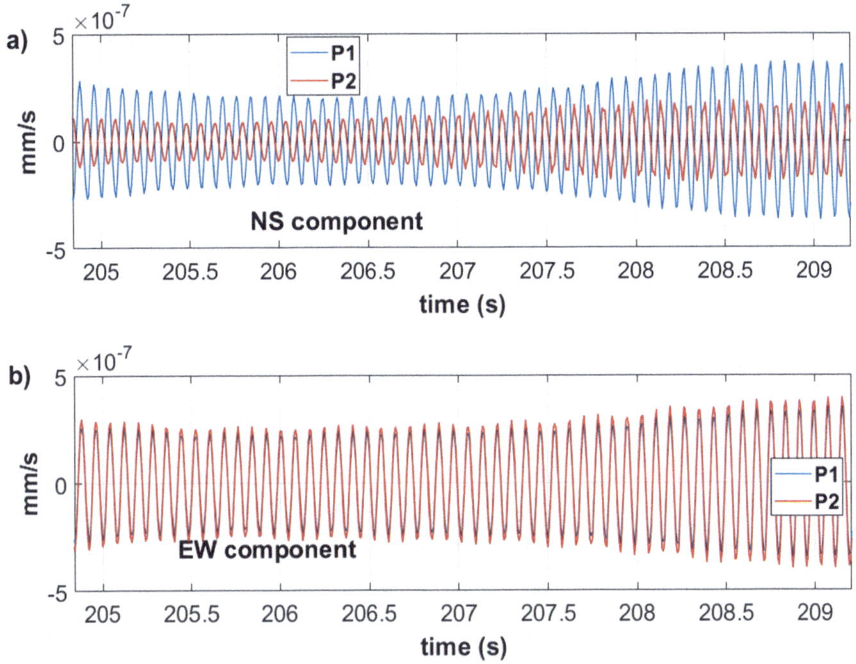

Fig. 5.21 Synchronous data: close up of the velocity data for the points P1 and P2 after the application of a narrow-band Gaussian filter centered at 12.4 Hz: **a** NS components (traces are out of phase by 180°), **b** EW components (traces are practically identical). The vertical scales are intentionally kept the same for both the plots so to clearly compare the amplitudes

Yamamoto 2004). According to such approach, the damping D can be defined through the computation of the following ratio (see Fig. 5.22):

$$D = \frac{1}{2}\frac{f_2 - f_1}{f_m} \tag{5.4}$$

where

f_m = frequency of the considered eigenmode (amplitude A_{max})
f_1 = frequency ($f_1 < f_m$) corresponding to the amplitude $A_{max}/\sqrt{2}$
f_2 = frequency ($f_2 > f_m$) corresponding to the amplitude $A_{max}/\sqrt{2}$.

Figures 5.22 and 5.23 present the computation of the damping for the 8.37 EW flexural mode and the 12.40 Hz torsional mode, respectively.

The obtained values (2.2 and 1.6%) clearly refer to the building behavior in the microtremor regime (during the data recording there was just a very mild breeze) and cannot be considered as representative of the behavior in case of the strong motion that occurs in case of earthquakes (we will come back to this point at the very end of this chapter).

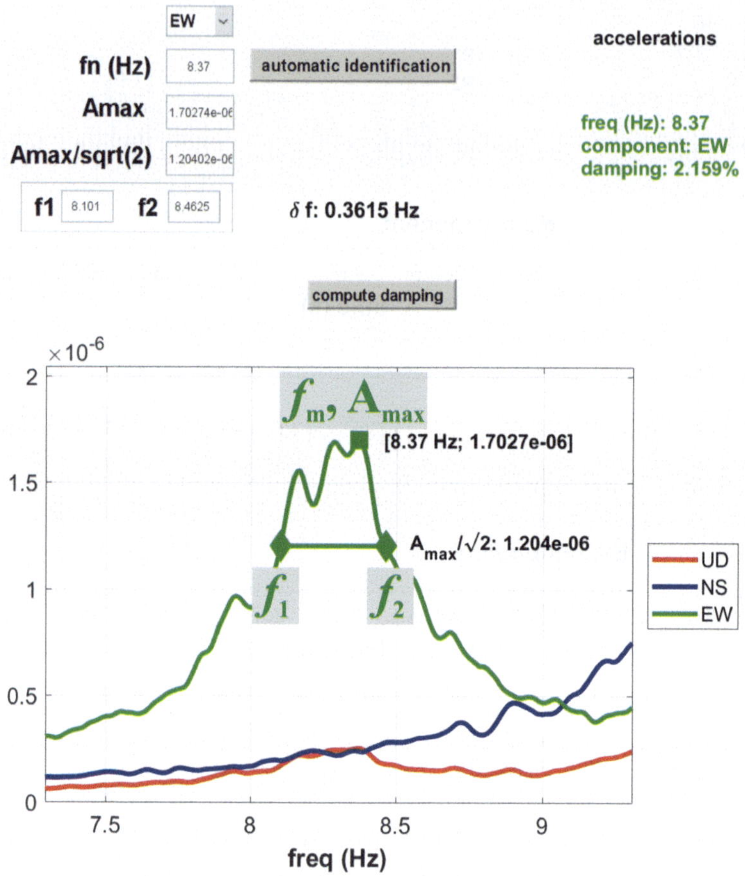

Fig. 5.22 Damping computation (classical $1/\sqrt{2}$ method) for the 8.37 Hz flexural mode (along the EW direction—the obtained value is 2.16%). Close up of the amplitude spectra for the three components around the f_m (8.37 Hz) frequency

Incidentally, the vibration data of this case study are available for downloading together with a couple of video animations useful to appreciate the horizontal motion for the 8.37 Hz flexural mode and the 12.40 torsional mode (see the link provided in the preface of the book).

Why the knowledge of the building behavior is crucial? We already emphasized that the very idea of a "special" site resonance frequency is somehow naïve and lacks of generality (see Fig. 3.16 and related text) since the actual site amplification is not represented by the HVSR curve. The consequence is straightforward: the idea that if one of the natural frequencies of a building is close to the HVSR peak frequency there is a special structural danger is quite pointless. If we go back to Fig. 3.16 and consider the P6 data (on the right panel) we can see that the actual (experimental) amplification at 7 Hz is about 10 while the HVSR value is below 1.

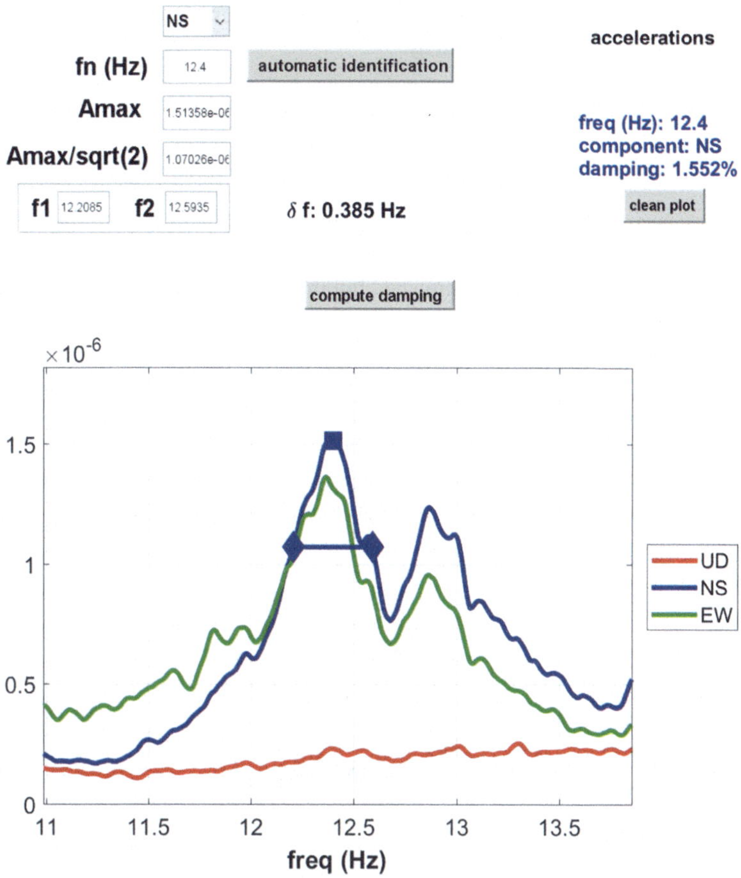

Fig. 5.23 Damping computation for the 12.4 Hz torsional mode (obtained value equal to 1.55%). Reported the value for the NS component only (for the EW component the value is nearly identical). Close up of the amplitude spectra around the f_m frequency (in this case 12.40 Hz)

The knowledge of the experimental eigenfrequencies is useful (actually necessary) in case the structural engineer needs to verify the correctness of the numerical model created in order to assess the structural integrity in case of wind loads and quakes.

Usually, for doing that, the structural engineer creates a Finite Element (FE) model of the structure but, in order to verify that all the properties of the numerous elements are properly defined, needs to compare the obtained results with the experimental evidences achieved through the analysis of the vibration data (see the next GHM case studies presented in the following sections).

5.2.2 Case Study #2: GHM Analysis of Non-synchronous Data for a 25-Storey Building

In cooperation with *Thomas M. Weber* (Synaxis AG Zürich, Switzerland)

and *Lorenz Keller* (roXplore gmbh—Switzerland)

The second case study refers to a 25-storey reinforced concrete building in Zürich (Westlink Tower, Zürich-Altstetten—Fig. 5.24). At the time of measurements, the main construction works were just finished but the interior fitting was not completed.

Due to logistical and security issues, it was only possible to fix three measurement points (P1, P2 and P3—Fig. 5.24) and, also because of the limited available time (construction activities could be stopped for not more than half an hour), it was not possible to record synchronous data. Data were then collected from a single 3C geophone moved to the three indicated points.

We should underline that, during the data acquisition, all the construction machines need to be switched off since the signals they inevitably create could be hard to identify/separate with respect to eigen vibrations of the building.

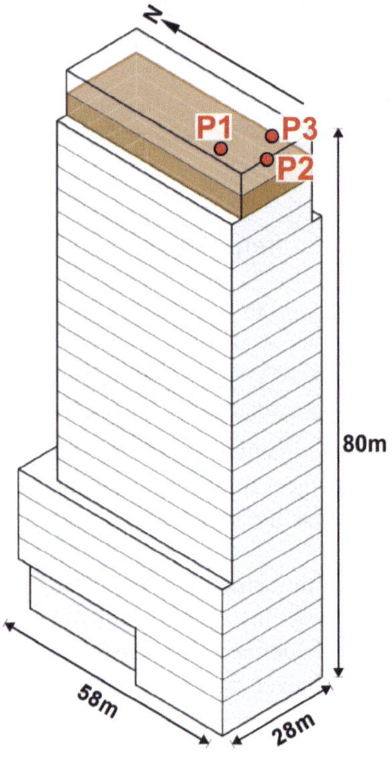

Fig. 5.24 Scheme of the Westlink Tower in Zürich with the three points where non-synchronous vibration data were recorded by means of a single 3C geophone (positioned in sequence at the three different points)

For all these reasons, recording time was 10 min for point P1 and only 5 min for points P2 and P3 (needless to say that, for such kind of tall buildings, longer recording times are highly desirable).

During the data acquisition weather conditions were quite good (no wind).

GHM analysis

The assessment of the amplitude spectra for the NS and EW components (not shown for the sake of brevity—details in Dal Moro et al. 2018) allowed the determination of three main (large-amplitude) eigenfrequencies at 0.59, 0.83 and 1.12 Hz and three further (smaller-amplitude) modes at 2.30, 2.69 and 3.47 Hz.

The GHM plots for the 0.83 and 1.12 Hz frequencies are reported in Figs. 5.25 and 5.26: the 0.83 Hz eigenfrequency clearly refer to a torsional mode while the 1.12 Hz vibration mode is flexural approximately (but not exactly) along the NS direction. The GHM plots for the other four modes are presented and commented in the mentioned article while Table 5.2 summarizes the data about the six identified modes.

As a matter of fact, for some of these flexural modes it is also possible to speculate about smaller torsional components and longer recording times are in general very useful to possibly identify spurious signals or various possible "contaminations" that can affect the data (and consequently the GHM plots). Furthermore, we should also consider that, in some cases, mixed modes are also possible.

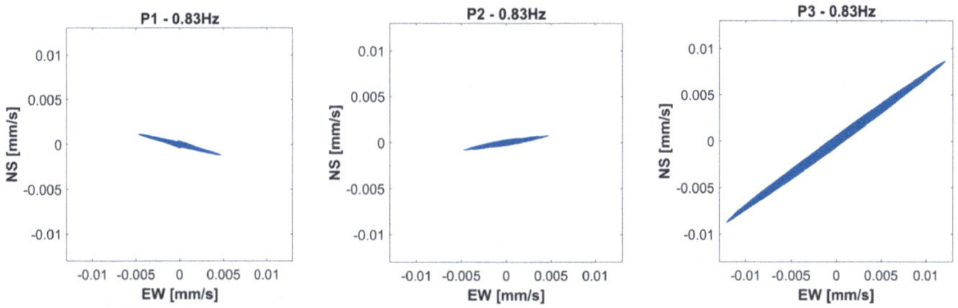

Fig. 5.25 Mode #2: 0.83 Hz GHM plots for points P1, P2 and P3. The mode is clearly torsional. After Dal Moro et al. (2018)

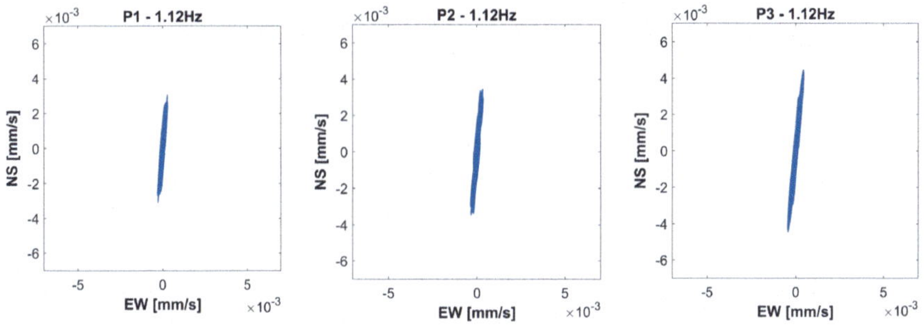

Fig. 5.26 Mode #3 (flexural): 1.12 Hz GHM plots for points P1, P2 and P3. From Dal Moro et al. (2018)

Table 5.2 Westlink Tower: summary table for the six vibration modes identified via GHM analysis

Mode	Frequency (Hz)	Type of mode
M1	0.59	EW flexural mode
M2	0.83	Torsional mode
M3	1.12	NS flexural mode
M4	2.30	EW flexural mode (smaller amplitude compared to M1)
M5	2.69	Torsional mode (smaller amplitude compared to M2)
M6	3.47	NS flexural mode (smaller amplitude compared to M3)

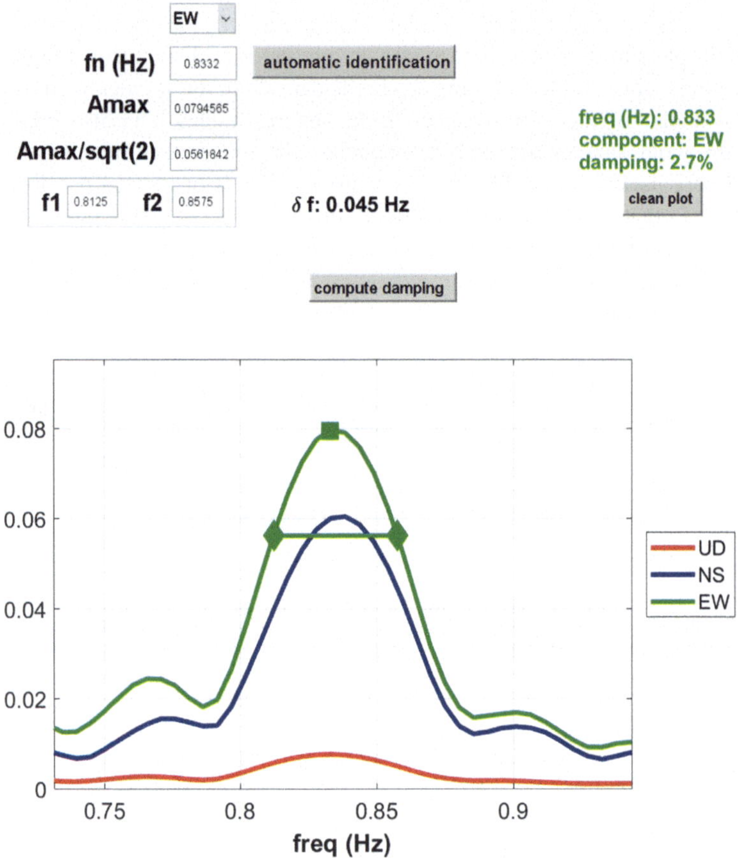

Fig. 5.27 Damping computation for the 0.833 Hz torsional mode (2.7%). Reported the value for the EW component only (for the NS component the value is nearly identical)

Table 5.3 Vibration modes of the Westlink Tower based on the finite element (FE) modeling and comparison with the results of the experimental GHM analysis

Mode	Frequency obtained via FE modelling (Hz)	Type of mode	Experimental frequency obtained via GHM technique (Hz)
M1	0.43	EW flexural mode	0.59
M2	0.60	Torsional mode	0.83
M3	0.83	NS flexural mode	1.12
M4	1.67	EW flexural mode	2.30
M5	1.92	Torsional mode	2.69
M6	2.38	NS flexural mode	3.47

Some further considerations can be put forward in particular about the 0.83 Hz torsional mode (Fig. 5.25). The directions and amplitudes at the points P1, P2 and P3, clearly indicate that the rotation axis lies between points P1 and P2. It is in fact obvious that the amplitudes in P1 and P2 are comparable but have specular directions and that, compared to P2, the amplitude in P3 is larger. This indicates that P3 is farther from the rotation axis and, consequently, has a larger amplitude compared to P1 and P2. Eventually, Fig. 5.27 reports the computation of the damping for the 0.83 Hz eigenmode (EW component).

Comparison with the results of the Finite Element modelling
The FE modelling of the Westlink Tower identified six eigenfrequencies quite close to the experimental data obtained via GHM analysis and also the mode shape (i.e. the type of mode) corresponds for all the six modes. Table 5.3 summarizes the FE and GHM results and provides the evidence that the parameters of the FE model were quite accurate. The small differences between the experimental eigenfrequencies and the values obtained from the numerical analysis can be attributed to the stiffening effects of non-structural components like facade elements and partition walls which were not incorporated in the FE model of the bearing structure (e.g. Su et al. 2005). Further elements possibly responsible for the small differences between measured and modeled eigenfrequencies are the assumed stiffness, the overall building weight and missing live loads (for a general discussion see Dal Moro et al. 2018 and references therein mentioned).

5.2.3 Case Study #3: A 13-Storey Reinforced Concrete Building

In cooperation with Ljuba Sancin

Dipartimento di Ingegneria e Architettura,
University of Trieste di Trieste (Italy)

The considered building is a 13-storey reinforced concrete frame with masonry infills realized in the 60s (Fig. 5.28). The staircase has concrete walls with very poor reinforcement. Although the reinforcement is poor, the concrete walls have a much higher stiffness than the r.c. (reinforced concrete) frame and consequently, if we do not consider the infills, the center of stiffness is likely very eccentric and far from the center of mass.

Some technical notes about the Finite Element modelling

A Finite Element (FE) model of the considered building (Fig. 5.29 reports two of the evaluated models) was defined in order to investigate how different modelling choices can influence the results. The structure was therefore modelled according to five different possible criteria and the results were compared in terms of natural periods, vibration modes and seismic forces.

The first model (N1) contains just the structural elements: r.c. columns and beams and the concrete walls of the staircase. The columns and the beams are modelled with *frame* elements, the walls with *shell* elements. In the second model (N2) there are the same elements, but the concrete walls are modelled as *frame* elements. This choice could be very convenient for a designer, since is computationally lighter compared to the first one. The third model (N3) is the same as the first one, but it has additional diagonals that represent the masonry infills, taken without considering the openings. The fourth model (N4) is very similar to the third with the only difference that the stiffness of the diagonals also accounts for the presence of openings (windows and balcony doors). The fifth model (N5) is the same as the second, but the *frame* elements representing the concrete walls are connected by rigid links that modify the behavior to the staircase.

Since the building codes do not specify how to consider non-structural elements, all these five models (and many more) are so-to-speak "legitimate" in order to evaluate (through numerical simulations) the seismic vulnerability of the building.

Table 5.4 summarized the main data about the first three natural periods (or eigenfrequencies) of each model and the corresponding participating mass in the two main directions (X direction is the long side of the building, Y direction is the short one—see Figs. 5.28 and 5.29). The table also reports the base shears for all the five models. As we can see, they differ quite a lot (e.g. up to 67% for model N3 compared to N1). It can also be noticed that the model N5 is quite similar to the model N1.

In any case the differences obtained while adopting different modelling choices are quite large and a comparison with experimental data is therefore necessary.

Experimental data: the GHM analysis

Ambient microtremor data were collected on two points of the building roof and data processed according to the GHM procedure.

Through the computation of the amplitude spectra (Fig. 5.30) we were initially able to determine a series of frequencies that were then used for the computation of the respective GHM plots. These latter are reported in Figs. 5.31 (first three modes —M1, M2 and M3) and 5.32 (successive three modes—M4, M5 and M6). Table 5.5 reports the main data about the identified eigenfrequencies: M1 and M4 are NS flexural modes, M2 and M5 torsional modes and M3 and M6 mixed modes.

If we compare these experimental evidences with the results of the FE modelling (Table 5.4) we can see that, in this case, the numerical models significantly differ from the observed data since the experimental eigenfrequencies are significantly higher compared to the ones obtained via FE modelling. This fact clearly requires a deeper investigation on the reasons of such discrepancies, which are possibly

Fig. 5.28 The third GHM case study: a 13-storey r.c. building (Gorizia—NE Italy)

related to the contribution of non-structural elements and the effects of larger accidental loads not considered during the numerical modelling.

A special comment is necessary to understand the 6.19 Hz peak frequency labelled with the S letter in the amplitude spectra shown in Fig. 5.30.

A careful evaluation of the characteristics of the amplitude spectra can immediately provide some doubts about its actual meaning but the definitive proof about its spurious nature can be obtained only by computing and comparatively assessing four more (mutually related) "objects":

(1) the spectrograms of the time series;
(2) the time series filtered with a Gaussian filter centered at 6.19 Hz;
(3) the GHM plot;
(4) the damping value.

Table 5.4 Results for the five FE models considered (N1–N5): first three natural periods (T), their participating mass along the X (UX) and Y (UY) directions and base shear (T_b) at DLS (damage limit state)

Mode	N1			N2			N3			N4			N5		
	T (s)	UX	UY	T (s)	UX	UY	T (s)	UX	UY	T (s)	UX	UY	T (s)	UX	UY
1	1.96	0.20	0.12	3.18	0.00	0.63	1.01	0.00	0.65	1.10	0.00	0.65	1.93	0.45	0.00
2	1.75	0.06	0.50	2.60	0.07	0.00	0.64	0.59	0.00	0.76	0.55	0.00	1.74	0.00	0.62
3	1.02	0.42	0.00	1.72	0.61	0.00	0.52	0.12	0.00	0.59	0.15	0.00	0.76	0.31	0.00
T_b (X) (kN)	1840.7			1444.5			3068.2			2568.6			1404.2		
T_b (Y) (kN)	2024.2			1039.2			2543.0			2371.4			1934.5		

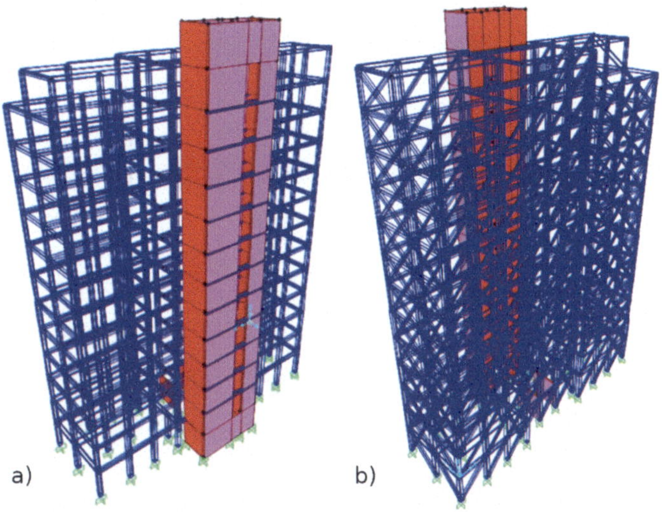

Fig. 5.29 Two FE models of the 13-storey building: **a** model N1; **b** model N2 (see text and summary Table 5.4)

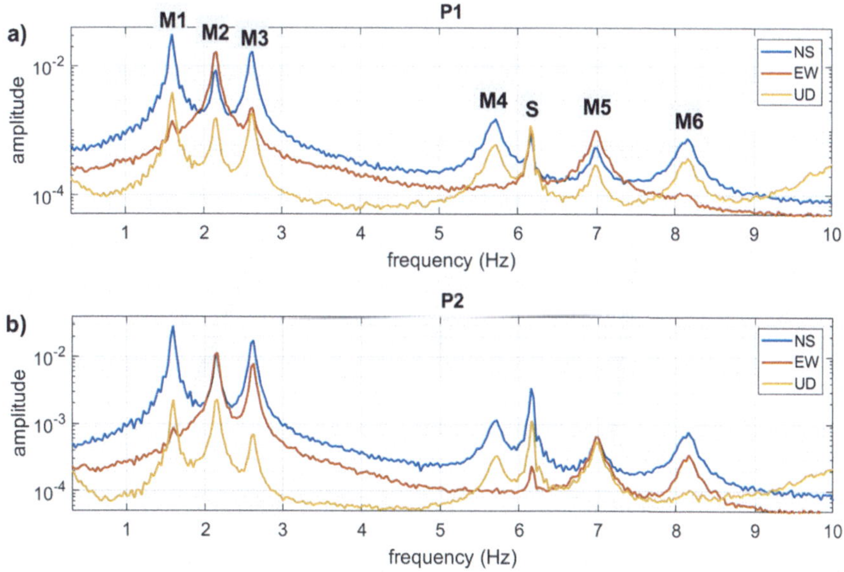

Fig. 5.30 Average amplitude spectra (log scale) for the point P1 (upper plot) and P2 (lower plot): highlighted the six identified modes. The "S" peak is a spurious signal not related to the building eigenmodes (see text)

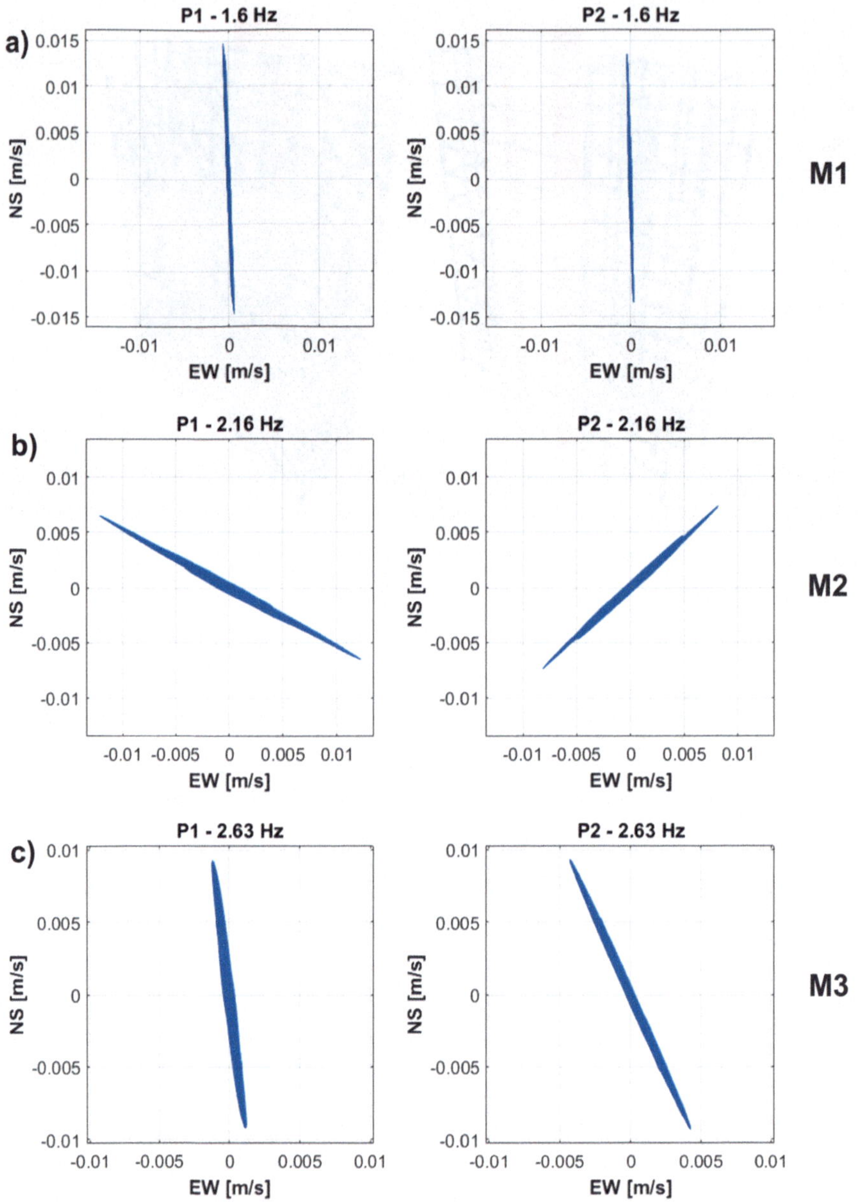

Fig. 5.31 GHM plots for the first three modes (see also Table 5.5): **a** 1.6 Hz flexural mode; **b** 2.16 Hz torsional mode; **c** 2.63 Hz mixed mode

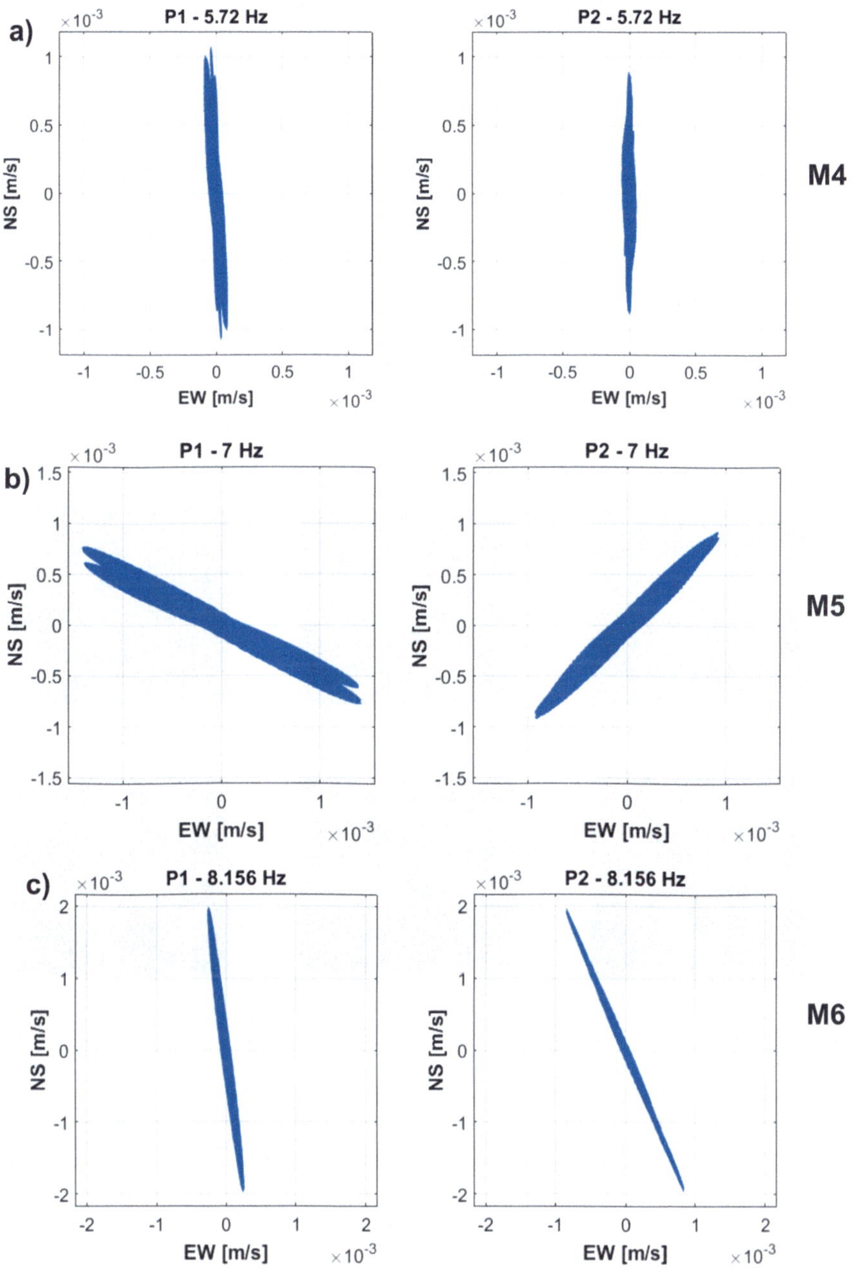

Fig. 5.32 GHM plots for the modes M4, M5 and M6 (see also Table 5.5): **a** 5.72 Hz flexural mode; **b** 7.0 Hz torsional mode; **c** 8.16 Hz mixed mode

Table 5.5 Modes identified via GHM: modes 4, 5 and 6 reproduce the same kind of behavior as the first three modes but at higher frequencies and with a lower amplitude (compare GHM plots shown in Figs. 5.31 and 5.32)

Mode	Frequency (Hz)	Period (s)	Type and comment
M1	1.60	0.6250	Flexural (NS axis)
M2	2.16	0.4630	Torsional
M3	2.63	0.3802	Mixed
M4	5.72	0.1748	Flexural (same as M1 but about one order of magnitude smaller)
M5	7.00	0.1429	Torsional (same as M2 but about one order of magnitude smaller)
M6	8.16	0.1225	Mixed (same as M3 but about one order of magnitude smaller)

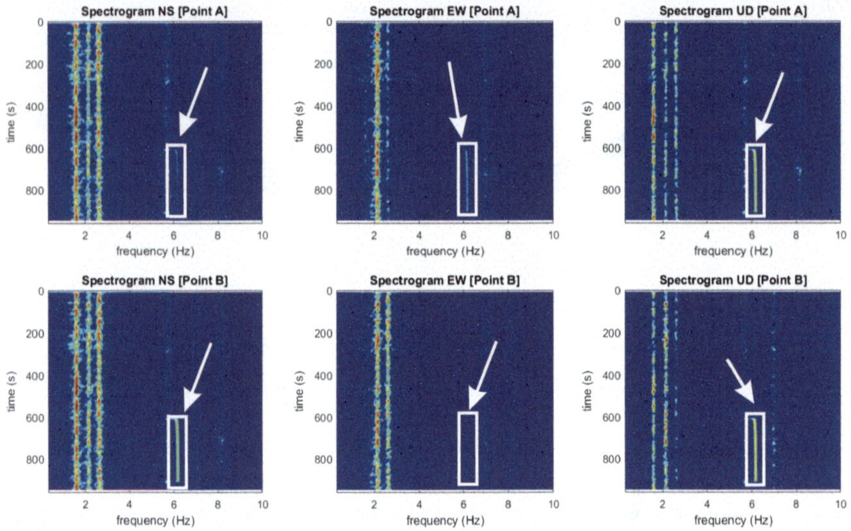

Fig. 5.33 Amplitude spectra over the time (i.e. spectrograms) for the three (NS, EW and vertical) components. Upper and lower plots report the data for P1 and P2, respectively. The presence of a spurious 6.19 Hz vibration between 600 and 900 s is extremely clear (it is a large-amplitude temporary and "artificial" signal likely produced by some household appliance). See also Fig. 5.34

Figures 5.33 and 5.34 show the spectrograms for all the three components for both the P1 and P2 points. It is clear that the 6.19 Hz signal is not permanent but appears and disappears all of a sudden (total duration is about 5 min) and therefore cannot be one of the eigenmodes of the building (which, since they are intrinsic charateristics of the structures, are permanent).

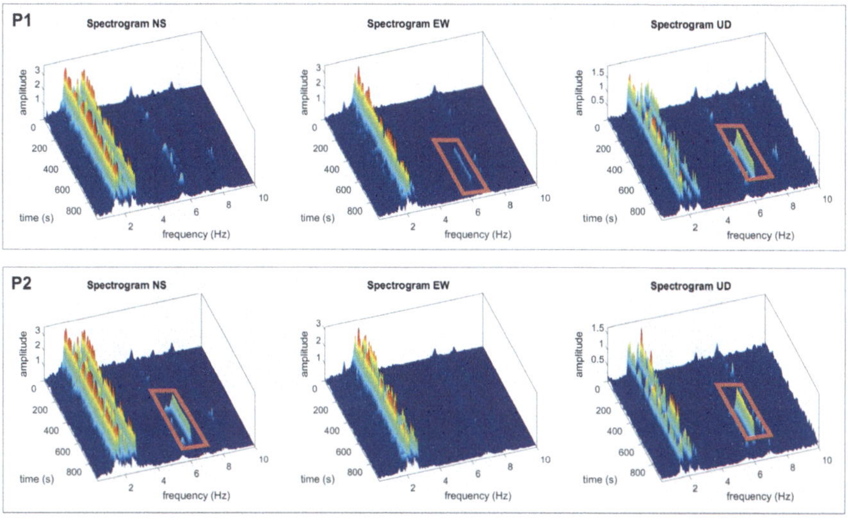

Fig. 5.34 Amplitude spectra over the time for the three components (upper plots P1, lower plots P2) according to a 3D perspective. The spurious 6.19 Hz signal is extremely clear (see text and compare also with Figs. 5.33, 5.35, 5.36 and 5.37)

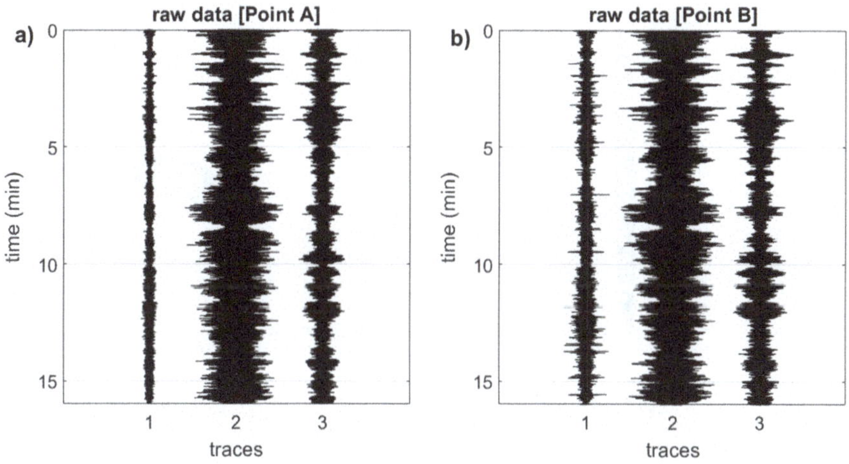

Fig. 5.35 Raw data for the point P1 (on the left) and P2 (on the right). Reported the vertical, NS and EW traces

If we have a look at the raw data (Fig. 5.35), nothing is apparent but once we apply a Gaussian filter centered at 6.19 Hz, the signal immediately shows up (Fig. 5.36).

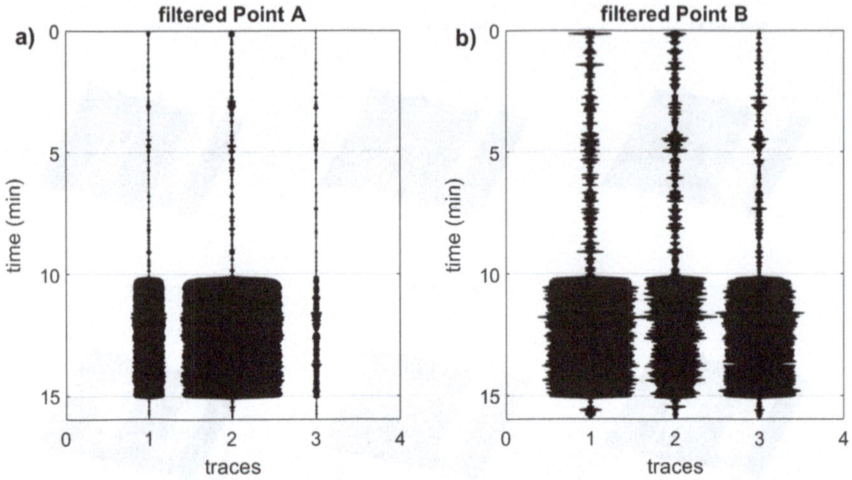

Fig. 5.36 Gaussian-filtered data at 6.19 Hz (vertical, NS and EW traces) for the two considered points (P1 on the left and P2 on the right): the spurious signal has a very distinctive start and end. Compare with the raw data in Fig. 5.35, where no evidence is apparent (the usefulness of the Gaussian filtering is apparent)

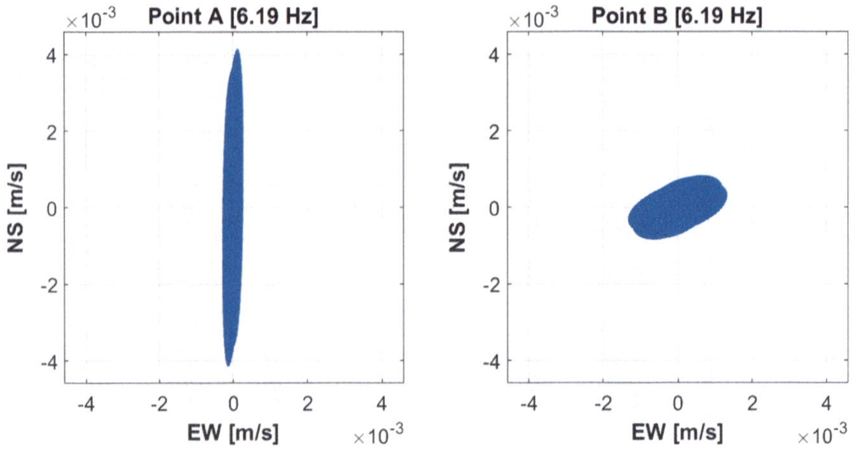

Fig. 5.37 GHM plot for the 6.19 Hz spurious signal. The motion at the two points is clearly pointless and cannot represent any building eigenmode

If we compute the GHM plot for this frequency (Fig. 5.37) we obtain something that cannot be explained in terms of flexural or torsional modes (video animation of the synchronous data at 6.19 Hz shows "meaningless" particle motion at P1 and P2);

Eventually we also computed the damping values for the eigenmodes and the S spurious signal. Figure 5.38 presents the damping computation for the first three modes (for the sake of brevity we present the data just about the Point 1, being the values for

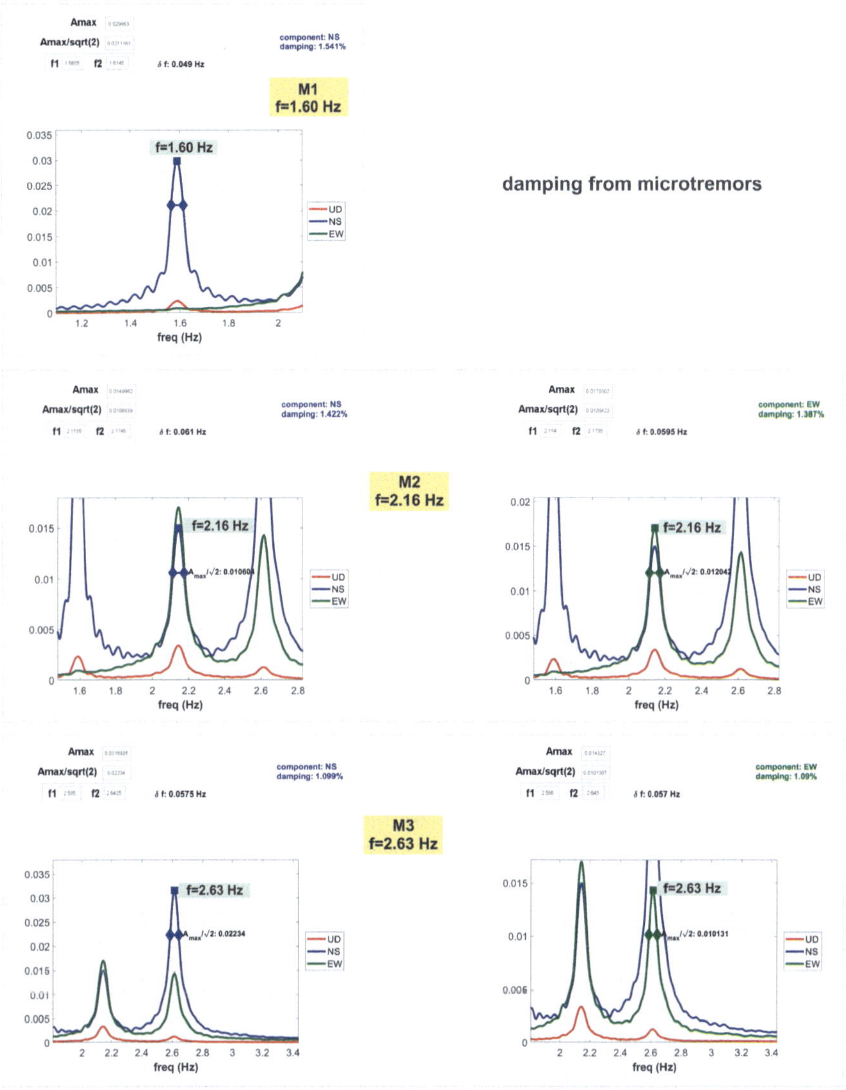

Fig. 5.38 Damping computation for the first three modes (measurement Point 1). From top to bottom: M1 (flexural mode—reported just the NS component), M2 (torsional mode—reported for both the NS [on the left] and EW [on the right] components), M3 (mixed mode—reported both the NS and EW data) (see Table 5.5). The values are around 1.5%, 1.4% and 1.1%, respectively

Point 2 nearly identical). If we compare the damping values for the eigenmodes (ranging from 1.1 to 1.5%—Fig. 5.38) with the value obtained for the S signal (0.4%—see Fig. 5.39) we obtain a further evidence about the "artificial" nature of such a signal which was likely produced by some household appliance or by the engine of some facility (this kind of spurious/industrial signals are always characterized by a much lower damping value compared to the ones of the real eigenmodes).

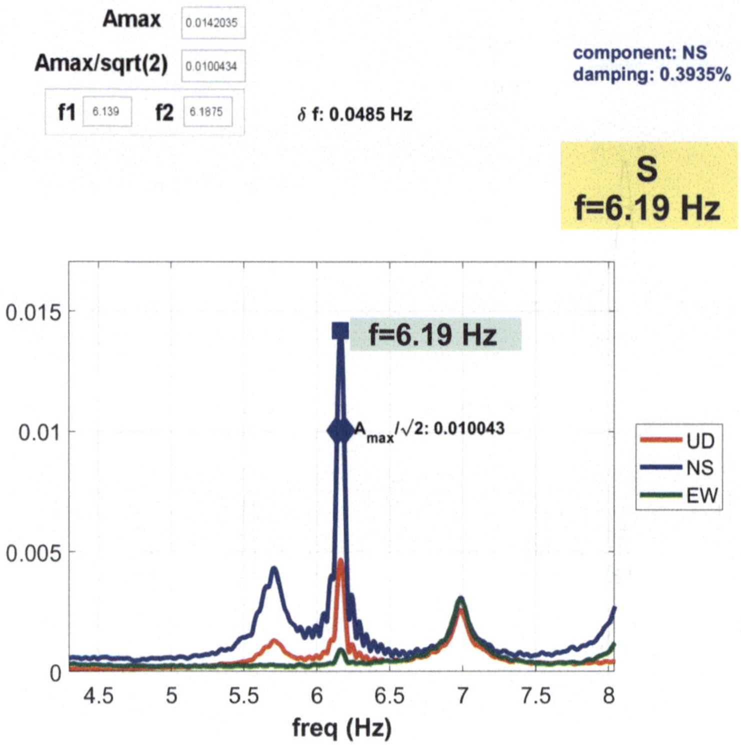

Fig. 5.39 Damping computation for the spurious S signal (see Figs. 5.33, 5.34, 5.35, 5.36 and 5.37) (for the sake of brevity we report only the computation for the NS component). Compared to the values obtained for the actual eigenmodes of the building (Fig. 5.38), the damping of the S signal is significantly lower (about 0.4%). This fact is apparent even through the simple visual comparison between the shape (very pointy) of the 6.19 Hz peak and the smoother shape of the adjacent signals at 5.72 and 7.0 Hz

Frequencies and *damping* factors as a function of the vibration amplitudes

Building vibrations are usually measured while considering ordinary microtremor data (i.e. small-amplitude vibrations), but what happens in the event of larger tremors such as those occurring during an earthquake?

In general, we observe an increment of the *damping* value (that can increase from 1 to 2% [in case of microtremors] to about 5%) and a decrease of the eigenfrequencies for a value indicatively equal to 10% of the values identified from the microtremor analysis.

These issues are described for instance in Michel et al. (2011) while an interesting and comprehensive series of real-world data is presented in Ceravolo et al. (2017) (see Fig. 5.40).

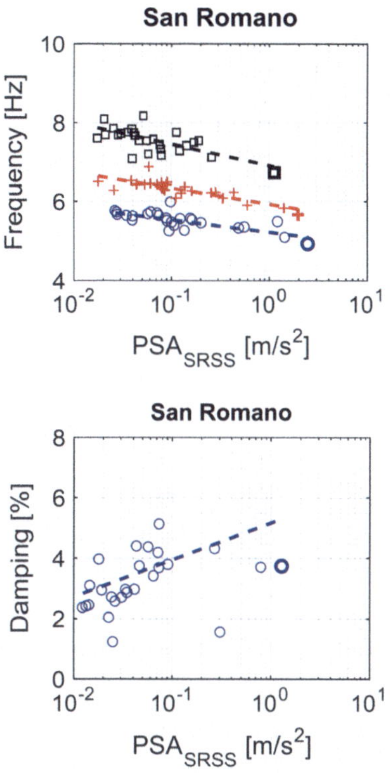

Fig. 5.40 City hall of the San Romani in Garfagnana town (Italy): variability of the frequency of the first three modes and changes of the *damping* value of the first mode as a function of the PSA (*Peak Structural Acceleration* recorded in the building). From Ceravolo et al. (2017)

References

Arakawa T, Yamamoto K (2004) Frequencies and damping ratios of a high rise building based on microtremor measurement. In: Proceedings of the 13th world conference on earthquake engineering, Vancouver, BC, Canada, 1–6 Aug 2004. Paper no 48

Athanasopoulos GA, Pelekis PC (2000) Ground vibrations from sheet pile driving in urban environment: measurements, analysis and effects on buildings and occupants. Soil Dyn Earthq Eng 19:371–387

Bonev Z, Vaseva E, Blagov D, Mladenov K (2010) Seismic design of slender structures including rotational components of the ground acceleration. In: Eurocode 8 approach. Proceedings of 14th European conference on earthquake engineering, Ohrid, Macedonia, vol 9, pp 6706–6713

Brownjohn JMW (2003) Ambient vibration studies for system identification of tall buildings. Earthq Eng Struct Dyn 32:71–95

Cagnetti V, Pasquale V (1979) The earthquake sequence in Friuli, Italy, 1976. Bull Seismol Soc Am 69:1797–1818

Ceravolo R, Matta E, Quattrone A, Zanotti L, Fragonara L (2017) Amplitude dependence of equivalent modal parameters in monitored buildings during earthquake swarms. Earthq Eng Struct Dyn 46:2399–2417

Charney FA (2010) Seismic loads: guide to the seismic load. Provisions of ASCE 7-05. American Society of Civil Engineers Press, Reston, p 231
Chopra AK (1995) Dynamics of structures: theory and applications to earthquake engineering. Prentice Hall, Upper Saddle River, NJ, 944 pp
Dal Moro G, Weber T, Keller L (2018) Gaussian-filtered Horizontal Motion (GHM) plots of non-synchronous ambient microtremors for the identification of flexural and torsional modes of a building. Soil Dyn Earthq Eng 112:243–245
Deckner F (2013) Ground vibrations due to pile and sheet pile driving—influencing factors, predictions and measurements. Licentiate thesis, Division of Soil and Rock Mechanics, Department of Civil and Architectural Engineering, School of Architecture and the Built Environment. KTH, Royal Institute of Technology, Stockholm, p 126
de Silva CW (2005) Vibration and shock handbook. CRC Press, Taylor & Francis Group, Boca Raton, p 1771
Garevski M (ed) (2012) Earthquakes and health monitoring of civil structures. Springer, Netherlands, 331 pp. ISBN 978-94-007-5182-8
Ghosh SK (2003) Seismic and wind design of concrete buildings. Kaplan Publishing, New York, 542 pp
Gupta AK (1990) Response spectrum method in seismic analysis and design of structures. Blackwell Scientific Publications, Cambridge, 170 pp
Hong LL, Hwang WL (2000) Empirical formula for fundamental vibration periods of reinforced concrete buildings in Taiwan. Earthq Eng Struct Dyn 29:327–337
Jongmans D (1996) Prediction of ground vibrations caused by pile driving: a new methodology. Eng Geol 42(1):25–36
Kim DS, Lee JS (2000) Propagation and attenuation characteristics of various ground vibrations. Soil Dyn Earthq Eng 19(2):115–126. https://doi.org/10.1016/S0267-7261(00)00002-6
Kouroussis G, Conti C, Verlinden O (2014) Building vibrations induced by human activities: a benchmark of existing standards. Mech Ind 15:345–353
Massarsch KR, Fellenius BH (2008) Ground vibrations induced by impact pile driving. In: 6th international conference on case histories in geotechnical engineering, Arlington, VA, 11–16 Aug
Michel C, Guéguen P, El Arem S, Mazars J, Kotronis P (2010) Full scale dynamic response of a RC building under weak seismic motions using earthquake recordings, ambient vibrations and modelling. Earthq Eng Struct Dyn 39:419–441
Michel C, Zapico B, Lestuzzi P, Molina FJ, Weber F (2011) Quantification of fundamental frequency drop for unreinforced masonry buildings from dynamic tests. Earthq Eng Struct Dyn 40:1283–1296
Naito Y, Ishibashi T (1996) Identification of structural systems from microtremors and accuracy factors. In: Eleventh world conference on earthquake engineering, Acapulco, Mexico. Paper no 770
Sato H, Fehler MC (2009) Seismic wave propagation and scattering in the heterogeneous earth. Springer, Berlin, Heidelberg. ISBN: 978-3-540-89622-7
Su RKL, Chandler AM, Sheikh MN, Lam NTK (2005) Influence of non-structural components on lateral stiffness of tall buildings. J Struct Des Tall Spec Build 14(2):143–164
Takada T, Iwasaki R, An DD, Itoi T, Nishikawa N (2004) Dynamic behavior change of buildings before and after seismically retrofitting. In: Proceedings of the 13th world conference on earthquake engineering, Vancouver, BC, Canada, 1–6 Aug 2004
Trifunac MD (2009) The role of strong motion rotations in the response of structures near earthquake faults. Soil Dyn Earthq Eng 29:382–393
Zembaty Z (2009) Rotational seismic load definition in Eurocode 8, part 6, for slender tower-shaped structures. Bull Seismol Soc Am 2009(99):1483–1485

Some Final Remarks and Recommendations

6

Partial knowledge is more triumphant than complete knowledge; it takes things to be simpler than they are, and so makes its theory more popular and convincing.

Friedrich Nietzsche—Human, All Too Human

G. Dal Moro, *Efficient Joint Analysis of Surface Waves and Introduction to Vibration Analysis: Beyond the Clichés*,
https://doi.org/10.1007/978-3-030-46303-8_6

Abstract

In the final chapter are summarized a series of practical advices about the efficient acquisition of field data. A general discussion about the joint analysis of several observables is provided together with a few more comparative examples and, in the very end, a synthetic list of the main facts illustrated in the book.

6.1 Summarizing

Most of the problems in seismic-data analysis are due to the fact that geologists without any specific and profound geophysical background believe that all the methodologies aimed at analyzing surface waves for the determination of the V_S subsurface profile are a sort of "geotechnical test". Such a word (*test*) comes from the geotechnical sector and has a very precise meaning. A *geotechnical test* is a series of well-defined (standardized) procedures that define a protocol that must be strictly followed so to obtain a final "number" that can be then compared to other "numbers" obtained while adopting exactly the same protocol.

This means that geotechnical tests have a universal (although relative) value: since the procedures to obtain a certain value are always the same, the obtained numbers can be compared. In order to perform a geotechnical test, is not necessary to be a professional: it is just necessary to follow the pertinent protocol since the obtained results depend on the correct execution of a series of standardized operations.

On the other side, this approach cannot be applied to geophysical surveys (acquisitions and analyses) and considering a geophysical survey as a geotechnical test is big (and extremely common) mistake.

In seismics (and, in general, in geophysics) nothing can be standardized. Both during acquisition and data processing, everything depends on personal (and to some degree arbitrary) choices and decisions. Potentially, the same data can be handled in different ways and, consequently, lead to different results.

Usually, when we fail to identify the correct subsurface model is not because the "methodology" is inaccurate, but because we did not record the necessary data, failed to understand them and/or performed erroneous choices during the analysis.

The data (i.e. the signals) necessarily depend on the subsurface model and the inversion/modelling procedures are aimed at reconstructing the subsurface properties by carefully analyzing them. The correctness of the retrieved model depends on the quality of our analyses.

The actual subsurface model is one (and unique). It is only through a careful joint analysis of different observables that it is eventually possible to converge towards the correct model.

It is a complex process that should be accomplished by an experienced professional (with specific background and skills) and not a mechanical application of "simple rules".

In general terms, the question to answer is: what are the properties of the subsurface materials?

The approach to follow must be fixed while considering the specific goals and site characteristics.

This is the point: the goal is not performing a "MASW" (or whatever) survey. The goal is the detailed reconstruction of the subsurface properties through a careful analysis of the data recorded according to the methodology that better suits the goals and local conditions: in this book we have seen several possible ways to record and analyze surface wave propagation and we have also seen that the standard MASW (vertical geophones and modal dispersion curves) can provide inaccurate results and is nowadays quite obsolete.

Let us now summarize main technical facts:

(1) If we consider the state-of-the-art about surface-wave analysis, the MASW acronym has completely lost any meaning since does not indicate anything precise. Everything can in fact be considered "MASW":

  (a) M stands for *Multichannel* but everything is necessarily multi-channel (for instance, HVSR and HS require the analysis of three traces and are therefore multi-channel);
  (b) *Surface Waves* (SW). It is not stated what kind of surface waves we are considering: Love waves? Rayleigh waves? Scholte waves? And which component of Rayleigh or Scholte waves? The vertical, the radial or both?
  (c) It is not clear what kind of analysis (A stands for *Analysis*) we are speaking about (interpreted modal curves? effective dispersion curves? FVS?).

For instance, the case study briefly illustrated in Appendix G can be surely called MASW but it should be clear that the Rayleigh-wave propagation is analyzed in a holistic manner through the FVS analysis of both the Z and R components (phase-velocity spectra) also including the RPM frequency-offset surface. The consequence is that the obtained results are necessarily much more reliable when compared to the "standard" analysis of the modal (picked, i.e. interpreted) dispersion curves of the vertical component of Rayleigh waves only.

On the other side, the HS approach based on the active data recorded by a single 3C geophone can also be called MASW since it is based on multi-channel (but single-offset) data (the three Z, R and T components) and is about surface-wave analysis accomplished while considering up to six observables (group velocity spectra of the Z R and T components together with the RVSR and RPM curves and, in case we also recorded passive data, the HVSR).

(2) The most recent methodologies require properly-designed and high-quality *acquisition systems*. The "meaning" of a set of seismic traces does not depend on the geophones or on the seismograph but on the specific *combination* of seismograph (the A/D conversion unit), cables and geophones (see for instance Figs. 1.17 and 1.18 and related text). The acquisition software (i.e. the software that controls the seismograph and that represents the way we "communicate" with the A/D unit) is very often the critical point and little care is paid to improve its functionalities. For instance, the way some seismographs manage the vertical stack (e.g. Figs. 1.32 and 4.2) or the quality check of passive data (e.g. Fig. 4.22 and A.5) is often neglected or cumbersome.

(3) In order to define the phase velocities we need multi-offset data (MASW, ESAC, SPAC, ReMi, MAAM etc.) while group velocities can be obtained by considering just a single trace/geophone (MFA/FTAN/HS). Of course, both the velocities bring the same kind of information about the subsurface materials (even if from two different "points of view").

(4) *Joint analysis* (a precise and quantitative process) should not be confused with a qualitative *data integration* (where ambiguous seismic data are interpreted and compared).

(5) The joint analysis of several observables is the only way to solve the non-uniqueness of the solution and the possible ambiguities of the data (see Sect. 1.3) and provide reliable V_S profiles.

6) While recording MASW (multi-offset) data, the number of geophones/ channels used is irrelevant if the velocity spectra are computed via phase shift (see e.g. Figs. 1.15 and 2.3 together with the related text).

(7) The phase shift method (used to obtain phase-velocity spectra from the field data) allows analyzing non-equally spaced data (Fig. 1.16 and related text).

(8) ESAC/SPAC/MAAM methodologies (i.e. the passive techniques used to determine the dispersive properties) have two major issues: (a) it severely suffers from the non-uniqueness problem (since the same *effective* curve can be the result of different mode combinations such a problem can be even more serious than with the active MASW data); (b) for common professional applications, we can deal with the vertical component of Rayleigh waves only (the analysis of the other components would require three-component data and more complex processing procedures).

(9) Combining Z-component MASW and ESAC/SPAC/ReMi/MAAM does not make much sense (it cannot be considered as a real joint analysis). Both are about the dispersion of the vertical component of Rayleigh waves. In the low-frequency range (say below 5 Hz) passive techniques provide clearer information but two facts should be considered: on one side ESAC/SPAC/ReMi methodologies require relatively-complex field operations and very large arrays; on the other side, low frequencies are easily analyzed by means of the HVSR (which is much easier to obtain).

(10) Horizontal geophones are useful to record Love waves, the radial component of Rayleigh waves, the SH-wave refraction travel times (and, if desired, the radial component of the P-wave refraction). Vertical geophones can be used

just for the acquisition of the vertical component of Rayleigh waves and for P-wave refraction surveys but, nota bene, P-wave velocities are strongly influenced by the saturation conditions and are therefore quite useless in all those situations where the water table is shallow (the layers below the water table are usually "invisible").

6.2 Miscellaneous Notes

The differences between a multi-channel acquisition aimed at defining the phase velocities (MASW, ReMi, ESAC, SPAC, MAAM etc.) and the data acquisition aimed at defining the group velocities (HS/MFA/FTAN) need to be underlined with respect to some practical consequences:

(1) for the definition of the group velocities (through active data) it is necessary that the first sample of the data exactly defines the "origin time". We need to work with the correct time (the "zero time" is the time of our hammering) exactly as for any refraction/reflection studies. No pre-trigger time must be applied;
(2) group-velocity analysis allows the definition of the average subsurface model between the source and the receiver.

On the other hand, the analysis of multi-offset data for the definition of the phase velocities is based on the correlation (*lato sensu*) between the different traces and the absolute time is not relevant. We just analyze the "slope" of a signal and this means that an inaccurate (or completely unknown) "zero time" is completely irrelevant. This is the reason why correlation techniques (that allow the determination of the phase velocities) are used in the analysis of passive data (ESAC, ReMi, etc.): the arrival time is not relevant because are considered just the "relationships" between the traces.

Figure 6.1 shows an example of field acquisition aimed at recording a dataset sufficiently comprehensive to allow a robust analysis and obtain therefore a reliable subsurface velocity profile. In case we have just one 3C geophone, we can initially deploy it at position#1 and record the active data (see HS methodology presented in Chap. 4). We then keep the geophone in the same position and record some passive data for the determination of the HVSR at that position. Successively, we move the 3C geophone in the position#2 and record a second passive dataset aimed at obtaining a second HVSR curve. During the data analysis, we will compare the two HVSR curves. If the two HVSR curves are similar, we obtain two important information:

1. Recorded data are consistent and, therefore, meaningful;
2. No major lateral variation is present.

On the other side, depending on the characteristics of the HVSR curves, major differences can be interpreted as representative of significant lateral variations or, in some cases, attributed to some failure in one of the two datasets. Passive data can in

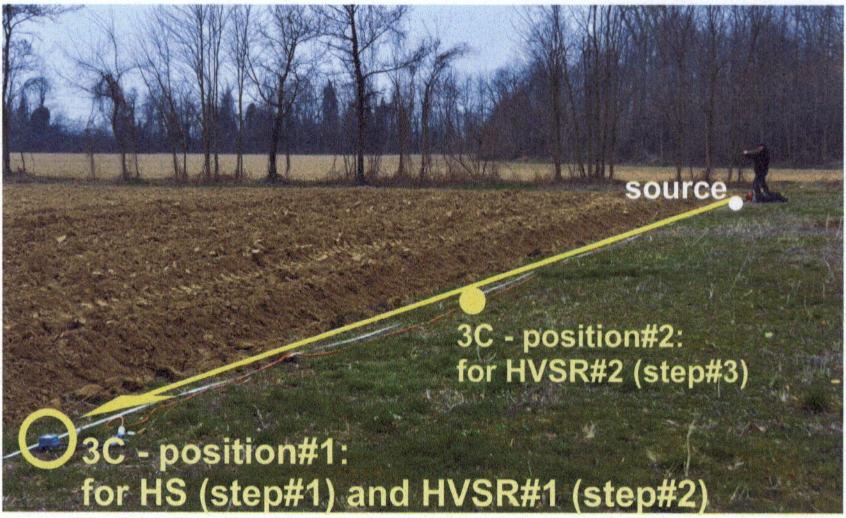

Fig. 6.1 Example of field operation for a HS + HVSR comprehensive acquisition in case we have just one 3C geophone. We initially deploy the 3C receiver (geophone) in the position#1 and record the active (HS) data. We then keep the geophone in the same position and record the passive data for the determination of the HVSR in that position (step#2). We then move the 3C geophone in the position#2 and record a second passive dataset aimed at obtaining a second HVSR curve (step#3). See text for further details and comments

fact sometimes fail for reasons that are not always easy to identify and, in that case, we should be able to understand that that specific dataset is meaningless and must be therefore rejected (this clearly requires both a robust theoretical background and a significant field experience).

In case we have two 3C geophones (and a suitable *acquisition system*), we can opt for a different field procedure that allows to save some time (Fig. 6.2). We first deploy one of the two geophones so to record the active (HS) data (step#1 in Fig. 6.2). Successfully (step#2), by connecting also the second 3C geophone, we can simultaneously record the passive data necessary to define the two HVSR curves (at the two locations). This way, we will save some time and obtain a single (passive) 6-trace dataset that can be used to define two HVSR curves. Needless to say that, in order to apply this kind of optimized acquisition procedure, the whole *acquisition system* needs to be properly designed (see Chap. 1).

Incidentally, the joint use of two 3C geophones also allows the acquisition of synchronous data useful for the analysis of the building vibrations (see Chap. 5) or, in some cases, the definition of the SSRn (Chap. 3).

As far as it concerns the dispersion analysis, throughout the book we highlighted that a single component (i.e. a single *observable*) is usually insufficient to adequately constrain the analysis and it is always recommended to perform multi-component joint analyses.

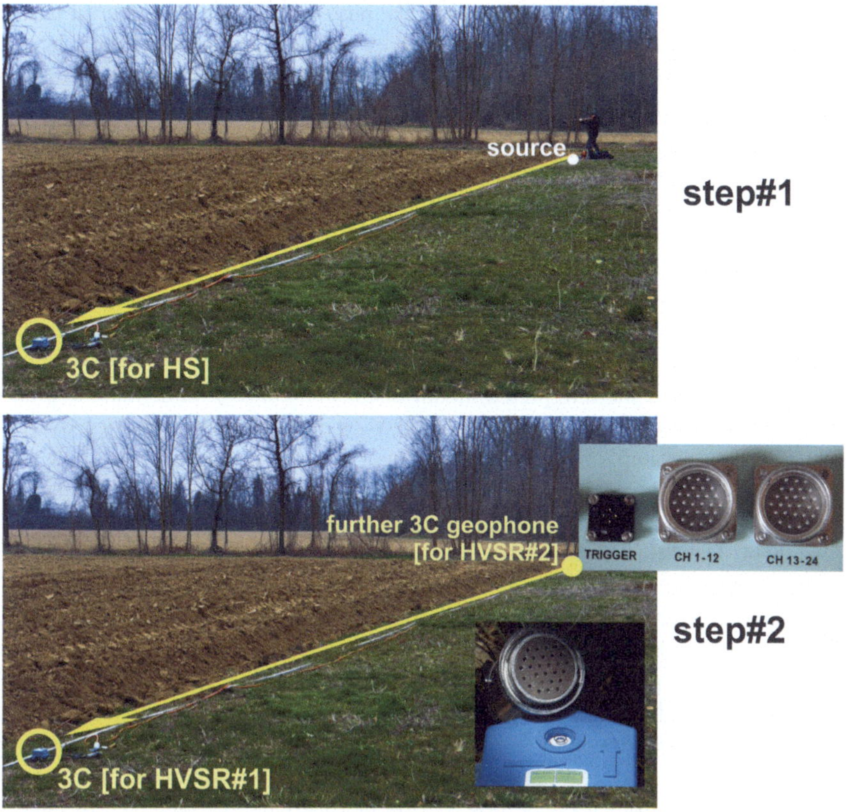

Fig. 6.2 Example of field operations for a HS + HVSR acquisition in case we have two 3C geophones. We place the first 3C geophone so to record the active (HS) data (step#1). We then keep that geophone in the same position and connect to the seismograph a second 3C geophone so to simultaneously record the data for the determination of two HVSR curves (at two different positions) (step#2). See text for details

The choice of the most appropriate observables depends on the characteristics of the site and on the specific goals. Often, in order to plan an efficient acquisition campaign for the exploration of a large area, it would be convenient to carry out a small preliminary survey aimed at obtaining an overall understanding of the area and identify the most effective approach necessary to unambiguously explore the entire area.

In general, thanks to their behavior (i.e. the simplicity of their velocity spectra that describe the dispersion) Love waves represent a very useful component that should be considered as a sort of safety belt for any type of work (see for instance case studies and synthetic datasets presented in Safani et al. 2005; Dal Moro 2014; Dal Moro et al. 2015, 2017).

By the way, it is not possible to define *the* perfect approach in universal terms, since it depends on both the specific objectives and complexity of the data. In this respect, a very important point should be underlined: a simple stratigraphy (i.e. site) can give rise to very complex (i.e. potentially misleading) data then cannot be solved through the analysis of a single component. It is therefore important to avoid embracing the naive idea that simple sites are necessarily associated to simple data while complex sites inevitably produce complex data.

The presence of higher modes is not a problem *per se*. If their presence is correctly understood and the analysis performed accordingly, higher modes are a major source of information but in case they are mistakenly interpreted, we inevitably obtain an erroneous subsurface model. As a matter of facts, in the standard MASW analysis (performed while considering the vertical-component of Rayleigh waves and the interpreted modal dispersion curves) mode misinterpretation is extremely common and the obtained V_S values are therefore often overestimated. The example reported in Fig. 1.10 is a typical example of single-component dataset (the vertical component of Rayleigh waves) that necessarily leads to an incorrect subsurface V_S model (more examples are reported in Dal Moro 2014; Dal Moro et al. 2015).

This is why, in case we intend to work with the standard multi-offset MASW approach, it is recommend the record also Love waves. The previous volume (Dal Moro 2014) is fundamentally about this critical point (several synthetic and field datasets are presented with the goal of discouraging from using just one component—see also guidelines in Appendix A).

But, as we saw throughout the book as well in the appendices, we can follow several approaches and go beyond the multi-offset MASW approach. In Fig. 6.3a are shown the penetrometer data (down to about 15 m) and the three HVSR curves (Fig. 6.3b) collected in an alluvial plain in the framework of a ordinary geotechnical study.

For such a study, we initially recorded the HS (active data) (step#1 in Fig. 6.2) and then, for the acquisition of the passive data necessary for the computation of two HVSR curves, deployed a second 3C geophone nearby the position of the seismic source. This way we recorded the 6-trace dataset for the computation of the HVSRs at the two extreme points of the HS array. In order to obtain a further/final (third) HVSR curve in the central position (where the penetrometer test took place), we recorded a further (passive) dataset while positioning one of the two geophones in the central point (position#2 in Fig. 6.1).

By comparing the three curves (Fig. 6.3a) we can obtain some information about possible lateral variations along the HS array. In this case, if we have a look at three HVSR curves (Fig. 6.3b), we realize that below 5 Hz some non-negligible variation is likely. Since in this case we were interested in just a single V_S 30 value (i.e. an average model), the joint analysis with the HS data (that necessarily represent an average value of the dispersive properties) was then accomplished while considering the mean HVSR curve obtained from the three experimental curves obtained at the three points.

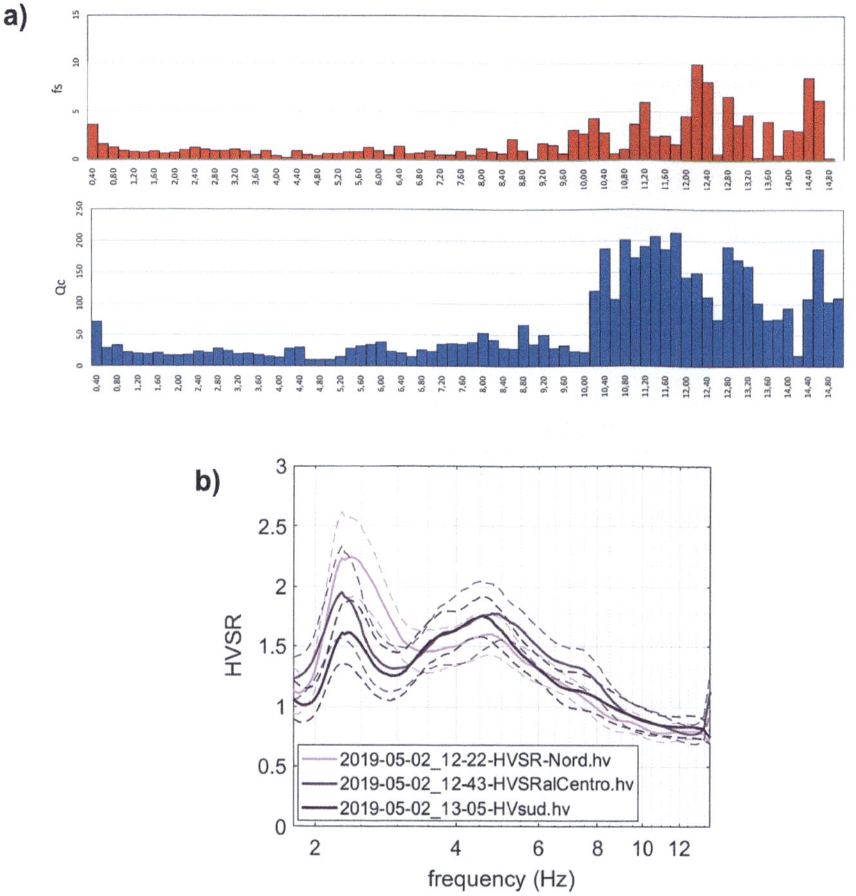

Fig. 6.3 Example of site characterization. Comparison between penetrometer and seismic data: **a** Q_C (tip resistance) and f_s (sleeve friction) data; **b** HVSR curves at three points (along an HS array). The central one corresponds to the location of the penetrometer test while the other two curves refer to two points 25 m away from it (one to the North and the other to the South)

Penetrometer data show quite clearly that silty layers characterize the first 10 m while deeper strata are composed of relatively dense sands. A more careful exam of the data reveals that between 5 and 10 m the sediments are slightly more competent compared to the first 5 m (very likely because of some amount of sand).

The joint analysis of the HS (group-velocity spectra of the Z and R components) and HVSR data is summarized in Fig. 6.4 and the obtained V_S profile is in excellent agreement with the penetrometer data (the silt-sand contact is perfectly identified and the velocities are those typical of the this kind of sediments).

Lateral variations are often little considered but can be quite large not only nearby the river beds or in mountainous and piedmont areas (where they are usually expected—see Fig. 6.5). Characterizing such areas can be complex and this is the

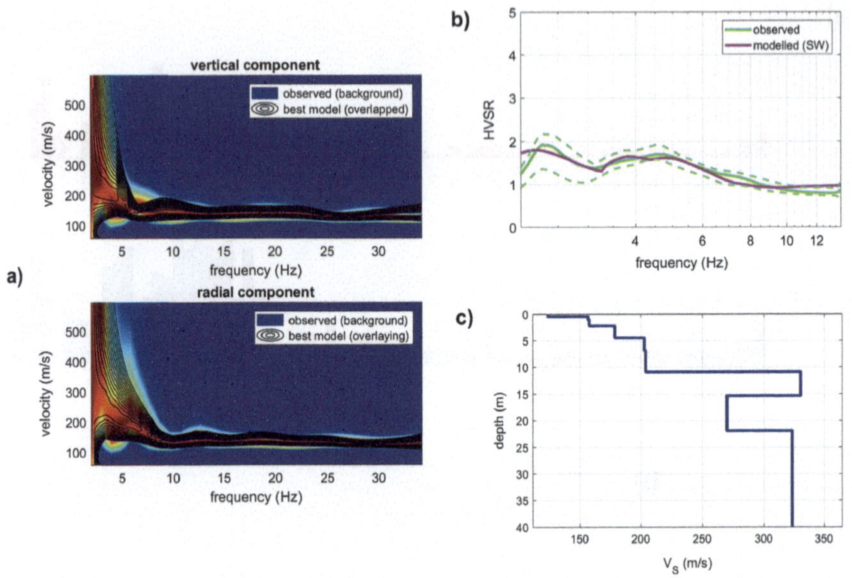

Fig. 6.4 Example of site characterization obtained by following the HS approach. Result of the joint analysis of the group-velocity spectra of the Z and R components of Rayleigh waves (**a**) and mean HVSR (**b**). Obtained V_S profile is shown in the lower right corner (c). The mean HVSR was obtained from the three curves reported in Fig. 6.3b

reason why we should never rely on a single HVSR (see basic guidelines in Appendix A and the case study in Appendix C, where a dramatic example of abrupt lateral variations in a soft-sediment perilagoon area is reported).

Figures 6.6, 6.7, 6.8 and 6.9 show a series of data collected in an area where thick peat layers (an abandoned meander of an ancient paleo-channel in NE Italy—Fontana et al. 2008) are occasionally present (captions provide all the information necessary to understand the shown data and compare them). Due to their "extreme" characteristics, peats are particularly interesting from the environmental and scientific points of view (their correct identification is critical also from the geotechnical point of view).

Shown data and analyses intend to highlight two main facts:

(1) peats are characterized by extremely-low shear-wave velocities and represent a serious geotechnical problem (e.g. Mesri and Ajlouni 2007; Boylan et al. 2011);
(2) different methodologies (in this case MASW, HS and MAAM) must provide nearly-identical results (if this does not happens, it means that the analyses are erroneous).

Fig. 6.5 Lateral variability along the bed of a river. Within a few meters, gravels, conglomerates and silt/clay sediments coexist

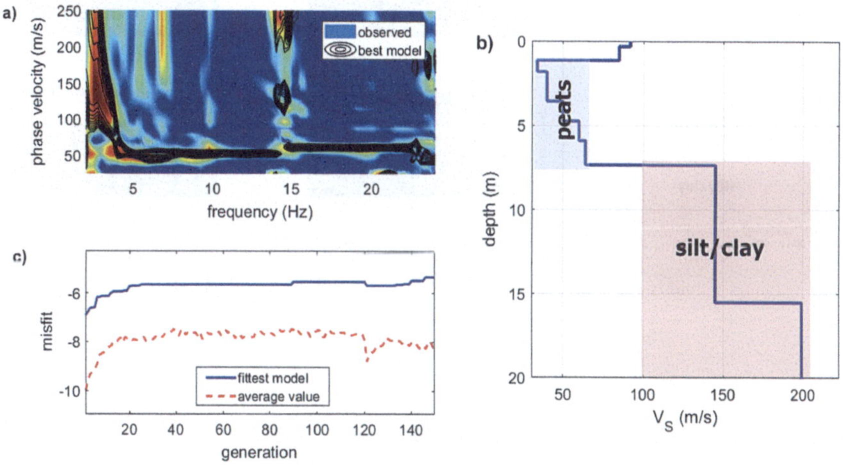

Fig. 6.6 A peat area: Z-component MASW analyzed according to the FVS approach (Chap. 2): **a** phase velocity spectra for the field data (background colors) and for the identified model (overlaying black contour lines); **b** identified V_S profile; **c** evolution of the misfit between the velocity spectrum of the field data and the velocity spectrum of the best model (for each generation/iteration)

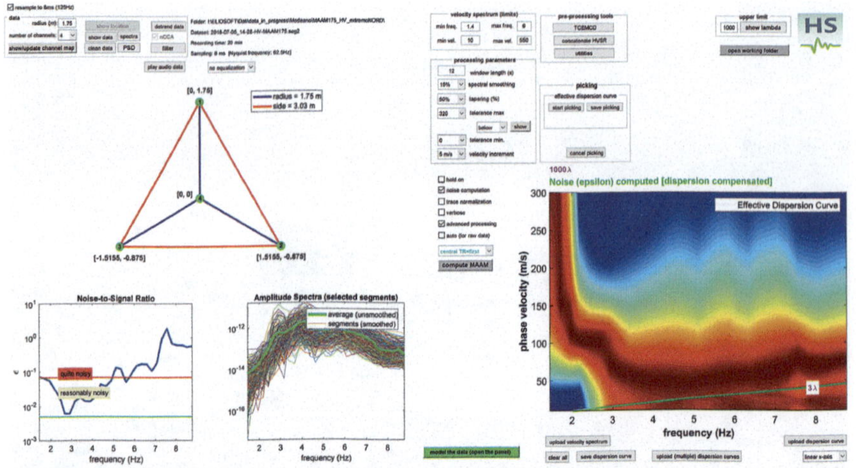

Fig. 6.7 The peat area: *Miniature Array Analysis of Microtremors* (MAAM). In the right corner is presented the phase-velocity spectrum identified using the passive data collected from the 1.75 m-radius array shown in the top left corner. The congruence with the phase velocities identified through classic MASW data (Fig. 6.6a) is excellent

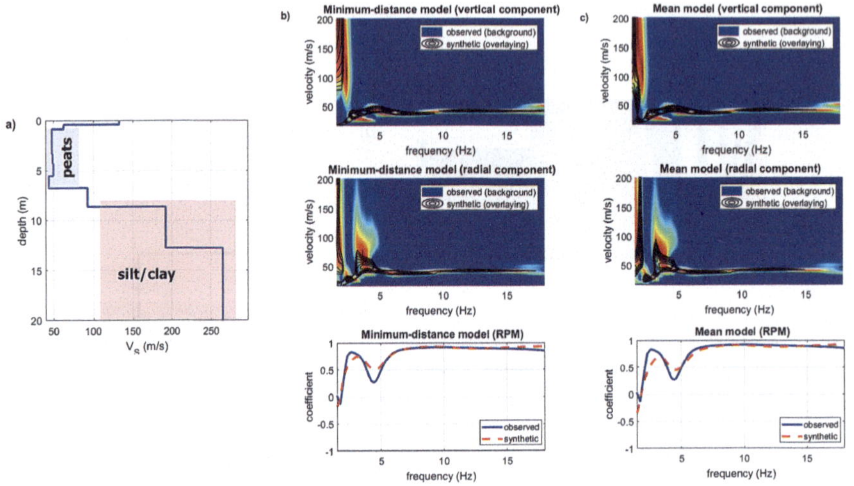

Fig. 6.8 The peat area: HS (Z + R + RPM) joint analysis. The identified model is practically identical to the one reported in Fig. 6.6 (obtained via FVS analysis of multi-offset MASW data): **a** minimum-distance V_S model; **b** field and modelled data for the minimum-distance model; **b** field and modelled data for the mean model (see Chap. 4 and Dal Moro et al. 2019)

Data shown in Fig. 6.9 attempt once again to go beyond the routine analysis so to be able to gain more and more insights about the material properties. In the upper and lower plots are reported the seismic traces and amplitude spectra of the Z and R

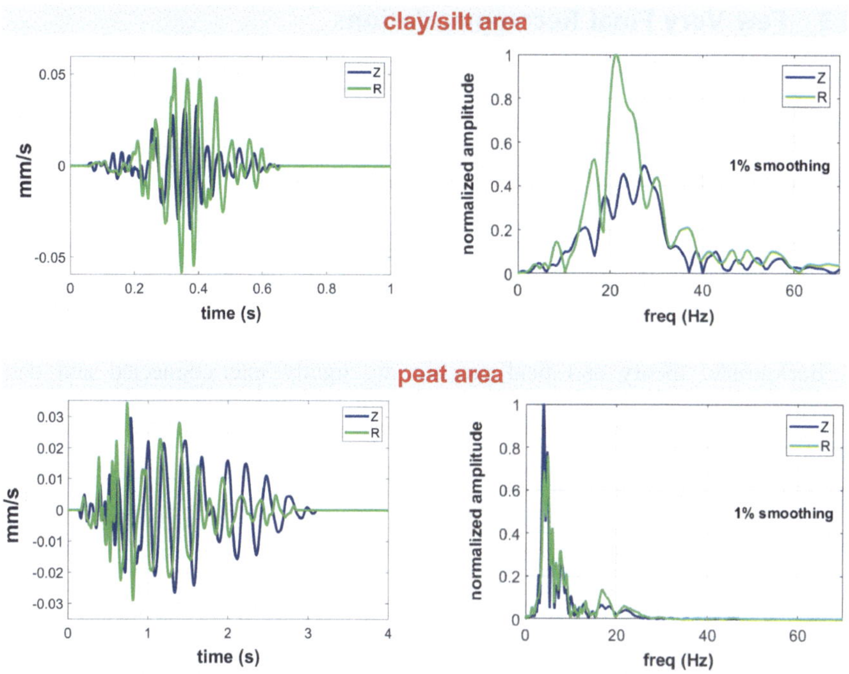

Fig. 6.9 Z and R active seismic traces (single-offset HS data) and amplitude spectra (normalized and 1% smoothed) for two sites around a paleo-channel in the Venetian-Friulian plain (NE Italy). The offset (42 m) and the source (vertical 8-kg sledgehammer, stack 20) were the same and it is therefore possible to compare the data. Upper panel: data for a "regular" clay/silt site. Lower panel: data for the peat site (see also Figs. 6.6, 6.7 and 6.8). In order to improve the readability, the seismic traces (left graphs) are plotted with two different time scales. Compared to the regular silt/clay site, at the peat site the velocities are significantly lower (see the time scale of the field traces) and the high frequencies (>10 Hz) are strongly attenuated

traces (HS data) recorded at two points around the considered paleo channel. Amplitude spectra are normalized by dividing the spectra by the maximum amplitude of the largest component (for the peat area the Z component, for the silt/clay are the R component). One site is a so-to-speak "regular" (ordinary silt/clay sediments) while the other site is characterized by a thick peat layer. Since the acquisition parameters were the same (offset 42 m and stack 20), we can compare the data and highlight two major facts:

(1) not surprisingly, Rayleigh waves are much slower in the peat area (see seismic traces in Fig. 6.9 [lower panel] and group velocity spectra in Fig. 6.8);
(2) in the peat area, attenuation is extremely massive and only frequencies lower than about 10 Hz "survived" while higher frequencies were completely cancelled out because of the intrinsic attenuation.

6.3 Few Very Final Recommendations

The topics, methodologies and examples presented in this volume intended to provide a wider scenario about surface wave analysis. Very often, both in the academic and professional sectors, surface waves are recorded and analyzed through popular and rigid procedures that are often not very efficient since cannot fully describe the characteristics of the particle motion.

If the considered observables are not sufficiently comprehensive (i.e. fail to describe the different aspects of surface-wave propagation), we risk to be unable to set up a properly-constrained inversion procedure thus obtaining a pointless subsurface model.

Background theory and field practice are tightly inter-connected and data acquisition and analysis should be accomplished only once the theoretical facts are clear.

Here a compact list of the main points highlighted throughout the book:

1. surface-wave analysis is not and cannot be considered and treated like a geotechnical *test*. Seismic data are *data*, i.e. they are necessarily *right* (if the acquisition is accomplished correctly). If the data analysis provides erroneous subsurface models is only because the analyses are based on mistaken data interpretations/understanding. In order to avoid that, a wide and sound theoretical background is necessary;
2. surface-wave acquisition and analysis is possible through several methodologies and we do not have to stick just to MASW, ReMi or HVSR techniques (which are pretty popular but often quite ineffective and/or insufficient to fully describe surface-wave propagation);
3. in case we are interested in the analysis of the phase velocities according to the active multi-offset MASW approach (multi-offset MASW), vertical geophones are not very useful since a small set of horizontal geophones allows to easily record both Love and Rayleigh (radial component) waves (see Chap. 1 and Appendix A);
4. for refraction studies we can surely use the 4.5 Hz geophones we need to record surface waves (we do not need to buy a set of high-frequency geophones for refraction studies and a further set of low-frequency geophones for MASW/ESAC/SPAC/ReMi/MAAM);
5. group velocities (MFA, FTAN, HS) are a useful (and very classical) method to analyze surface-wave propagation as much as phase velocities (MASW, ESAC/SPAC, MAAM, ReMi) (applied geophysicists are often little familiar with group velocities because of the lack of fair relationships with the seismological community, where group velocities are the routine tool);
6. quantitative *joint analysis* has nothing to do with (qualitative) *data integration*;
7. joint inversion is necessarily a *compromise*: the misfit of each single considered observable cannot be the minimum possible since what we are pursuing is the best model capable of explaining (in a "reasonable" but quantitative way) *all*

the observables (which can be influenced by noise and inhomogeneities in a different manner);

8. about MASW (multi-offset) acquisition: geophone distance does not significantly influence the quality of the phase velocity spectra computed according to the phase shift approach (for ordinary arrays of some tens of meters, 12 geophones are more than sufficient). In other words, *more* does not mean *better*: 12 horizontal geophones provide much more information (Love waves and radial component of Rayleigh waves) then 24 or 48 vertical geophones (that are useful just to define the vertical component of Rayleigh waves);
9. as shown in Chap. 4, the well-known *less is more* expression (from the pre-eminent modernist architect Mies van der Rohe) applies very well to surface wave analysis since the mindful use of a 3C geophone allows the determination of up to six observables that can be used to set up a very-well constrained inversion procedure (e.g. Sect. 4.3);
10. in the light of these improved techniques, the MASW acronym has completely lost its meaning since it is unable to identify any well-defined methodology (see Sect. 6.1);
11. shear-wave velocities are useful for several geotechnical applications and surely not only for the quasi-bureaucratic issues related to the several seismic-hazard national and international codes;
12. HVSR is not a straightforward estimate of the site amplification and the very idea of *resonance frequency* often does not apply since several frequencies can undergo some kind of amplification and, in some cases, a very wide frequency range is actually subject to large amplifications (with a trend quite different from the shape of the HVSR curve—see Sect. 3.1.5);
13. vibration data (for monitoring the vibration produced at a construction site or to characterize the eigenmodes of a building) can be recorded with the same equipment necessary to collect the seismic data;
14. an efficient *acquisition system* should be designed by properly taking into account all the state-of-the-art techniques (see the concerns presented in this volume for instance about the polarity issues, the active data stack, the data quality check etc.).

The more you know, the less you need

Yvon Chouinard

References

Boylan N, Long M, Mathijssen FAJM (2011) In situ strength characterisation of peat and organic soil using full-flow penetrometers. Can Geotech J 48(7):1085–1099

Dal Moro G (2014) Surface wave analysis for near surface applications. Elsevier, Amsterdam, The Netherlands, 252 pp. ISBN 978-0-12-800770-9

Dal Moro G, Moura RM, Moustafa SR (2015) Multi-component joint analysis of surface waves. J Appl Geophys 119:128–138

Dal Moro G, Al-Arifi N, Moustafa SR (2017) Improved holistic analysis of Rayleigh waves for single- and multi-offset data: joint inversion of Rayleigh-wave particle motion and vertical- and radial-component velocity spectra. Pure Appl Geophys 175:67–88. https://doi.org/10.1007/s00024-017-1694-8 (open access)

Dal Moro G, Al-Arifi N, Moustafa SR (2019) On the efficient acquisition and holistic analysis of Rayleigh waves: technical aspects and two comparative case studies. Soil Dyn Earthq Eng 125. https://www.sciencedirect.com/science/article/pii/S0267726118310613

Fontana A, Mozzi P, Bondesan A (2008) Alluvial megafans in the Venetian-Friulian Plain (north-eastern Italy): evidence of sedimentary and erosive phases during Late Pleistocene and Holocene. Quatern Int 189:71–90

Mesri G, Ajlouni M (2007) Engineering properties of fibrous peats. J Geotech Geoenviron Eng 133:850–866

Safani J, O'Neill A, Matsuoka T, Sanada Y (2005) Applications of love wave dispersion for improved shear-wave velocity imaging. J Environ Eng Geophys 10(2):135–150

Appendices

G. Dal Moro, *Efficient Joint Analysis of Surface Waves and Introduction to Vibration Analysis: Beyond the Clichés*,
https://doi.org/10.1007/978-3-030-46303-8

Appendix A: Basic Guidelines for Surface-Wave Data Acquisition

The present appendix is not meant to be an exhaustive guide to surface-wave acquisition. What we need to consider during field operations (as well as during data processing) can be fully understood only through a balanced synergy between a solid theoretical background and a long field practice.

Nota bene: indeed, data acquisition is a highly theoretical action since it is necessary to consider a huge number of facts that can influence the signal generation and the wave propagation (and recording). Field conditions are always different and it is often difficult to fix the details of a survey in advance since some decisions can be taken only once we are on the spot and consider all the possible issues.

***MASW* (multi-offset) acquisitions: Rayleigh and Love waves**

During a standard multi-offset (MASW) acquisition aimed at the determination of the phase velocities from active data, there are two fundamental parameters to fix with respect the geometry: the *minimum offset* (i.e. the distance between the source and the first receiver) and the *geophones spacing*.

In order to investigate the deepest layers, it is important that the array is as long as possible (the array length is usually limited by the actual room available), so the field operations must be performed while keeping in mind a very simple fundamental idea: let us take advantage of all the available room.

As we have seen in Sect. 1.5, the geophone spacing is quite irrelevant for the determination of the phase velocity spectrum so it must be set depending on the available room. A simple example can clarify this point.

Let us imagine that we have a 12-channel seismic cable and the available room is 75 m. In this case, the array can be fixed with a geophone spacing equal to 6 m and a minimum offset of 9 m, so to exploit all the available room.

The minimum offset should be large enough to avoid all the near-source phenomena that can have a pernicious effect on the computed velocity spectra. But the minimum offset depends also on the source (the more powerful, the larger the minimum offset). In very general terms, while using a common 8 or 10 kg sledgehammer, it is recommended a minimum offset equal to 5 or more meters but in case of more powerful sources (e.g. a weight drop or a thumper) such a value could be larger. Of course, the first point to avoid is the saturation (clipping) of the first trace(s) (see later on).

G. Dal Moro, *Efficient Joint Analysis of Surface Waves and Introduction to Vibration Analysis: Beyond the Clichés*,
https://doi.org/10.1007/978-3-030-46303-8

Table A.1 Summary table of the fundamental acquisition parameters during a multi-offset (standard MASW) acquisition for near-surface applications

Minimum offset (mo): distance between the source and the first geophone	5–20 m
Geophone spacing (dx)	*Key point*: the length of the array must be as long as possible (let us take advantage of all the available room). If the available space is for instance 62 m and you have 12 geophones, you can fix the geophone distance equal to 5 m and a minimum offset of 7 m Please, also consider that data do not need to be equally spaced (see Sect. 1.5)
Geophones	*Rayleigh waves*: vertical and/or horizontal geophones *Love waves*: horizontal geophones (see Sect. 1.2 and the figures that illustrate how to orientate the geophones) For active (multi-offset MASW) data, we can proficiently use just 12 horizontal geophones that enable us to record both Love and Rayleigh (radial component) waves (to jointly invert) Eigen frequency: 4.5 Hz (useful also for refraction studies—see Sect. 1.8)
Record time/length	2 or 4 s are usually sufficient (the complete surface wave trend must be entirely recorded even at the very last channels/traces, so it actually depends on the length of the array and on the type of sediments). For details, see Dal Moro (2014)
Number of channels/geophones	12 channels (but often even less) are sufficient (see Sect. 1.5) The crucial point is anyway the total length of the array, possibly not less than say 50 m, much better 70–90 m. The length of the array influences the lowest frequency that can be soundly analyzed and, consequently, the investigated depth (the longer the array, the deeper we go)
dt (sampling period)	0.001 s (1 ms) is more than enough (2 ms are also fine) Higher sampling rates (sampling frequencies) are completely useless (compare the Nyquist frequency and the frequency range we are interested in for common geological applications)
Stack	It depends on several facts: length of the array, background noise, efficiency/power of the seismic source and overall quality of the equipment In general terms, five shots are the minimum value but in case of data acquisition in noisy urban areas, such a value can raise up to 10 and more
Source	Rayleigh waves can be produced by a vertical-impact source (VF—Vertical Force) such as a very classical sledgehammer or weight drop, as well as by an explosive force In order to produce high-quality Love waves we need a shear-source (see Fig. A.1 and Chap. 1) A 5-kg sledgehammer is usually unable to produce enough energy in the low-frequency range and a 10-kg version is often recommendable (8 kg can represent an acceptable compromise)
Notes	No AGC (*Automatic Gain Control*) No filter Keep the same gain for all the channels Avoid saturation/clipping of the traces (see later on)

Fig. A.1 Data acquisition for SH-wave refraction/reflection travel time analysis and/or Love-wave dispersion analysis: compare the axis of the wooden beam used as shear-wave source and the orientation of the horizontal geophones (their axis is perpendicular to the array). See also Sect. 1.2

The main aspects to consider for obtaining good multi-offset MASW dataset are summarized in Table A.1.

HS data

As shown in Chap. 4, HS is a methodology aimed at analyzing the propagation of surface waves while using the (active) data recorded by a single 3C geophone. As we know, Rayleigh waves are produced by a vertical (or explosive) force while in order to obtain Love waves we need to deal with an horizontal force (Figs. A.1, 1.37 and 1.38). This means that, in case we want to record and analyze both Rayleigh and Love waves, we need to record two separate datasets: one obtained with a vertical force (VF) and the other with a horizontal force (HF).

In order to optimize the data processing, we can then re-assemble the traces that are relevant from the two datasets. The VF data are clearly meant for the analysis of Rayleigh waves and the relevant traces are therefore the vertical (Z) and radial

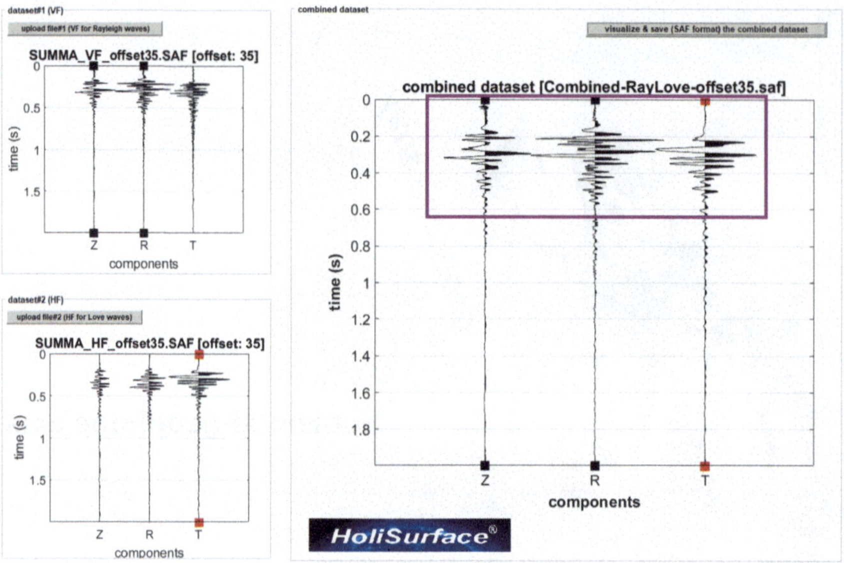

Fig. A.2 An example of HS data combination for the analysis of both Rayleigh and Love waves. On the left side are reported the data obtained while using a vertical force (VF) (upper plot) and a horizontal force (HF) (lower plot). From the VF data we select the first two traces (i.e. the vertical Z and radial R components of Rayleigh waves), while from the HF data we select the third trace (i.e. the transversal T trace which contains the information about Love waves). We then obtain a single file (traces shown on the right panel) that contains the information both about Rayleigh and Love waves. Also notice that after about 0.7 seconds the traces do not contain anything relevant for our purposes (surface-wave analysis) and data can be therefore cut in order to keep only the relevant part of the traces

(R) ones. On the other side, Love waves generated by means of a horizontal force (HF) are recorded by the transversal (T) trace. This means that we can select the traces as shown in Fig. A.2 so to obtain a single file that contains the information about both Rayleigh (on the first two traces—Z and R) and Love waves (the third—T—trace).

Acquisition parameters follow the same rules as for the multi-offset MASW methodology (see Table A.1). Of course, in this case we are dealing with only one offset which should be set by considering that, roughly speaking, the investigated depth is approximately equal to half or 2/3 of the offset (for further details see Dal Moro et al. 2019).

A 5-kg sledgehammer is usually unable to produce large-.amplitude low frequencies and we should prefer the 10-kg version (8 kg can represent a reasonable compromise).

The correct orientation of the 3C geophone is clearly crucial and, in order to meet the most common format, it should be such to obtain traces in the following sequence: Z, R and T (see data shown in Fig. A.2).

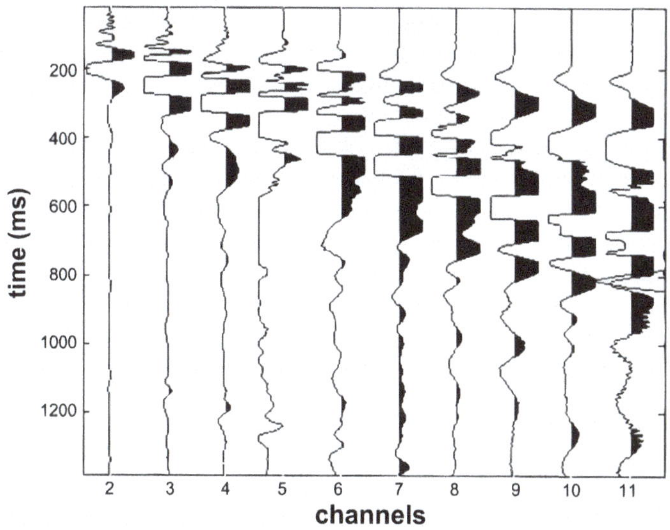

Fig. A.3 Multi-offset dataset characterized by a severe problem regarding the trace saturation. The amplitude is too large for the dynamic range of the seismograph and traces are therefore clipped/saturated. In order to avoid this (see e.g. Fig. A.4), gain must be reduced and/or the power of the source should be reduced

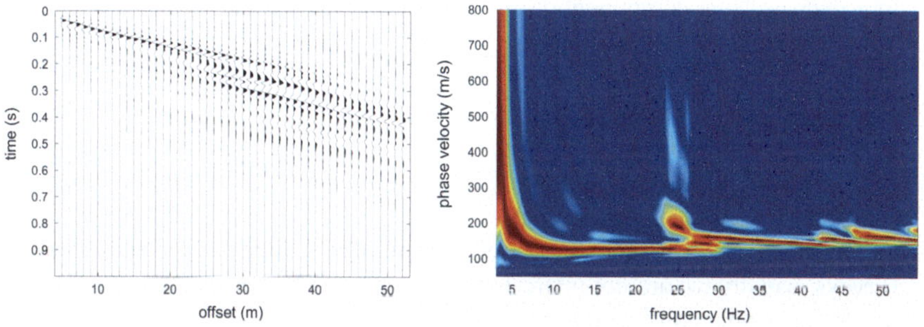

Fig. A.4 Example of data from an urban area (seismic traces on the left and phase-velocity spectrum on the right). Please, notice that, in spite of the fact that we are in a noisy urban area and on a gravel path of a city park, the seismic traces are anyway quite clean. This quality was obtained thanks to a large number of stacked shots (*stack* 15). Seismic traces on the left are normalized (each trace is divided by its maximum value so to obtain a series of traces whose maximum amplitude is 1 for all the traces)

Gain

Some seismographs require to set a gain value aimed at obtaining traces with a good Signal-to-Noise ratio. If this is the case of your seismograph, remember that the gain should be the same for all the channels (so not to alter the amplitude decrease along the array and have the chance to analyze the attenuation).

In case you want to analyze surface-wave dispersion, it is important to avoid trace saturation/clipping (Fig. A.3). Saturation occurs when the amplitude of the generated waves is too large for the dynamic range of the seismograph. This can happen because of the combination of several possible facts: the gain of the seismograph is too high, its dynamic range is too small, the source is extremely powerful and/or the offsets (i.e. the distance between the geophones and the source) is small.

The quality of the recorded traces should be similar to the one of the data shown in Fig. A.4 (recorded in a noisy urban area): no trace is clipped and the signal-to-noise ratio is quite good even for the distant offsets (such a result was obtained thanks to the *stack* [shown traces are the result of the stack of 15 shots]).

Of course, in case we have just a couple of clipped traces, there is no serious problem and we can surely proceed with the analysis of the dataset.

ReMi, ESAC and SPAC (for Rayleigh waves)

Introductory note: SPAC (*SPatial AutoCorrelation*) is, so-to-speak, a special case of the ESAC (*Extended Spatial AutoCorrelation*) approach. From the practical point of view, the difference is that while SPAC requires a rigid circular geometry, ESAC can be applied to any (bi-dimensional) array and is therefore much more flexible (and the field operations much simpler).

Although some very advanced acquisition and processing techniques allow the analysis of Love waves also from passive data, passive methodologies are usually used to record and analyze only Rayleigh waves and this can represent a serious limitation because of the consequent (and inevitable) ambiguities we face while dealing with just one component/observable (see Chaps. 1 and 3).

Few general points:

1. *Sensors*: vertical 4.5 Hz geophones;
2. *Record time*: 10–30 min (depending on the characteristics of the site, length of the array and goals of the survey);
3. *Sampling interval*: 16 ms (62.5 Hz) are usually sufficient (for this kind of surveys, usually we are not interested in frequencies higher than about 30 Hz—see Nyquist-Shannon theorem);
4. the geophone array should be as long as possible because, somehow similarly to the principles that rule the MASW approach, this kind of methodologies are fundamentally based on the "wavelength principles": the minimum frequency that can be soundly detected and analyzed depends on the length of the array. Dealing with small arrays (say smaller than 60 m) makes no sense since we would not be able to retrieve reliable information about the low frequencies;
5. About ReMi: the linearity of the array does not allow handling the directivity of the passive signals (that can come from any azimuth). The interpretation of the obtained phase-velocity spectra is therefore subjective and unclear (this is way in the last few years such approach has been replaced but the ESAC

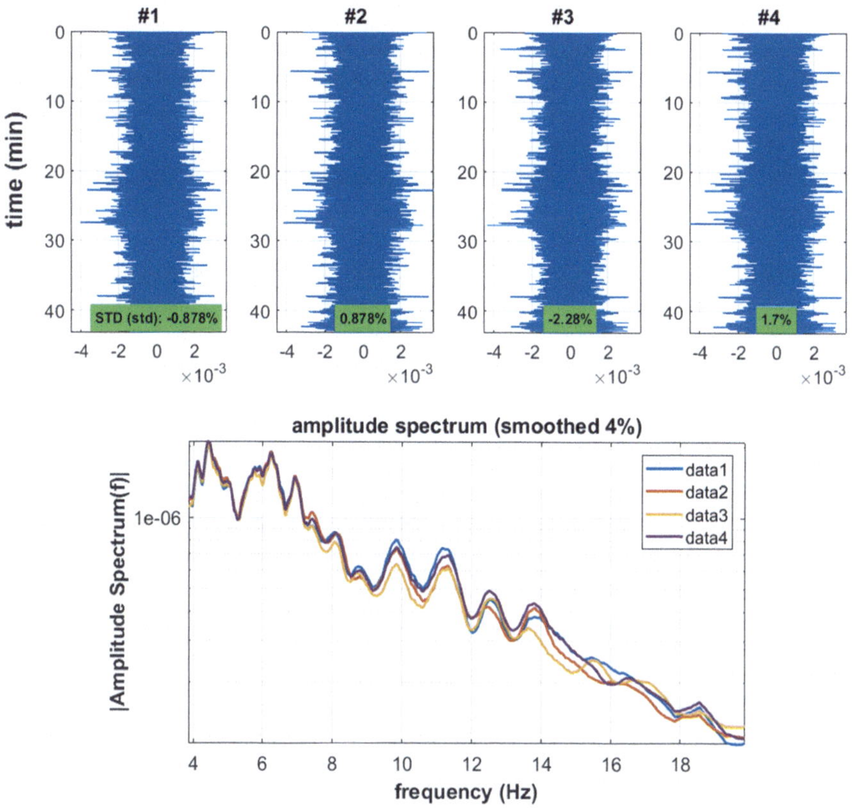

Fig. A.5 Passive dataset from a MAAM acquisition (triangular geometry, radius 1.8 m). The overall quality is assessed by computing the amplitude spectra (lower plot) and the standard deviations (STD) of the four traces. The deviations of the STD values from the median value (green boxes in the upper plots) provide a quantitative assessment of the similarity of the four traces. In case the percentage deviations from such a median values are small (say less than 10%) and the amplitude spectra are similar, data are likely suitable for the MAAM methodology. Further details in Dal Moro (2019)

methodology that, thanks to its mathematics and the bi-dimensional array, provides phase velocity spectra that do not have to be interpreted since the effective dispersion curve is represented by the maximum correlation—see Dal Moro 2014 and the ESAC examples reported in Chap. 3).

HVSR

As shown in Chaps. 3 and 4, for the determination of the HVSR we need a properly-calibrated 3C geophone (which can be used also for the acquisition of HS and vibration data).

Very few synthetic recommendations:

1. *record time*: it depends on the minimum frequency we intend to investigate (details in AA.VV.—SESAME, 2005) and, for near-surface applications, can vary from about five up to 30 min. In very general terms, 5 min can be used to define the HVSR down to about 1 Hz while for analyzing frequencies of about 0.1 Hz, we need to record at least 30 min;
2. do not ever rely on just one dataset: record at least two datasets at two different points (few meters away one from the other—see for instance Figs. 3.21, 3.22, 6.1 and 6.2 and related text) and compare the obtained curves;
3. *sampling rate*: 16 ms (62.5 Hz) is sufficient (just consider the Nyquist-Shannon theorem and the fact that for geological and engineering applications usually we are not interested in frequencies higher than about 20 Hz) [we can surely use a higher sampling rate and decimate the data during the processing, so to reduce the computation times];
4. *data format*: since we are dealing with three different components (vertical, NS and EW), depending on the *acquisition system* (i.e. on the combination between seismograph, seismic cable and 3C geophone—see Chap. 1) the sequence of the three traces can be different (it can be NS, EW, vertical or EW, NS, vertical and so on). Please, consider that the most common (recommended) format/sequence is: UD (vertical), NS and EW. In any case, in order to compute the HVSR, it is fundamental to know the actual format produced by your own *acquisition system*.

MAAM

Radius: it depends on the frequency range we intend to investigate and on the type of sediments. In general terms (and roughly speaking), in order to investigate the 4–20 Hz frequency range in unconsolidated sediments (silt, clay, gravels), the radius can vary between 2 and 4 m;

Record time: 15–30 min;

Sampling interval: 8–16 ms are sufficient (just consider the Nyquist-Shannon theorem and the fact that, in general, we are not interested in frequencies higher than about 30 Hz);

Among the methodologies considered in the present book, the *Miniature Array Analysis of Microtremors* (MAAM) is probably the trickiest one since requires very high-quality data. Data quality is obtained thanks to a high-quality acquisition system and through careful field operations (the geophones must be *perfectly* vertical and the ground coupling carefully checked).

Passive data acquisition can be anyway problematic because, usually, there is no easy way to assess the data quality during the acquisition (please, notice that active data can be easily and quickly checked through a simple visual check of the traces).

During the acquisition of the passive data to use for the MAAM, it is advisable to have a QC (Quality Check) software that computes the standard deviations of the traces and the amplitude spectra for a small test dataset. From the operational point of view, we can record a 2-min dataset and run a test similar to the one shown in Fig. A.5. If the standard deviations and amplitude spectra of the recorded traces are sufficiently similar (since we are dealing with microtremor data and the geophones are quite close to each other, the amplitude of the traces must be necessarily very similar), we can then proceed with the definitive (longer) data acquisition.

File naming: a key point (often completely neglected)

A file name should contain all the information necessary to understand the actual meaning of the data. File names like for instance *First_Data.seg2* or *Cairo1.segy* do not help in understanding what those data/files are about.

How can we choose a file name so to help us (or our colleagues) to manage the data properly and easily?

It is important to provide information both about the type of source and geophones as well as about the geometry of the acquisition.

In the following some suggestions which, of course, can be personalized.

MASW (multi-offset data)

In this case, we can adopt the nomenclature widely explained in Dal Moro (2014) and briefly recalled in Sect. 1.2. A few examples will clarify the point:

ZVF_dx5_mo5.seg2

vertical (Z) geophones (for analyzing the vertical component of Rayleigh waves) and vertical force (VF), geophone distance 5 m and minimum offset (i.e. the distance between the source and the first geophone) 5 m.

RVF_dx4_mo8.seg2

horizontal geophones set radially (R—radial component of Rayleigh waves) and vertical force (VF), geophone distance 4 m and minimum offset 8 m.

THF_dx4_mo8.seg2

horizontal geophones set transversally (T component, i.e. Love waves) and horizontal force (HF), geophone distance 4 m and minimum offset 8 m.

In case of non-equally spaced data, a possible solution can be the clear indication of the offsets (the offset is the distance between the source and the geophone). Two examples:

ZVF_off10_13_17_23_29_35_40_47_58.seg2
THF_offset7_11_19_27_31_39_47_59.seg2

HS

In this case, the receiver is a 3C geophone so we need to clarify only two things: the kind of source (VF or HF) and the offset. For instance:

VF_off50.seg2 or HF_off28.saf

In case, for any reason, we decide to swap the position of the source and receiver, we could for instance modify the previous file names this way:

VF_off50_reverse.seg2 and HF_off28_reverse.saf

Needless to say that the 3C geophone must be properly oriented and that the trace sequence should be also clear (see Chap. 4). Therefore, in case our acquisition system does not follow the standard format (vertical, radial, transversal), it can be important to clarify the actual meaning/sequence of the 3-trace dataset. For instance:

VF_off60_T_R_Z.seg2 or HF_off50_Z_T_R.seg2

MAAM

The file name should contain (at least) two information: the radius and the position of the central geophone (usually the central geophone is the first or the last trace). Examples of clear file names are therefore:

MAAM_radius3m_centralFIRST.seg2

MAAM_rad2m_CentralGeophoneLast.seg2

HVSR

Depending on the equipment, it can be important to clearly state the meaning of the trace sequence (which depends on the combination of geophone, seismic cable and seismograph). In any case, it would be important to include some hint about the geophone orientation with respect the local conditions (see Chap. 3). For instance:

HVsite1_UD_NS_EW_GeographicNorth.seg2 (in case we used the Geographic North to orientate the geophone)

or

HVsite2_UD_EW_NS_NorthParallel2ValleyAxis.seg2 (in case, being in a mountain valley, we aligned the NS direction of the 3C geophone with the axis of the valley)

Appendix B: An Urban Park: Multi-component (Z + R) and Multi-offset Rayleigh Wave Joint Analysis also Together with the RPM Frequency-Offset Surface

The data of this case study were collected in a NW-Italy urban area (city of La Spezia), where several further data were collected in order to compare the results from different methodologies (see Dal Moro 2019, 2020a).

In general terms, the site is characterized by about 15 m of soft sediments covering a thick sequence of gravel-like materials.

Here we present only the multi-component (Z + R) multi-offset data acquired along a pathway covered with a stiff layer of crushed gravel (Fig. B.1) of a public park (acquisition parameters are reported in Table B.1).

The Z and R traces (i.e. the vertical and radial components of Rayleigh weaves) were processed in order to obtain three observables (Fig. B.2):

(1) the phase velocity spectrum of the ZVF component;
(2) the phase velocity spectrum of the RVF component;
(3) the Rayleigh-wave Particle Motion (RPM) frequency-offset surface (see Chap. 4 and in particular Sect. 4.7).

Data presented in Fig. B.2 put in evidence that down to about 5 Hz the phase velocities are quite small (around 130 m/s) while below such a frequency an abrupt velocity increase is apparent. At the same frequency, the Rayleigh-wave Particle Motion changes (Fig. B.2c): while at higher frequencies the motion is fundamentally retrograde (correlation value equal to about +1), below 5 Hz the particle motion becomes prograde (correlation value equal to about −1). These two facts are clearly mutually related and the V_S profile obtained through the joint inversion of the three considered observables provides the reason of it (see results presented in Fig. B.3).

Although the Rayleigh wave particle motion is the result of several parameters and it is impossible to predict its behavior (even very simple subsurface models can excite prograde motion), in some cases it was observed that prograde motion is

G. Dal Moro, *Efficient Joint Analysis of Surface Waves and Introduction to Vibration Analysis: Beyond the Clichés*,
https://doi.org/10.1007/978-3-030-46303-8

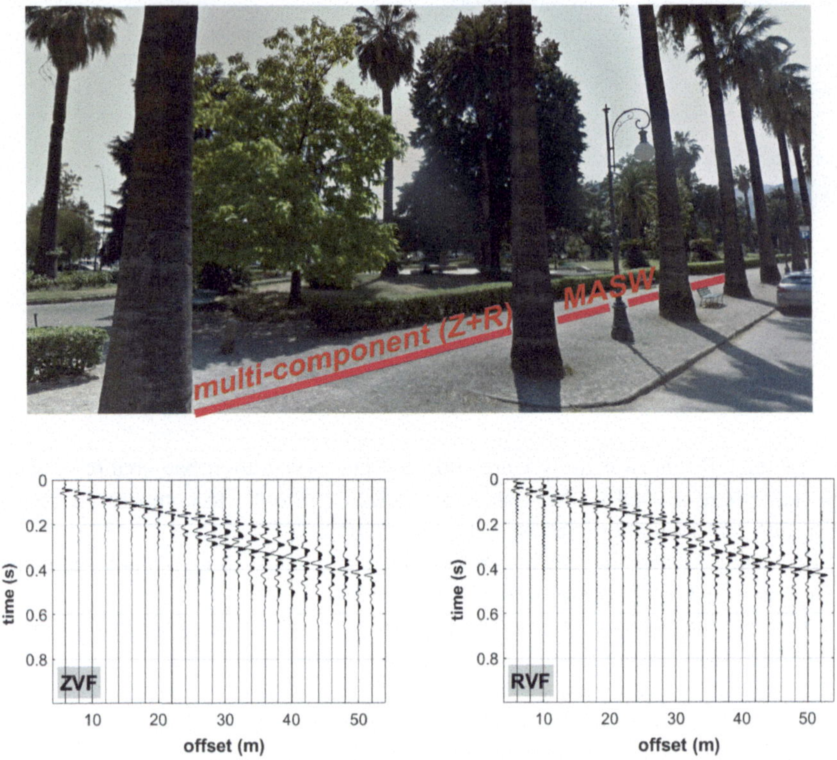

Fig. B.1 The NW Italy urban site where we recorded the vertical (Z) and radial (R) components of Rayleigh waves according to a traditional (MASW style) multi-offset setting

Table B.1 Acquisition parameters for the ZVF (vertical geophones) and RVF (horizontal geophones set radially—see Fig. 1.7) traces shown in Fig. B.1

Sampling interval	1 ms (1000 Hz)
Acquisition length	1 s
Minimum offset	6 m
Geophone spacing	2 m
Number of channels	24
Stack	15

As source was used a standard 8 kg sledgehammer (vertical force—VF)

simply the result of an abrupt increase of the shear-wave velocities (Tanimoto and Rivera 2005; Malischewsky et al. 2008; Dal Moro et al. 2017; Dal Moro 2019).

The present data/site is an example of it: the change from retrograde to prograde at 5 Hz is due to the contact between the superficial silt sediment and the gravels at a depth of about 15 m (see V_S profile shown in Fig. B.3d).

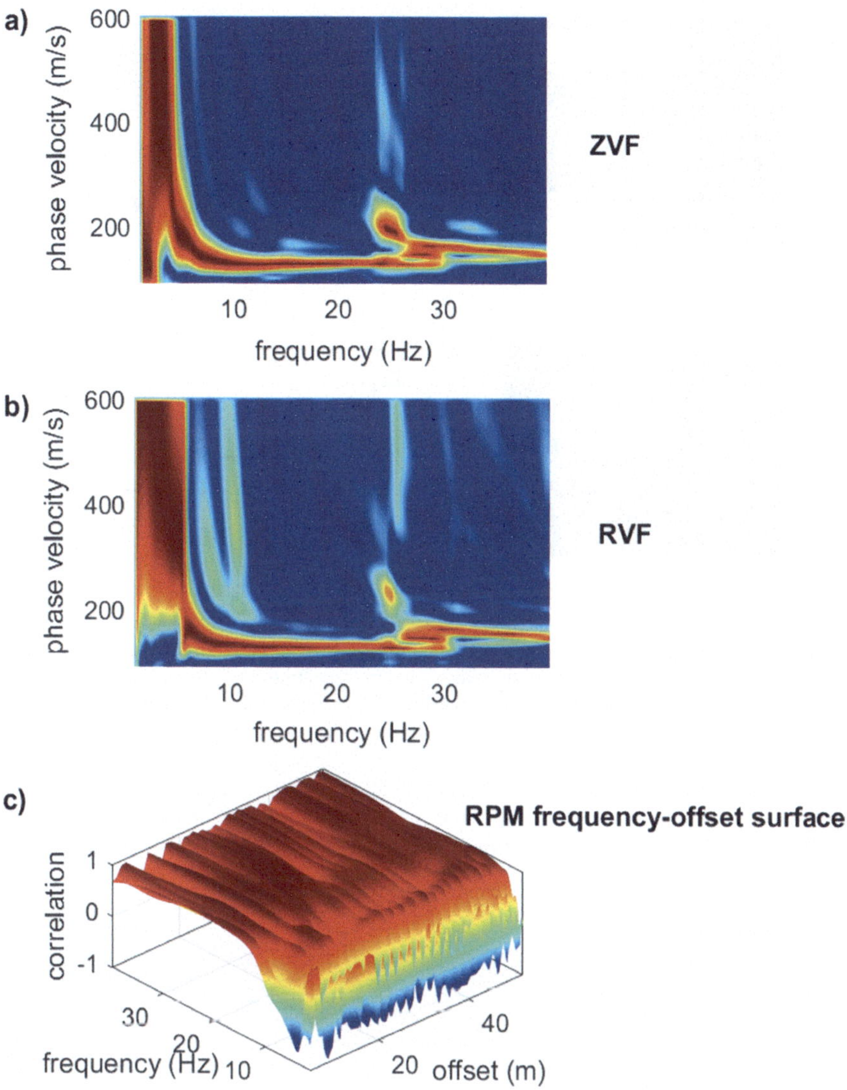

Fig. B.2 The three observables obtained from the Z and R seismic traces shown in Fig. B.1: **a** phase-velocity spectrum of the Z component; **b** phase-velocity spectrum of the R component; **c** RPM frequency-offset surface

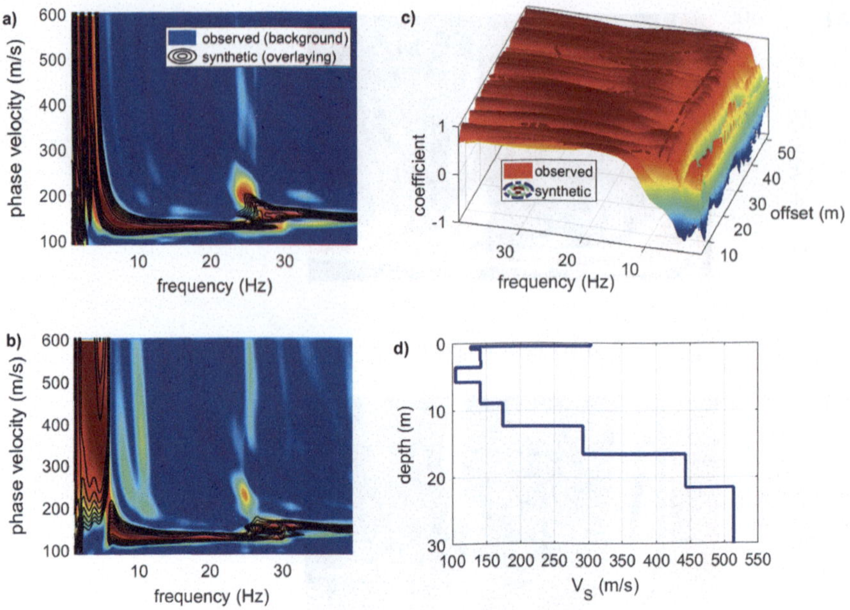

Fig. B.3 Result of the joint inversion of the Z and R phase velocity spectra (FVS approach) together with the RPM surface: **a** phase-velocity spectrum of the ZVF component; **b** phase-velocity spectrum of the RVF component; **c** RPM surface; **d** retrieved V_S profile

The results obtained through the analysis of these multi-offset and multi-component data can be compared with the subsurface model presented in Dal Moro (2019) and obtained while analyzing the HS, MAAM and HVSR data.

Appendix C: Large Lateral Variations in a Soft-Sediment (Perilagoon) Area

The Venetian–Friulian Plain (NE Italy) is a large piedmont basin that, in the Southern part (along the Adriatic coast), is filled with hundreds of metres of unconsolidated sediments and characterized by two lagoon systems: on the West the Venetian and on the East the Grado-Marano lagoon (Fontana et al. 2008). In the Grado-Marano peri-lagoon area, a series of seismic data were collected in order to characterize the sediments from the geotechnical point of view (e.g. Dal Moro et al. 2015; Dal Moro and Puzzilli 2017) and obtain some insight about the origin of the paleo dunes that characterize the eastern part of the Grado lagoon.

In this case study, we are considering some of the HS and HVSR data collected about 3 km south from the ancient Roman city of Aquileia (Gallo et al. 2014), over an area locally known as *Beligna*.

The DTM (Digital Terrain Model) shown in Fig. C.1 clearly shows an elongated geomorphological feature that a surface recon reveals dominated by coarse sands (it must be underlined that the surrounding area is otherwise characterized by silty sediments). Site#1 is approximatively at the center of the considered geomorphological feature while site#2 at the western extreme.

We recorded both resistivity and seismic data (Fig. C.1): one ERT (Electrical Resistivity Tomography) line was set up perpendicularly to a prospective residual dune (the sandy elongated feature), while HS and HVSR seismic data were collected at two points along the ERT line. Of course, in order to minimize the effects of the lateral variations, the HS data were recorded with the source and the receiver aligned with the sandy elongated dune (i.e. in the N45E-S45W direction).

Since we were interested in obtaining information also about the depth of the bedrock (i.e. the thickness of the sedimentary cover), we considered the HVSR down to 0.2 Hz.

Data and results of the joint (HS + HVSR) analysis are summarized in Figs. C.2, C.3, C.4 and C.5 (captions reports and underline several details) and provide a very consistent image of the shallow subsurface conditions. In fact, ERT and seismic data identify a central area (site#1) characterized by stiff (V_S of about

G. Dal Moro, *Efficient Joint Analysis of Surface Waves and Introduction to Vibration Analysis: Beyond the Clichés*,
https://doi.org/10.1007/978-3-030-46303-8

450–500 m/s) and high-resistivity (over 500 Ωm) material overlying slightly-softer (V_S of about 250 m/s) and low-resistivity (less than 1 Ωm) layers.

Very soft sediments are instead present in the western area (site#2), where the superficial stiff sand wedges out: here, the sedimentary sequence is characterized by V_S values around 100 m/s in the first 7 m and the resistivity decreases to about 1 Ωm. The presence of such thick (7 m) soft-sediment cover over harder sediments (V_S values increase to 200 m/s at 7 m and approximately 500 m/s at about 15 m of depth) is responsible for the distinctive H/V peak at about 2.5 Hz (Fig. C.5c) which is not present at site#1 where harder sandy sediments are present already at the surface (also compare the seismic data in Figs. C.2 and C.3 and the photo in Fig. C.6a). The V_S profiles obtained for site#1 (in the middle of the resisual dune) and site#2 (where the superficial stiff sand fades away and very soft sediments are present) are shown together in Fig. C.5f and can be compared to the ERT section shown in Fig. C.1 (the agreement is apparent).

Also in the light of paleontological, ichnological and geomorphological data, such a well-defined stiff sandy feature can be interpreted as the "root" of a paleo dune that was almost completely cancelled during the heavy reclamation works that took place in order to expand the agricultural areas (a couple of residual dunes are still present in the eastern areas).

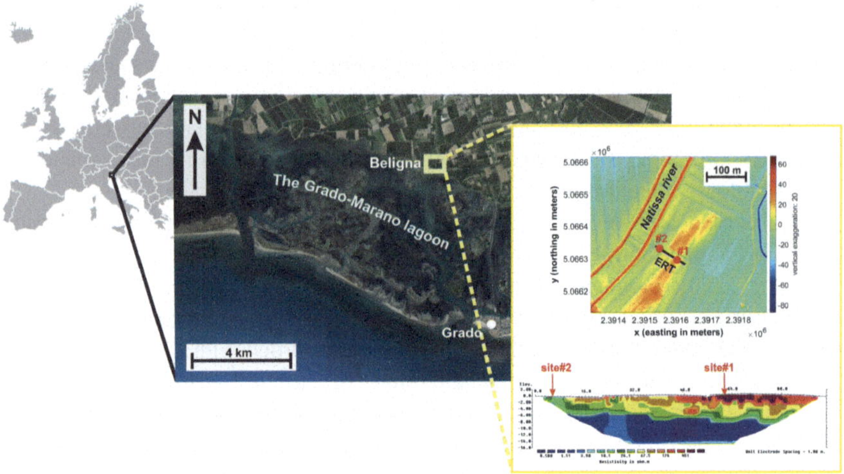

Fig. C.1 The study area (not far from the ancient Roman city of Aquileia—NE Italy). Reported the location and, on the right, the DTM (Digital Terrain Model) map with indicated the location of the ERT section (reported in the lower-right corner) and the two points (site#1 and #2) were HS and HVSR data were recorded. As can be seen, site#1 is right in the middle of the elevated sandy geomorphological feature (a prospective paleo dune), while site#2 is at its extreme western border (were the sandy sediments that characterize site#1 fade away and become a "regular" silty soil). Please, notice that, in order to emphasize the topography, in the DTM we applied a 20× vertical exaggeration

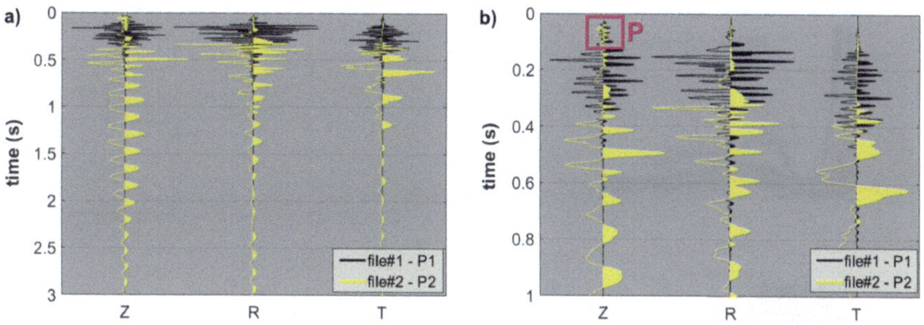

Fig. C.2 Comparison of the HS active seismic traces (offset 35 m) recorded at the site#1 [black lines] and #2 [yellow lines] (see map in Fig. C.1): **a** 3 s traces (Z, R and T components); **b** close up of the data (shown only 1 s). A simple visual check of the seismic traces provides an instantaneous evidence of the large differences between the shallow sediments at the two considered points. The signals represented by the yellow traces (site#2, just about 50 m from site#1) are clearly significantly slower compared to the black traces/signals recorded at site#1. In other words, compared to the traces recorded at site#2, the HS traces at site#1 immediately provide the evidence that the shallow sediments at site#1 are markedly stiffer (surface waves are significantly faster). A closer examination of the first arrivals along the Z trace at site#2 reveals also the presence of the P-wave refraction due to the shallow water table (see high-frequency arrivals at about 0.035 s [box P] and Dal Moro et al. 2015)

Fig. C.3 Comparison of the HVSR curves at site#1 and #2. While the 0.4 Hz peak is apparent in both the datasets, the 2.5 Hz peak is present only at site#2. The joint analysis with the HS data (see and compare data and analyses summarized in Figs. C.4 and C.5) provide the explanation of such a large difference

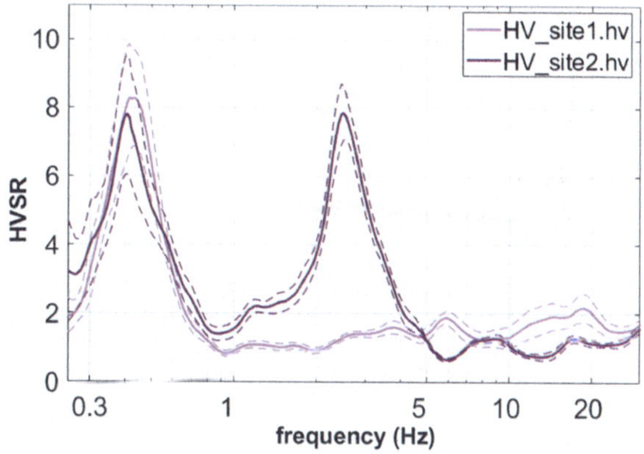

The very low-frequency HVSR peak at about 0.4 Hz (clearly visible in both the HVSR curves) is easily attributed to the deep bedrock (see V_S profiles presented in Figs. C.4d and C.5d and compare with the data obtained from borehole seismics about 8 km southeast of the Beligna area and presented in Della Vedova et al. 2015).

This case study shows that, through the joint analysis of HS and HVSR data (which require very simple field operations), it is possible to characterize a complex area (where significant lateral variations occur).

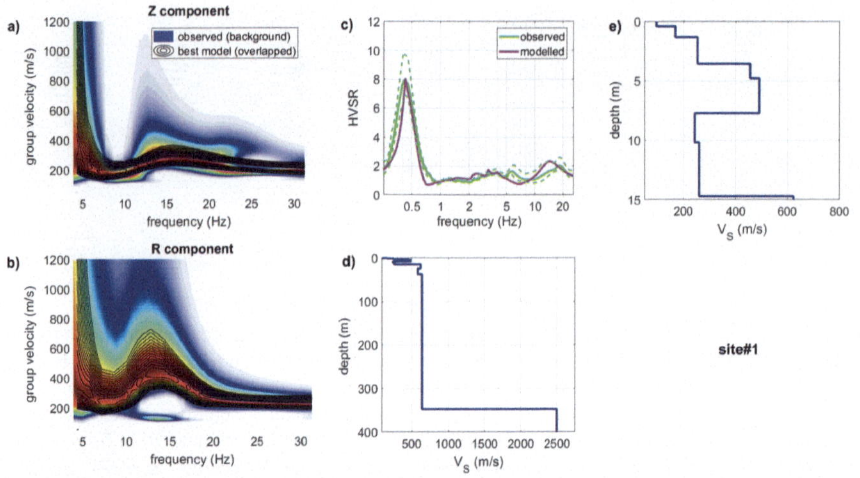

Fig. C.4 Site#1 (see map in Fig. C.1): V_S profile obtained from the joint analysis of the HS (Z and R components) + HVSR seismic data. **a** Group velocity spectra of the Z component; **b** group velocity spectra of the R component; **c** HVSR curves; **d** obtained V_S profile (down to 400 m); **e** close up of the shallow part (first 15 m) of the obtained V_S profile. The HVSR peak at about 0.4 Hz is due to the deep bedrock at a depth of about 350 m

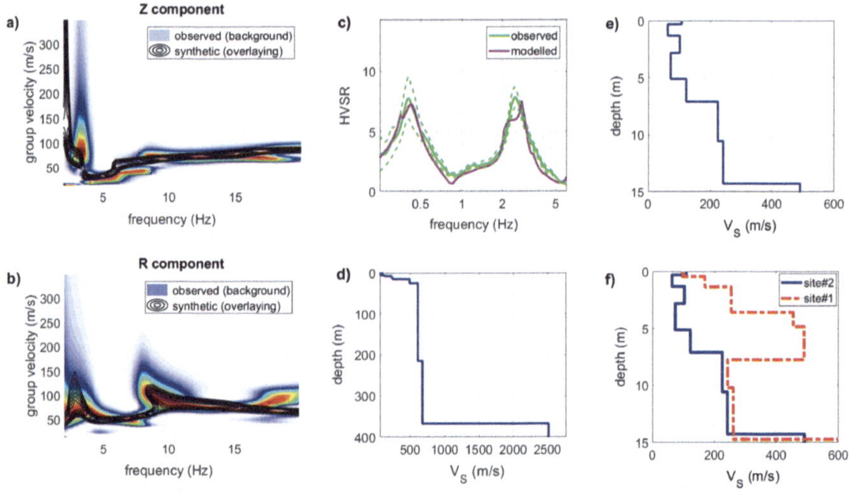

Fig. C.5 Site#2 (see map in Fig. C.1): V_S profile obtained from the joint analysis of the HS (Z and R components) + HVSR seismic data. **a** Group velocity spectra of the Z component; **b** group velocity spectra of the R component; **c** HVSR curves; **d** obtained V_S profile (down to 400 m); **e** close up of the shallow part (first 15 m) of the obtained V_S profile. The HVSR peak at about 2.5 Hz is due to the contact between the deeper and stiffer sediments with the superficial very soft sediments (very low V_S values); **e** comparison of the V_S profiles at site#1 and site#2 (first 15 m)

Fig. C.6 Two photos of the sediments/samples that can be found on the surface in the study area: **a** pieces of cemented sand at site#1 (Fig. C.1); **b** bigger and stiffer samples with evidences of tubercles and shafts (detailed ichnological analyses will be available in a study currently in progress)

It should be strongly underlined, that the goal was not related to the seismic-hazard assessment (too often shear-wave velocities are considered only with respect to site amplification studies), but for the exploration of a complex area and the collection of data useful to cast some light to its geological and geomorphological evolution.

Appendix D: Seismic and Geological Bedrock in a NE Italy Prehistoric Site

Pozzuolo del Friuli is a municipality in NE Italy about 9 km south of Udine. The area is known among the archaeologists for some important evidences of the iron and Bronze Age as well as, in some areas, of the Neolithic era.

From the geological and geomorphological point of view, the area is quite interesting because its hilly landscape (in the middle of the otherwise flat Friulian plain—Fontana et al. 2008) is the result of complex tectonic phenomena that brought to surface the geological *bedrock*.

In the framework of the microzonation study of the area, a series of seismic data were collected also with the goal of comparing the results that can be obtained through different methodologies:

- Multi-component MASW (phase velocities analyzed according to the FVS approach)
- *HoliSurface* analysis (group velocities of the Z and R components together with the RPM frequency curve)
- HVSR data (jointly with the above-mentioned dispersion data)

The photo in Fig. D.1 shows the contact between the Miocene sandstone and the Pleistocene conglomerates and, in order to define the local seismic response it was necessary to define the shear-wave velocities of such outcropping formations.

Site#1: the Miocene sandstone

In the area shown in Fig. D.2 there is a plateau where a first seismic dataset was recorded. In the point shown in Fig. D.2 sandstones dominate, while further north-west, a few meters of conglomerates cover the Miocene sandstones (see the section about site#2).

At site#1 we recorded multi-component (Z+R) MASW, HS data and two microtremor datasets for the definition of two HVSR curves (at two points about 25 m one from the other—see HVSR data in Fig. D.3).

The two microtremor datasets used for the determination of the HVSR curves were carried out with the aim of verifying the congruence of the two curves (therefore of the overall consistency of the data) and the absence of strong lateral variations (see Chap. 3). The second dataset was then acquired considering a very limited recording time (about 4 min). In this way, it is possible to verify only the

G. Dal Moro, *Efficient Joint Analysis of Surface Waves and Introduction to Vibration Analysis: Beyond the Clichés*,
https://doi.org/10.1007/978-3-030-46303-8

Fig. D.1 Pozzuolo del Friuli (Udine—Italy): the contact between the Miocene sandstones (below) and the Pleistocene conglomerates (above). What are the respective shear-wave velocities?

"high" frequencies (here of paramount interest) since, at lower frequencies, the second HVSR is necessarily ill-defined (see Fig. D.3).

As demonstrated by the unsmoothed amplitude spectra (plotted with linear scales—Fig. D.3c) and coherence functions (Fig. D.3b), the HVSR "peak" at about 1.5 Hz (Fig. D.3a) is clearly related to some industrial facility. Such a signal is actually present in most of the NE Italy area (see Chap. 3 and Dal Moro 2020b).

Multi-component (Z + R) MASW + RPM frequency-offset surface

Figure D.4 presents the field traces and phase velocity spectra of the Z and R components, together with the RPM frequency-offset surface.

The joint inversion of these three observables produced the results shown in Fig. D.5 (remember that the larger the number of "objects" considered in the inversion process, the smaller the uncertainty/ambiguity of the identified subsurface model).

From these results we can see that these sandstones (Fig. D.2) are characterized by V_S around 400 m/s, quite far from the value necessary to associate such a

Fig. D.2 Site#1 (south-east of the plateau): outcropping sandstone

formation to the *seismic bedrock*. The velocity inversion (confirmed also by the analyzes presented in the next section) may be easily explained by levels richer in marl component or (very daring hypothesis) by weaker levels related to tectonic phenomena (the area is extremely complex from the tectonic point of view and very strong lateral variations occur).

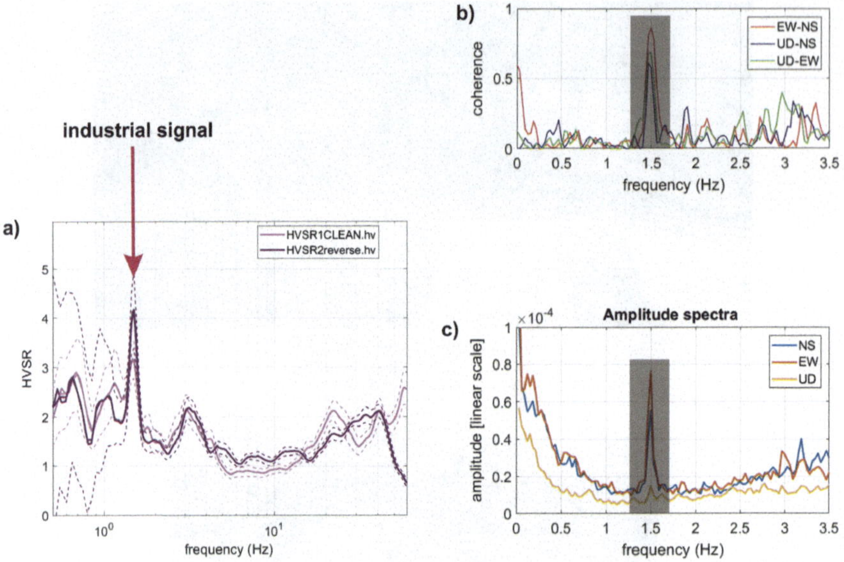

Fig. D.3 Site#1: comparison of the HVSR curves from two points 25 m away from each other (one close to the small escarpment shown in Fig. D.2 and the other 25 m away from it, South-East). In the low frequencies, the second curve is badly defined (see large standard deviations in the **a** plot) because obtained from the analysis of a 4-min dataset (see recommendations about the length of the passive data necessary to define the HVSR curve summarized in the Appendix A). The signal at about 1.5 Hz highlighted by the gray rectangle is apparently spurious/industrial (see large values of the coherence functions and unsmoothed amplitude spectra [**b** and **c** graphs computed by considering the HVSR#1 dataset])

HS [*HoliSurface*®] + HVSR

As already underlined throughout the book, in the HS approach we do not consider the phase velocities (as in the MASW, ReMi, MAAM or ESAC techiques) but the group ones. For this reason, a single trace (for each component) is sufficient (see the traces reported in Fig. D.6 for the three Z, R and T components). For simplicity, in this case we considered only the Z and R components, together with the RPM frequency curve and, in order to investigate deeper levels, the mean HVSR curve obtained after having removed the industrial components (see results summarized in Fig. D.7). Since in this case we considered also the HVSR curve, the investigated depth is larger compared to the active multi-offset data considered in the previous section (see analysis presented in Fig. D.5). In any case, if we compare the V_S profile shown in Figs. D.5 and D.7, the agreement is apparent and this demonstrates the consistency of the two methodologies (the multi-offset and multi-component technique and the single-offset multi-component HS approach).

Site#2: the Pleistocene conglomerate

Some tens of meters NE from site#1, a conglomerate layer covers the Miocene sandstone (Fig. D.8).

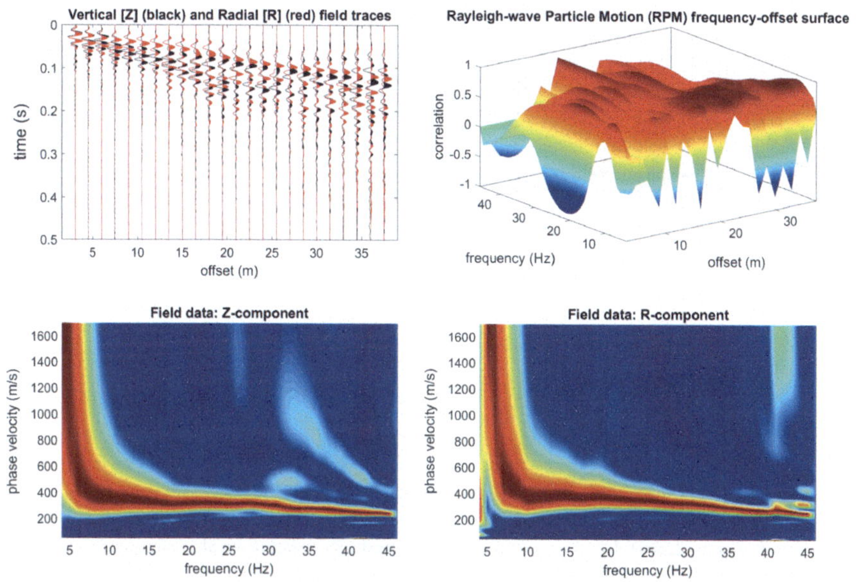

Fig. D.4 Site#1: field traces (Z and R components), RPM frequency-offset surface and (below) phase velocity spectra for the Z and R components

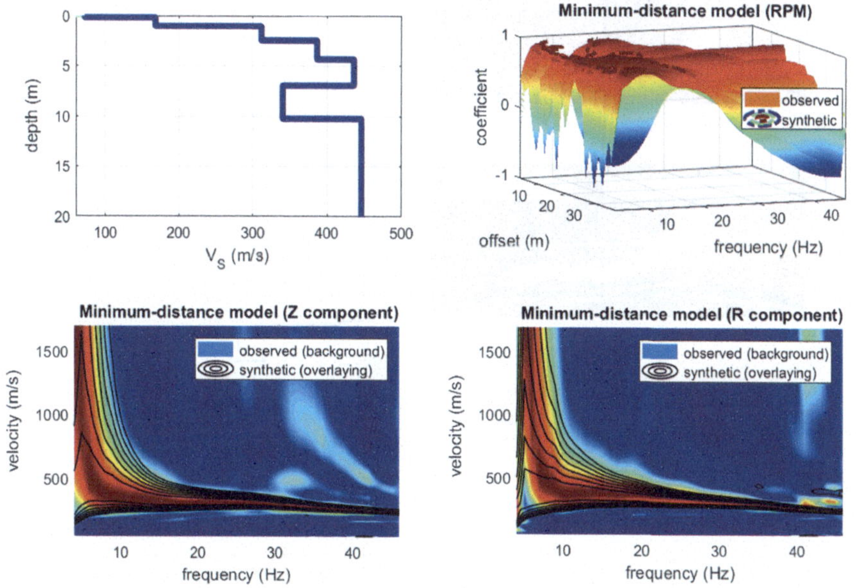

Fig. D.5 Site#1: result of the joint inversion of the Z and R phase-velocity spectra (FVS approach) together with the RPM surface. As usual: background colors represent the field data while overlying contour lines the synthetic data of the identified V_S model (shown in the upper left plot)

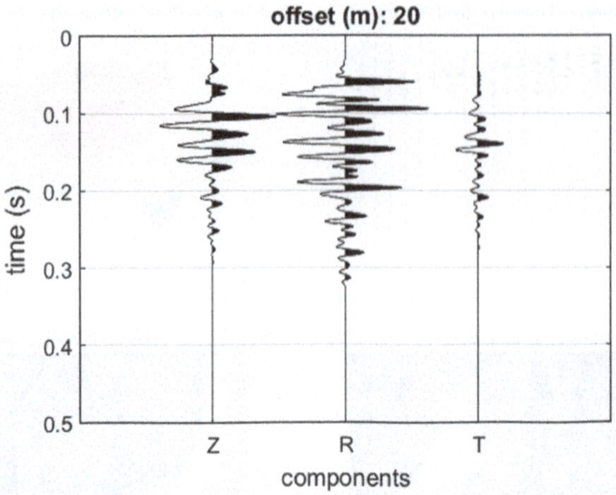

Fig. D.6 Site#1: HS data (Z, R and T traces—offset 20 m). Z and R traces are used to define the respective group-velocity spectra and define the RPM frequency curve (the T trace—Love waves—are here not considered). In order to increase the investigated depth we also analyzed the HVSR curve. Results of the joint analysis of the computed observables shown in Fig. D.7

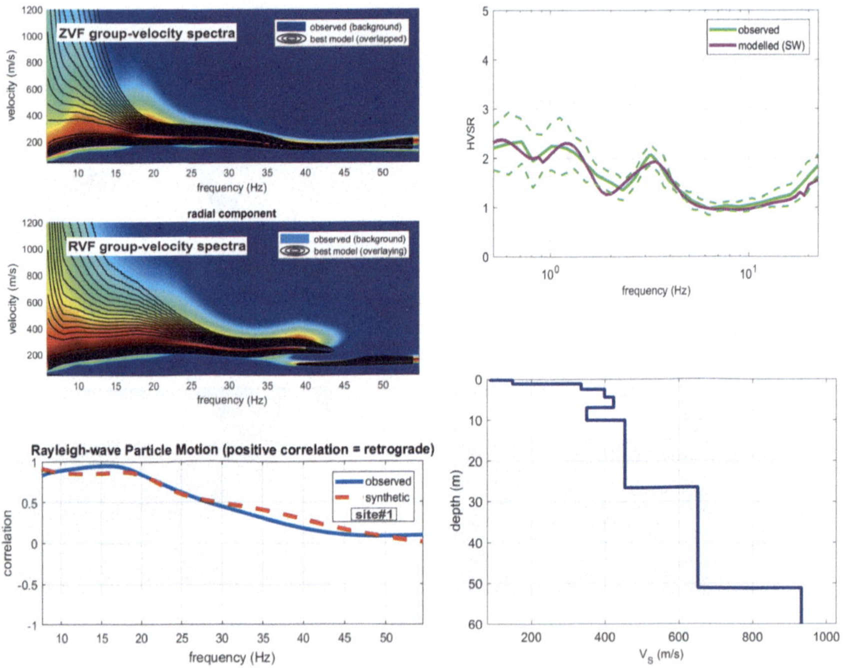

Fig. D.7 Site#1: result of the joint analysis of the Z (vertical) and R (radial) group-velocity spectra together with the RPM and HVSR curves

Fig. D.8 Site#2: about 50 m north-east from the site#1 (Fig. D.2), the Pleistocene conglomerates cover the Miocene sandstone

Fig. D.9 Site#2: comparison of the HVSR curves obtained from the microtremor data gathered at two points about 20 m distant from each other. The overall consistency is apparent: data are then reliable and no significant lateral variation is present

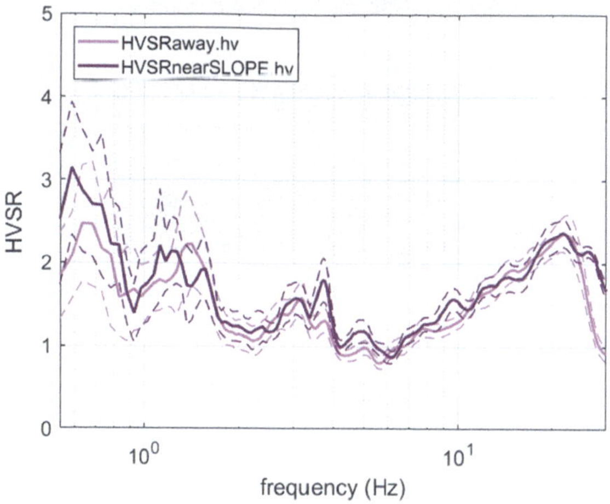

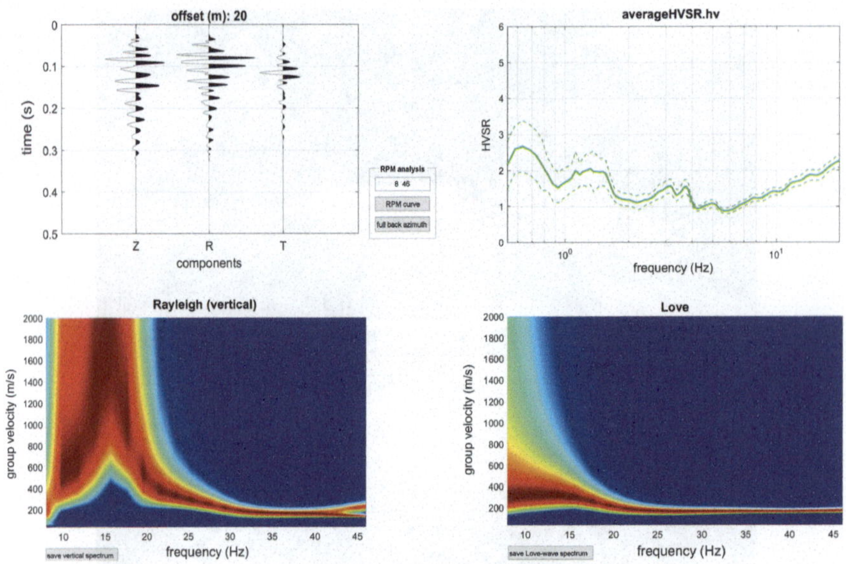

Fig. D.10 Site#2: active HS traces (Z, R and T components—offset 20 m) and mean HVSR curve. In the lower panel are shown the group-velocity spectrum for the Z (on the left) and T (on the right) components. The result of the joint analysis of these three observables is shown in Fig. D.11

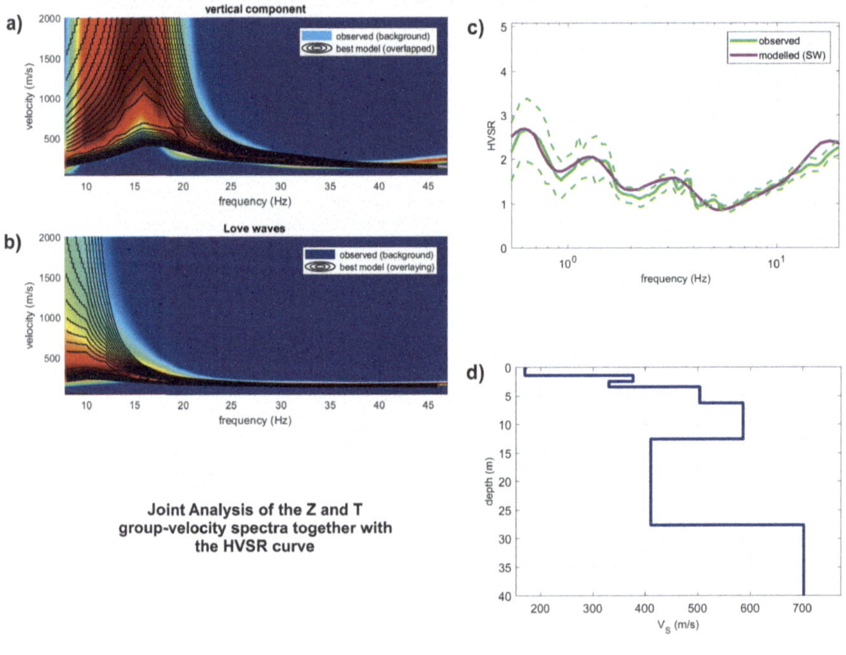

Fig. D.11 Site#2—joint analysis of the group-velocity spectra of the Z (a) and T (b) components together with the HVSR curve (c). The identified V_S model is shown in the bottom-right corner (d)

Fig. D.12 Outcropping sandstones not far from the site#2 (V_S is around 400 m/s)

Similarly to the work accomplished at site#1, we collected several active and passive datasets but, for brevity, we will only show the HVSR and HS data and analyses. The two HVSR curvs (from two points 20 m away from each other) are shown in Fig. D.9 and the overal consistency is apparent (this means that the subsurface conditions are quite homogenous). Moreover we can notice the absence of the 1.5 Hz industrial components that was present at site#1 (Fig. D.3) [site#1 data were recorded in the morning while site#2 data at lunchtime].

The data about the Z and T components (i.e., the radial component of Rayleigh waves and Love waves) are presented in Fig. D.10 together with the average HVSR curve obtained from the two considered HVSR curves (see Fig. D.9). The result of the joint inversion of these three observables is shown in Fig. D.11, where the overall good agreement is apparent.

Few final remarks

Data and analyses presented in this brief case study show once again that working with the phase velocities (standard multi-offset MASW technique) or with the group velocities (single-offset, multi-component HS approach) is necessarily equivalent. We verified that the shear-wave velocities of the Miocene sandstones and Pleistocene conglomerates do not reach the value considered to define the seismic bedrock (usually fixed in 800 m/s). In the considered area, such a value is reached only at a depth of approximately 50 m. The V_S values associated to the superficial part of the Miocene sandstones and to the Pleistocene conglomerates are between about 300 and 500 m/s (the Pleistocene conglomerates appear characterized by slightly-higher shear-wave velocities).

Appendix E: HS and Microtremor Data: A Small Example of Comparative and Comprehensive Analysis (Joint Inversion and SSRn)

The data considered for the present case study were recorded at the foot of a small and isolated hill in the Friulian alluvial plain (NE Italy—Fontana et al. 2008), where a cretaceous limestone sequence raises up and creates an elongated hilly with the main axis along the N60E-S60W direction (Fig. E.1).

In order to define the stratigraphic sequence and the depth of the limestone bedrock at a meadow in the Northern part of the hill (Fig. E.1), we recorded both HS (offset 25 m) and microtremor data.

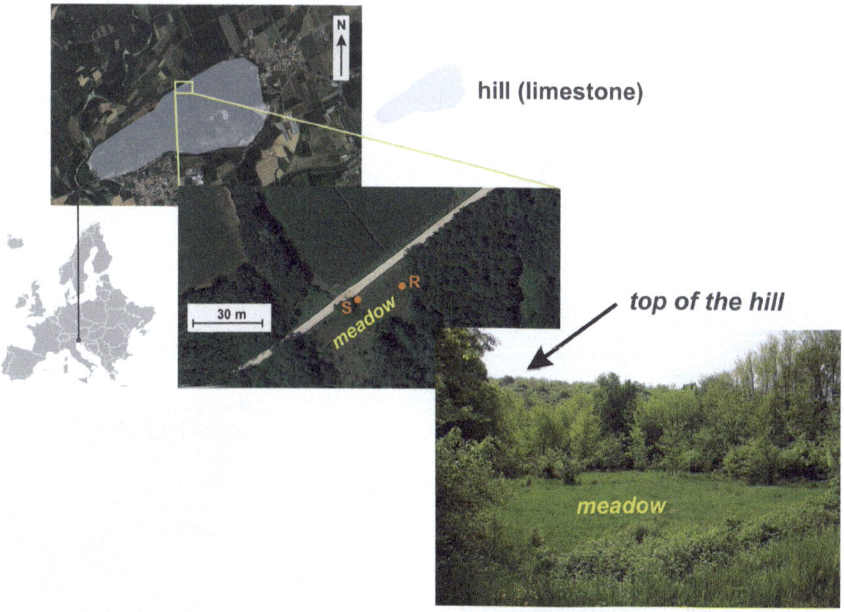

Fig. E.1 The study area: a meadow at the foot of a limestone hill. The points S and R represent the positions of the source (standard 8 kg sledgehammer) and receiver (3C geophone) for the acquisition of the HS data. How deep is the limestone *bedrock*?

G. Dal Moro, *Efficient Joint Analysis of Surface Waves and Introduction to Vibration Analysis: Beyond the Clichés*,
https://doi.org/10.1007/978-3-030-46303-8

Furthermore, thanks to the microtremor data simultaneously recorded with an identical equipment over a limestone outcrop in the wood (about 150 m away from the meadow), we also defined the SSRn curve (see Chap. 3).

Figure E.2 presents the active HS data (Z and R traces and group-velocity spectra) as well as the HVSR curve that were jointly inverted with the results shown in Figs. E.3 and E.4.

A similar example of HS + HV joint analysis is presented for instance in Dal Moro and Puzzilli (2017) (such a paper can be easily downloaded from the internet). In that particular case, part of the microtremor data were recorded while an excavator was at work and, in order to be exploited to define a meaningful HVSR curve, a radical data cleaning was necessary.

Since for the present case the depth of the *bedrock* revealed not too deep (about 15 m—Fig. E.4), we also tried to consider the active data only and performed a joint inversion of the two group-velocity spectra together with the RPM frequency curve. The result of the Z + R + RPM joint inversion is shown in Fig. E.5: the *bedrock* is identified at the same depth but, of course, the investigated depth is necessarily smaller with respect to the opportunity offered by the HVSR curve (that allowed to constrain the V_S profile much deeper—a possible approach to define the investigated depth while analyzing group velocities is illustrated in Dal Moro et al. 2019).

In order to explore a bit more the site and the data, we also considered the microtremor data collected over the limestone outcrop (reference rocky site) that can be found in the wood, about 150 m away from the meadow and compare them

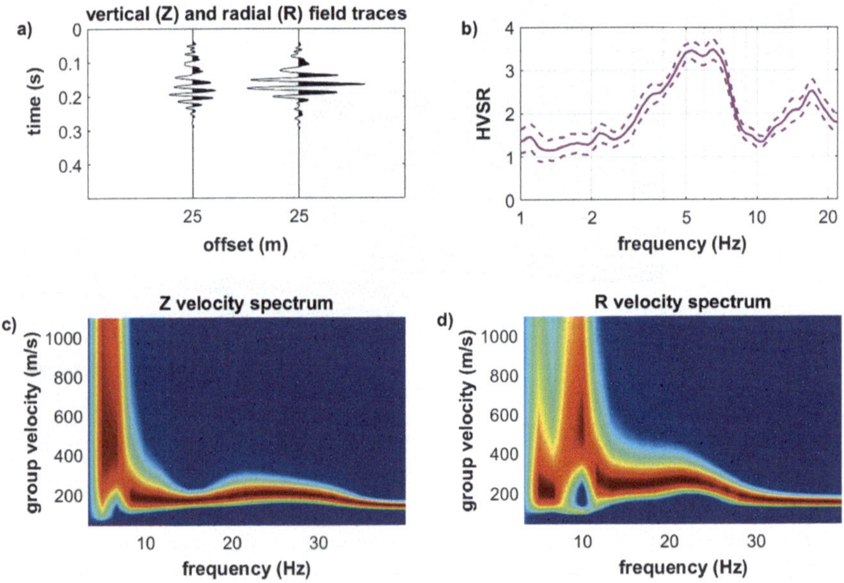

Fig. E.2 The HS and HVSR data: **a** field traces (offset equal to 25 m) of the Z and R components (VF source); **b** HVSR curve; **c** group velocity spectrum of the Z component; **d** group velocity spectrum of the R component

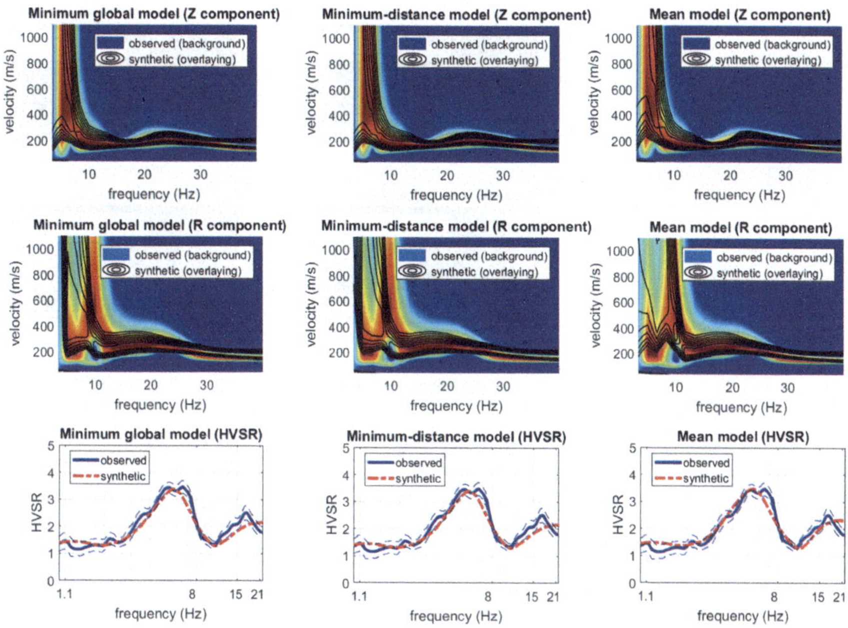

Fig. E.3 Result of the joint inversion of the active (HS) and passive (HVSR) data. The overall agreement for all the three considered observables is apparent. Shown the data pertaining to the three most important models. From left to right: the model with the minimum global misfit, the minimum-distance model and the mean model computed from all the final Pareto models (details in Dal Moro et al. 2019). It must be strongly underlined that when the inversion process is properly setup, the three models must be very similar (see the V_S profiles shown in Fig. E.4)

Fig. E.4 The V_S profiles of the three best models reported in Fig. E.3: the fact that they are pretty similar proves that the inversion process was properly setup

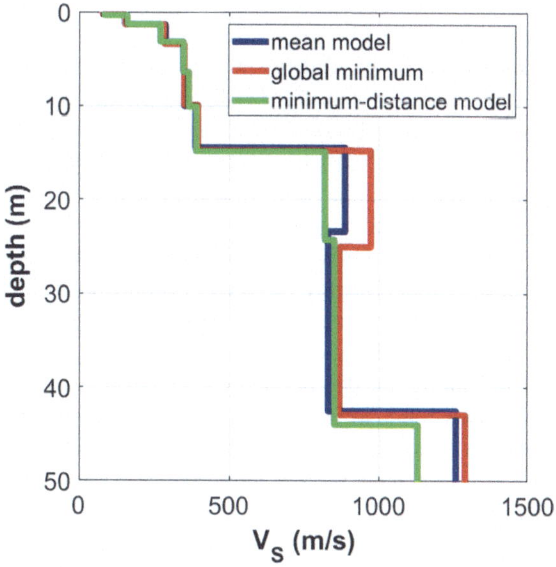

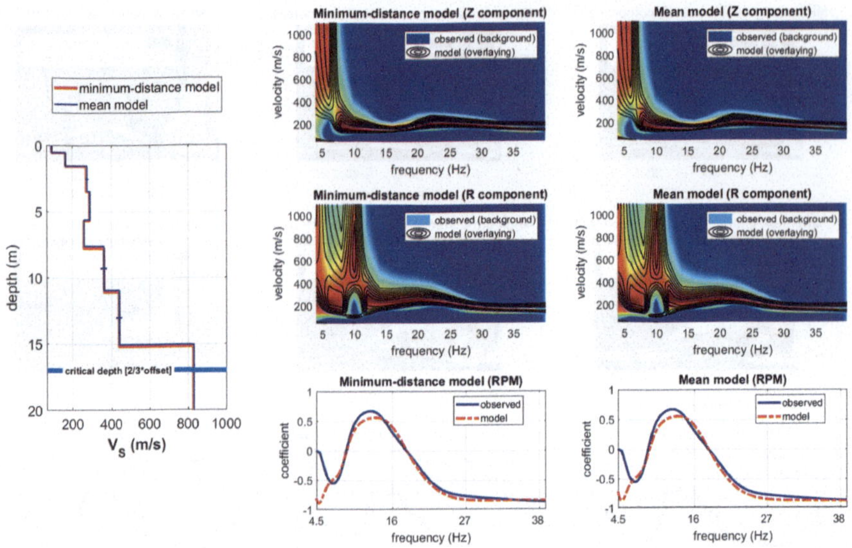

Fig. E.5 Inversion of the active HS data only: joint analysis of the Z and R group-velocity spectra together with the RPM frequency curve

with the microtremor data recorded in the middle of the meadow and used for the joint analysis presented in Figs. E.2 and E.3. In Fig. E.6 is shown the particle motion along the horizontal plane after having band-passed the data in the 0.5–32 Hz frequency band: the amplification of the horizontal motion in the middle of the meadow (where the *bedrock* is present at a depth of about 15 m) is apparent. In Fig. E.7 are shown and compared the HVSR (in the middle of the meadow) and SSRn curves computed while considering three different ways of computing the H component (quadratic, geometric and arithmetic mean—see Chap. 3). It is clear that the difference between different ways of defining the H component is irrelevant but the difference between the HVSR and SSRn curves is quite large.

As commonly observed by the author of this book and by several other authors (e.g. Perron et al. 2018), the SSRn appears significantly higher than the HVSR but it must be underlined that none of these curves can be considered as representative of the actual site amplification. In general terms this latter is in fact represented by the SSR (that we can define only through the analysis of quake data) although, as for instance pointed out in Olsen (2000), the peak velocity amplification is a specific characteristic of the considered earthquake and is not a persistent and invariable characteristic of the site.

Incidentally, as also pointed out in Diagourtas et al. (2001), one of the reasons why HVSR tends to underestimate the actual site amplification is that, contrary to one of the basic assumptions implicit in the so-called Nakamura's method, the vertical component is also amplified (see amplitude spectra and spectral ratios presented in Fig. E.8).

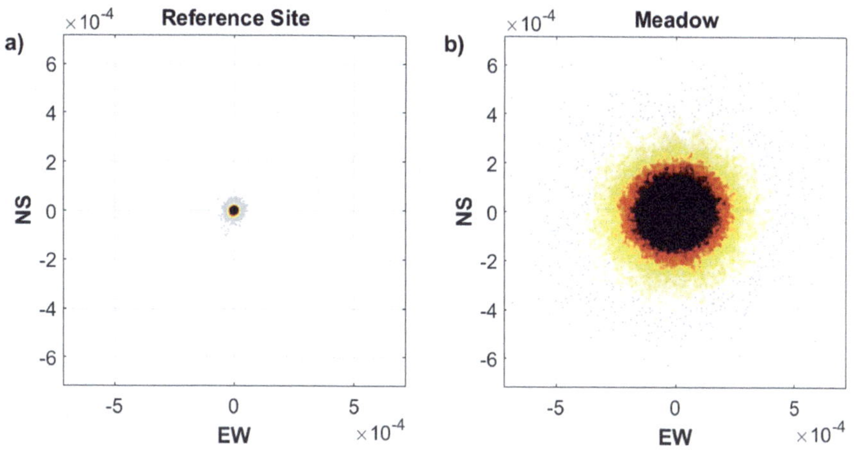

Fig. E.6 Comparison of the particle motion (density functions) over the horizontal plane at the reference rocky site (left panel) and in the middle of the meadow (right panel) (data in mm/s). The amplification is apparent (see also Fig. E.7)

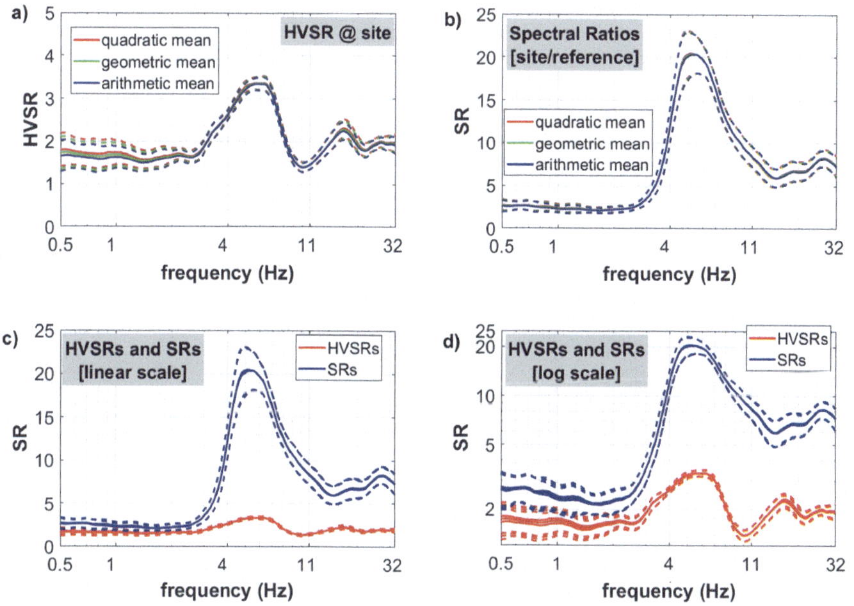

Fig. E.7 Comparison between the HVSR in the middle of the meadow and the SSRn curves: **a** HVSR (in the meadow) computed according to three different definitions of the H component (quadratic, geometric and arithmetic mean—see Chap. 3); **b** SSRn curves computed according to the same three definitions of the H component; **c** comparison of the HVSR and SSRn curves (linear scale); **d** comparison of the HVSR and SSRn curves (logarithmic scale). See text for comments

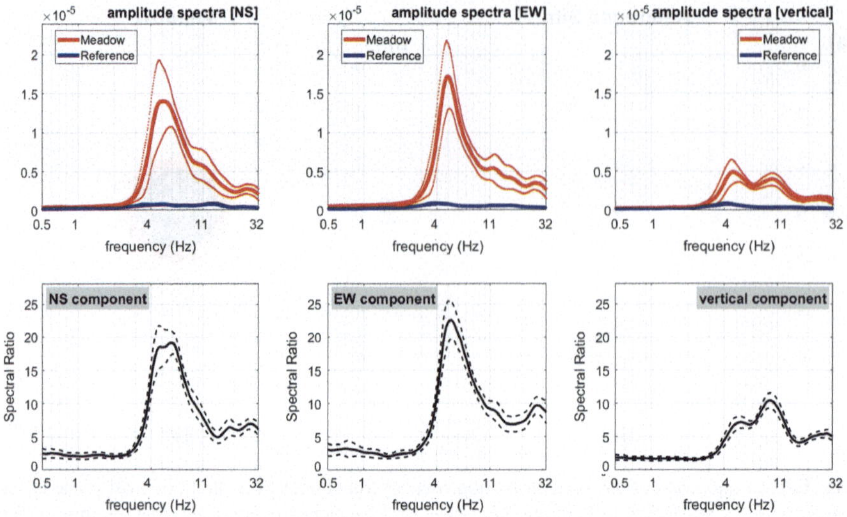

Fig. E.8 Analysis of the microtremor data for the definition of the SSRn: amplitude spectra (upper panel) and spectral ratios (lower panel) for the three components for the two sites (the rocky reference site and the meadow at the foot of the hill). The apparent and large amplification of the vertical component (right column) is one of the reasons why HVSR cannot be considered as a straightforward expression of the actual site amplification

This means that, in general, HVSR is surely useful to better constrain the deep V_S structure but cannot be used as a straightforward and reliable assessment of the site amplification since it systematically tends to underestimate it.

Appendix F: Example of HS for a Hardly-Accessible Site

The considered site is located in NW Italy (not far from the city of La Spezia) and is characterized by a sub-outcropping bedrock and significant access difficulties due both to the topography (the area is quite steep) and to the presence of lush vegetation (Fig. F.1). It is clear that these logistical conditions prevented from using classic multichannel (multi-offset) MASW techniques and we then proceeded with the HS methodology.

The collected HS data are shown in Fig. F.2 (offset: 27 m, source: 8-kg sledgehammer) and the presence of a certain amount of scattering and complex wave phenomena is apparent (field operator recorded the data according to a quick-and-dirty mood while considering that, as also supported by the general site conditions and by some penetrometer data, the bedrock had to be quite shallow).

With the goal of comparing different possible processing options, in the following are shown the results obtained by following three possible approaches:

1. analysis of the Radial (R) component of Rayleigh waves only (Fig. F.3);

Fig. F.1 The site is located over a steep flank and it is not easy to reach also because of the lush vegetation

G. Dal Moro, *Efficient Joint Analysis of Surface Waves and Introduction to Vibration Analysis: Beyond the Clichés*,
https://doi.org/10.1007/978-3-030-46303-8

2. analysis of the Love waves only (Fig. F.4);
3. joint analysis of the radial (R) component of Rayleigh waves + Love waves (Fig. F.5).

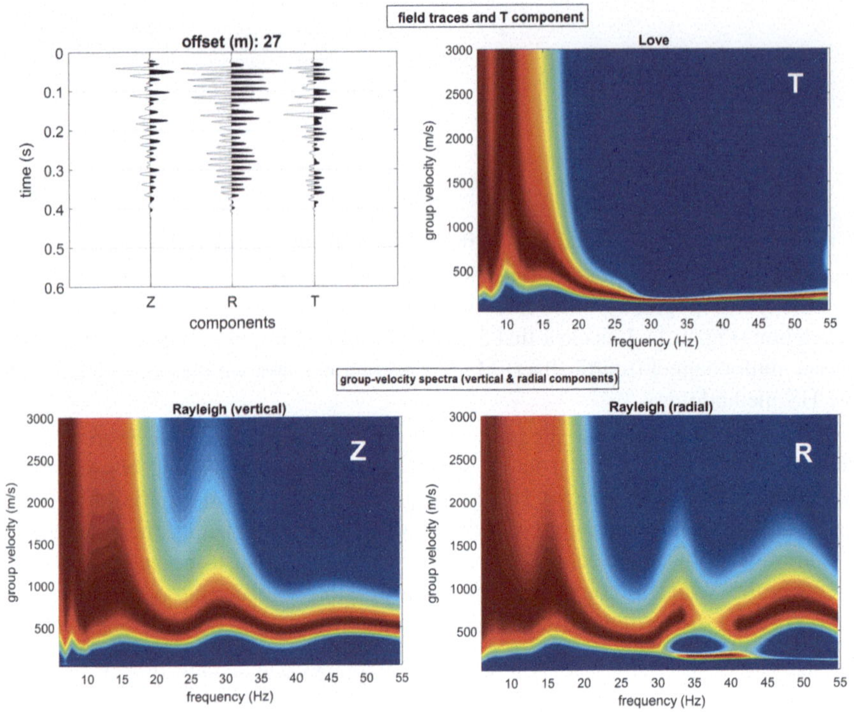

Fig. F.2 HS active data (offset 27 m): shown the three recorded traces (Z, R and T components) and the respective group-velocity spectra

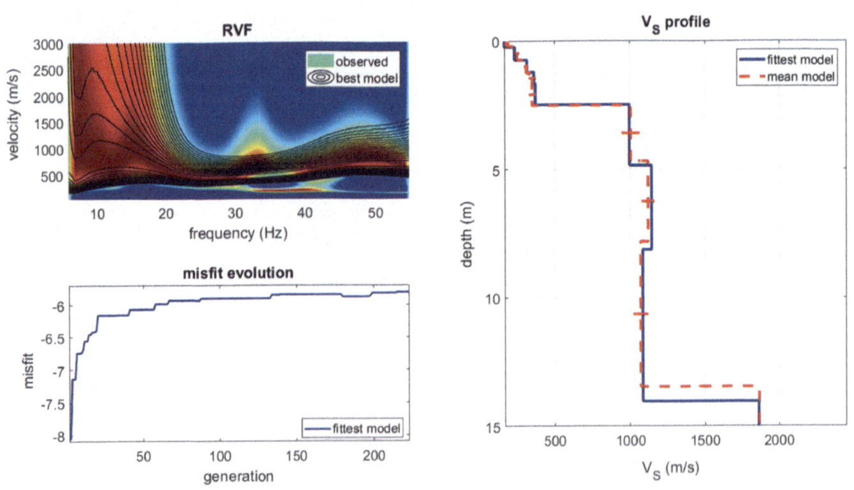

Fig. F.3 Single-component FVS analysis of the radial component of Rayleigh waves (RVF). Considering that the offset was 27 m, the V_S profile is shown only down to 15 m

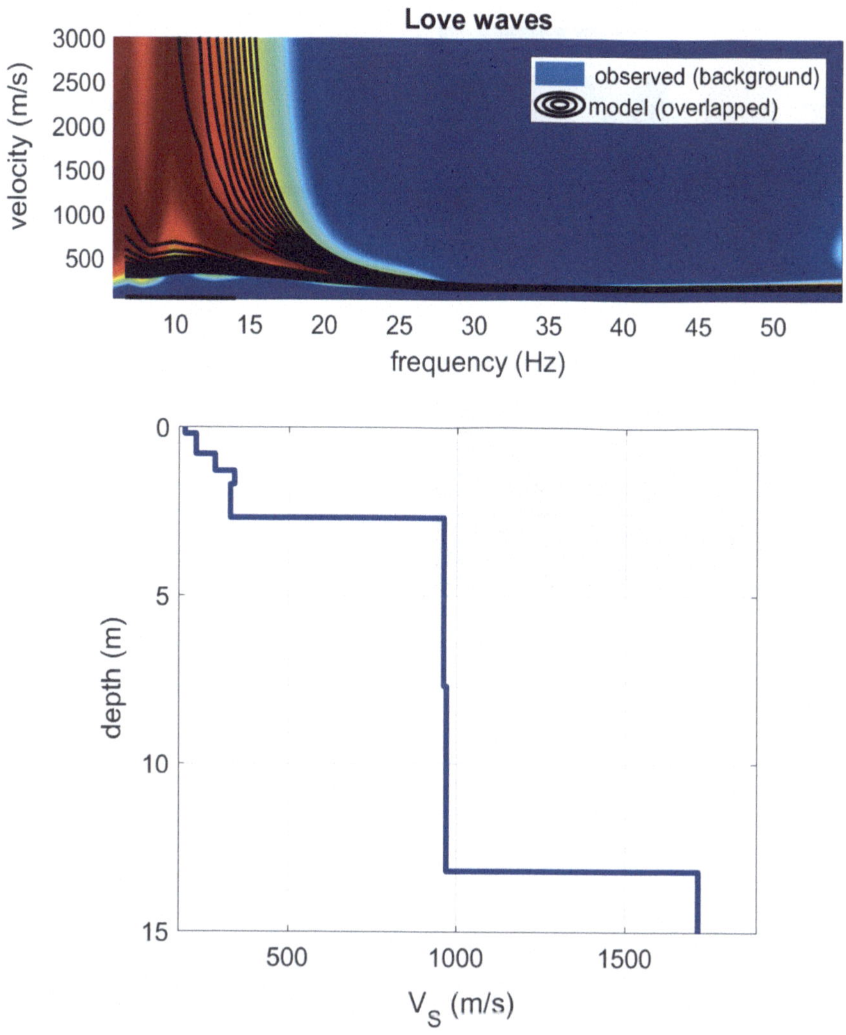

Fig. F.4 Single-component FVS analysis of the THF component (Love waves)

Incidentally, a similar comparative study is presented in Dal Moro and Puzzilli (2017), where a series of analyses are performed in order to attempt to assess and analyze the (active and passive) data recorded in a very noisy area dominated by soft sediments (silt, clay and sand).

By comparing the presented results (Figs. F.3, F.4 and F.5), we can appreciate the general congruence of the obtained V_S profiles which, also in agreement with the available penetrometer data, identify the bedrock at a depth of about 3 m.

Considered the offset of the HS data (27 m) we might wonder whether we actually reached the notorious 30 m of depth required by most of the building codes worldwide.

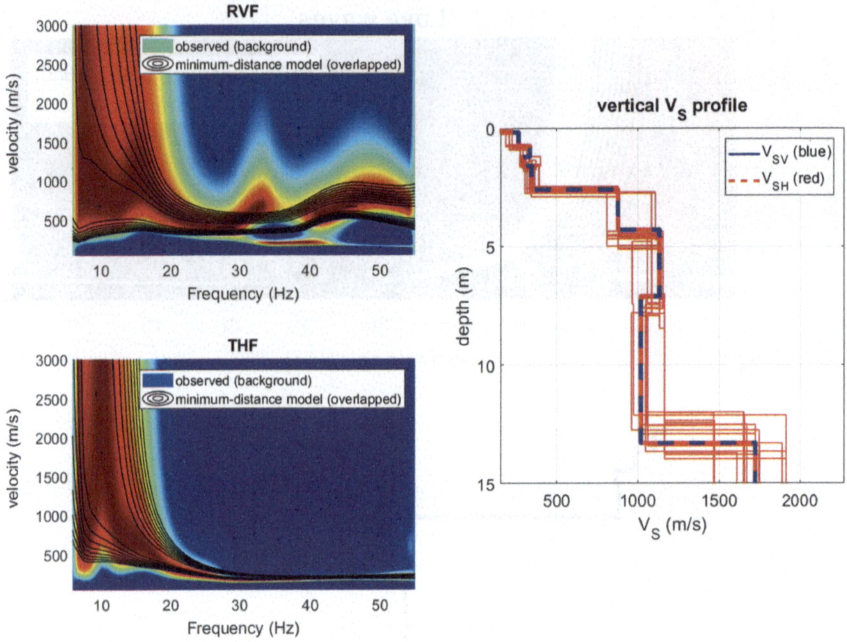

Fig. F.5 Joint FVS inversion of the radial component of Rayleigh waves (RVF) and Love waves (THF)

The estimation of the penetration depth from dispersion data is surely possible but necessarily approximate and questionable. Everything depends on the way we define the sensitivity and, strictly speaking, is based on the analysis of the eigenfunctions and, more specifically, their trend with depth (the subject was briefly discussed in Chap. 4 while describing the HS methodology). On the other side, for routine analyses (especially in the professional sector where productivity is a key issue), some simpler and quicker approach is necessary. We can therefore consider the hybrid approach described in Dal Moro et al. (2019) and in Chap. 4 (Sect. 4.5). For the present case, if we consider the $\lambda/3$ approximation and 9 Hz as the minimum reliable frequency (this is actually quite optimistic), thanks to the high V_S values we obtain an investigated depth of about 40–50 m, which is surely too optimistic but provides anyway an idea of what we can reasonably expect.

We should anyway also consider that:

(a) in general terms, in order to increase the investigated depth, we could also add/use the HVSR;

(b) in cases like the present one, we could also simply use the common sense (in geological terms): is there any valid geological reason to imagine that significant velocity inversions occur at layers deeper than 15–20 m (which is the depth surely investigated)? In simple terms: is it possible that below the

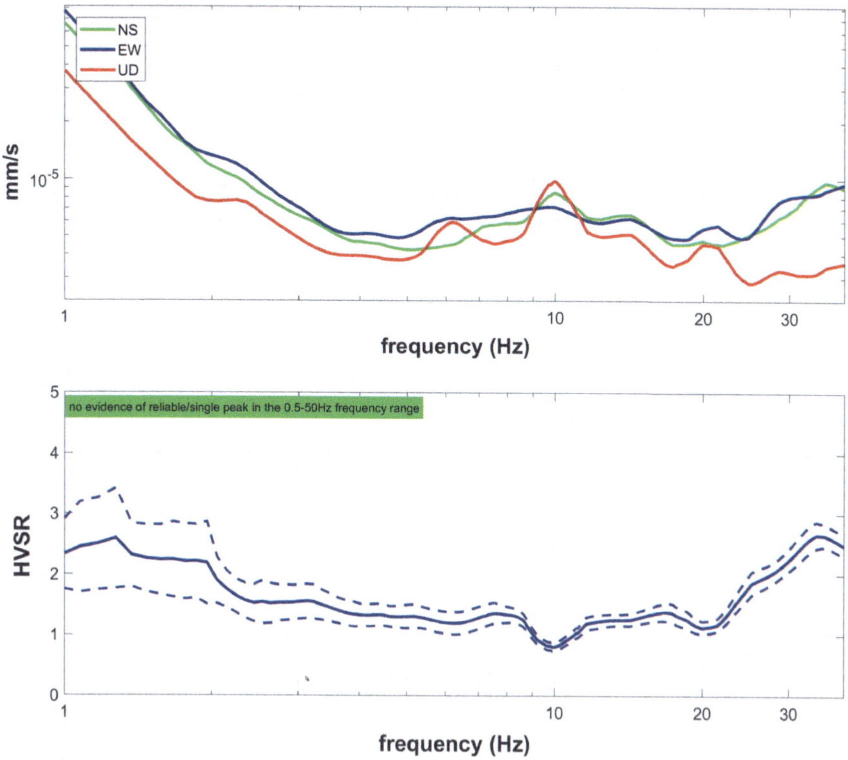

Fig. F.6 Mean amplitude spectra (upper plot) and HVSR (lower plot). See also data reported in Fig. F.7 and related comments

bedrock identified by the HS data (and confirmed by the penetrometer data) some softer/weaker layer is present?

A couple of microtremor datasets were also recorded and Figs. F.6 and F.7 present the main data about one of them. A series of industrial signals are apparent (see also Chap. 3): the two "valleys" at about 10 and 20 Hz in the HVSR curve (Fig. F.6) are clearly due to two of them. This is the evidence that an industrial signal does not necessarily produce a "positive" (i.e. >1) artefact (or "peak"). In case the amplitude of the vertical component is larger than the horizontal ones, the H/V spectral ratio decreases (see for instance the amplitude spectra and the HVSR at 10 and 20 Hz). In this case, we should call this sort of features "industrial valley" or "industrial hallow" (surely not "industrial peak"). Why all the other industrial signals (Fig. F.7) do not modify (pollute) the HVSR shown in Fig. F.6? They do not modify the HVSR for two reasons. The most important one is that the amplitude along the vertical and horizontal axes are quite similar and, therefore, the H/V ratio does not change; the second reason is that the applied *smoothing* inevitably tends to smooth out small industrial components (see Chap. 3 and Dal Moro 2020b).

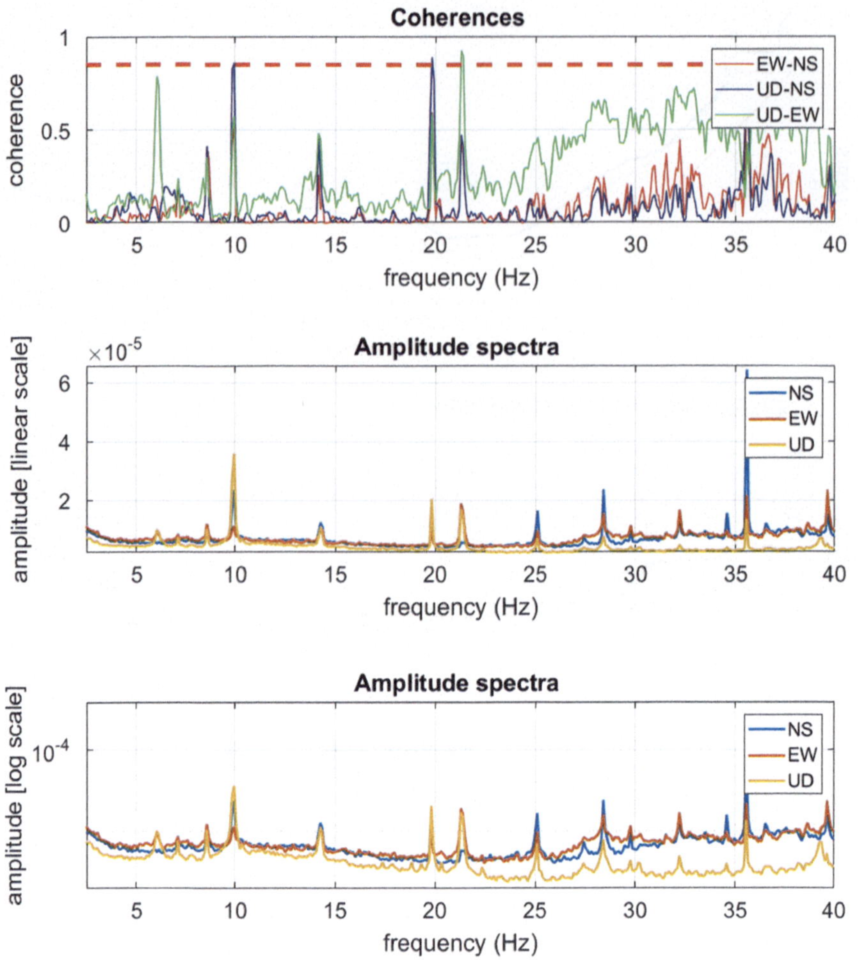

Fig. F.7 Coherence functions (upper plot) and amplitude spectra reported with linear (central plot) and logarithmic scales (lower plot). The presence of several (about 12–14) industrial signals is apparent. Fortunately, only two of them (at 10 and 20 Hz) somehow influence the H/V spectral ratio shown in Fig. F.6

Appendix G: 2D V_S Section of an Urban Area from the Multi-offset Holistic Analysis of Rayleigh Waves (Multi-offset Z + R + RPM)

In cooperation with Lorenz Keller—*roXplore (Amlikon-Bissegg—Switzerland)*, www.roXplore.ch

The present case study illustrates the acquisition and analysis of active multi-offset and multi-component data according to the Z + R + RPM approach (we therefore consider the data of Rayleigh waves only).

The data were acquired in the historic center of a Swiss town considering the classical multi-offset approach (which means that geophones are placed at gradually increasing distances from the source as in the typical so-called MASW approach). Needless to say that, since we were working along a urban street, we were dealing with an asphalt cover (Figs. G.1 and 1.12).

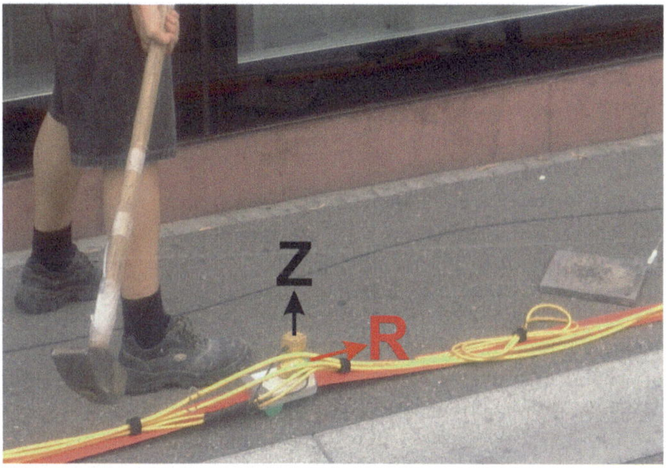

Fig. G.1 Multi-offset data acquisition while using a landstreamer (pulled by a vehicle). Note that, along the landstreamer, at each recording point there are two geophones: the vertical one (Z) and the horizontal (R) set with the axis radial to the array. See also Figs. 1.11 and 1.12

G. Dal Moro, *Efficient Joint Analysis of Surface Waves and Introduction to Vibration Analysis: Beyond the Clichés*,
https://doi.org/10.1007/978-3-030-46303-8

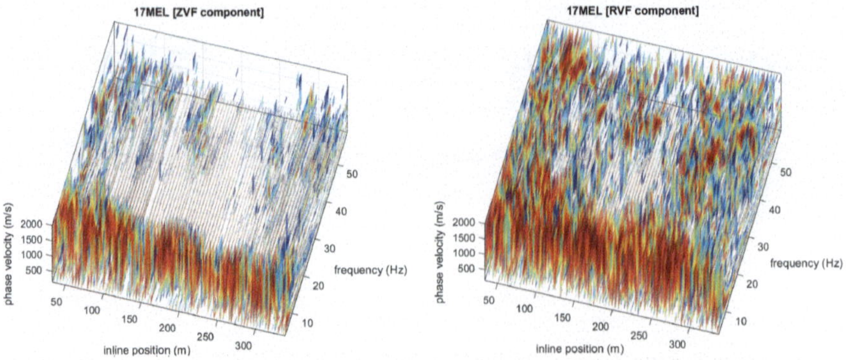

Fig. G.2 Phase velocity spectra of the Z and R components in a 3D perspective for the 78 shots (raw data) along the 340 m line. On the left the Z component, on the right the R component. The quality of the component R is slightly lower for the reasons reported in the text

The goal was to reconstruct the 2D profile along a 340 m line, so to be able to verify the subsurface model and properly plan the operations necessary for a *Horizontal Directional Drilling* (HDD). A 44-channel land-streamer was set up with 22 vertical and 22 horizontal (4.5 Hz) geophones (Fig. G.1). The land-streamer was pulled by a vehicle along the trace of the profile we needed to investigate.

This acquisition setting (recording of both the vertical and radial components of Rayleigh waves) is such that, for each shot point, we can obtain the data necessary to implement the Z + R + RPM inversion (see Chap. 4 and Appendices B and D). From the two components of Rayleigh waves we can in fact compute both the two velocity spectra as well as the RPM frequency-offset surface and, from the joint analysis of these three *observables*, we can therefore set up a well-constrained inversion process.

Figure G.2 shows the phase velocity spectra for both the Z and R components. The quality of the R-component spectra is in general slightly lower for three main reasons:

(1) a stiff layer at surface (asphalt and partly concrete slab) reduces the amplitude of the radial component of Rayleigh waves (this is the main reason why, in case of a superficial stiff layers, for frequency higher than about 7–8 Hz the HVSR curve tends to decrease—see Dal Moro 2014, Fig. 4.13 and related text);
(2) being in an urban environment, the background noise is relatively high;
(3) due to economic reasons (reducing the acquisition effort and related costs), the vertical stack was fixed to only 4.

The vertical stack can be a critical point (especially when dealing with the acquisition of body waves for refraction and reflection studies) and should be set considering the overall situation and, in particular, the noise level of the investigate area (an urban or industrial area is necessarily problematic from this point of view) and the length of the array (data from long arrays will inevitably suffer from higher attenuation and, consequently, require more care in the acquisition parameters—included the stack).

Considering the length of the array (46 m), the maximum investigated depth can be cautiously (*a priori*) estimated in about 20 m (more than sufficient for the specific goals). Actually, the penetration depth can be determined solo *a posteriori*, by estimating the maximum wavelength λ (associated to the lowest considered frequency through the $\lambda = v/f$ relationship) and by diving it by 2 or, more cautiously, 3 (details in Dal Moro 2014).

Figures G.3 and G.4 present the Z + R + RPM data for two of the 78 considered multi-component datasets. We must consider and assess the general excellent agreement between the field and the synthetic data of the model identified considering all the three observables, i.e. the phase velocity spectra of the Z and R components (analyzed according to the FVS approach) and the RPM surface.

The consistency of all these three observables indicates that the result, here the V_S model, is of high quality and robust from the point of view of reliability.

In Fig. G.5 we can observe that the *bedrock* is sub-superficial at the beginning of the line (depth of about 5 m) and gradually deepens to about 10–12 m towards the end. At the very end of the line, bedrock suddenly deepens, in agreement with a series of geological evidences and considerations. As a matter of fact, the analysis

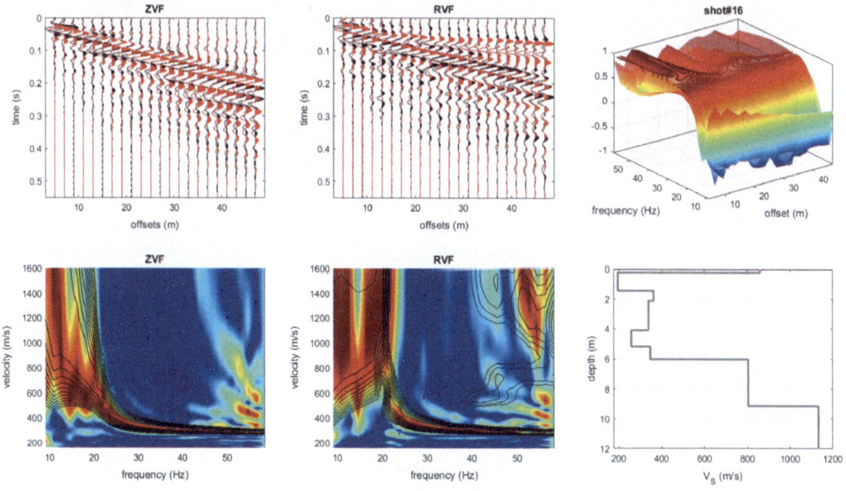

Fig. G.3 Example of Z + R + RPM (multi-offset) analysis for one of the 78 multi-component datasets recorded along the investigated line (about 340 m long—here the shot#16). Upper plots: traces of the Z and R components and the RPM frequencies-offset surface. Lower plots: the respective phase velocity spectra and the final V_S model. From the analysis of all the 78 multi-component datasets we obtained the 2D shear-wave velocity section shown in Fig. G.5

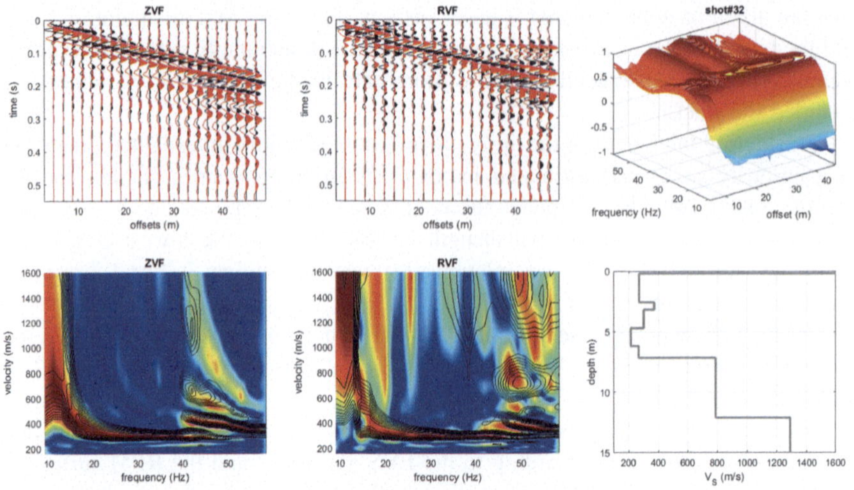

Fig. G.4 Further example of Z + R + RPM (multi-offset) joint analysis (shot#32). Same data as in Fig. G.3

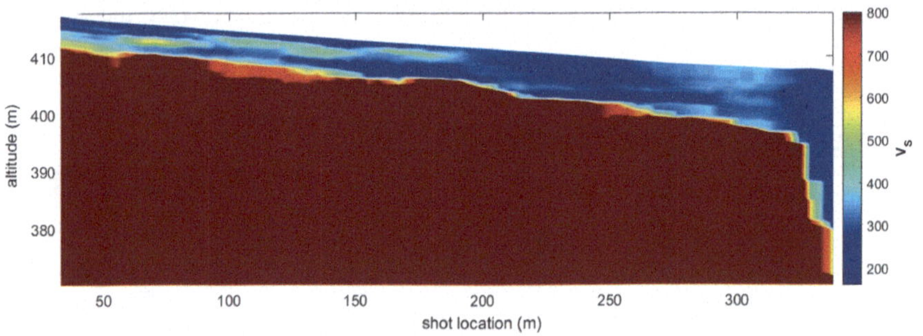

Fig. G.5 2D shear-wave velocity section from the joint analysis (Z + R + RPM) of 78 datasets considered (Figs. G.3 and G.4 report the data and analyses for two of the 78 shots)

of the data of the final part of the line required special care (for the sake of brevity, the details are here omitted).

Main lessons learned:

- surface waves are very useful far beyond the infamous V_{S30} parameter (i.e. are useful not only in microzonation studies);
- working on asphalt is not necessarily a problem (for many professionals, an asphalt cover is often seen as an incomprehensible bugbear—see also Sect. 1.13—Paranoia#1: the asphalt cover);
- in order to obtain a reliable subsurface model (free from major ambiguities), carrying out joint analyzes is crucial.

In fact, if we work with just one single component (or *observable*) we obtain a solution that *might* be correct but can be also seriously wrong; on the other side, if we work according to a joint (multi-component) approach we can set up a more constrained inversion process that necessarily leads to a more reliable subsurface model (see Fig. 1.8 and related text).

Appendix H: Identification and Automatic Removal of Industrial Signals in the Horizontal-to-Vertical Spectral Ratio

The microtremor data considered for this brief case study is characterized by a relatively-complex HVSR curve with a real peak (due to a stratigraphic contact between shallow soft silty sediments and a stiff gravel layer) contaminated by a spurious peak induced by an industrial component.

After the removal of the transient events, we processed the residual dataset (16.4 min long) by considering a standard 20 s window and by applying a 10% smoothing. Figure H.1 presents the obtained amplitude spectra and the HVSR curve.

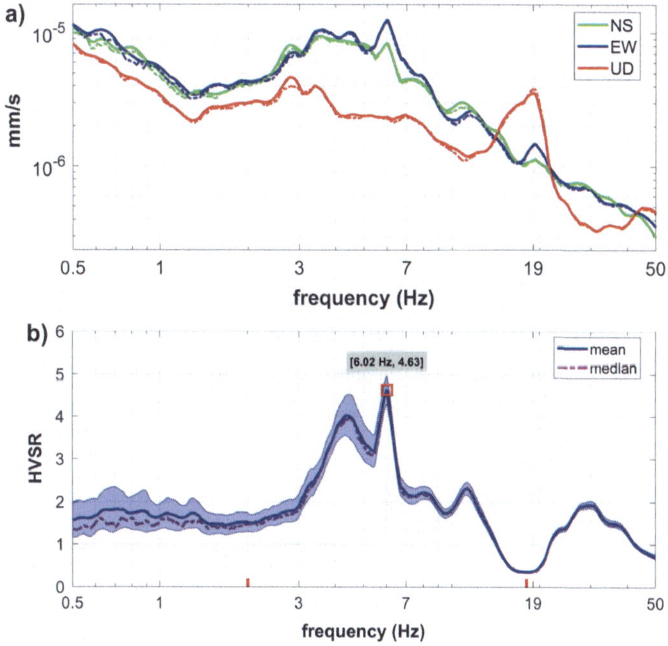

Fig. H.1 Amplitude spectra for the three components (a) and HVSR curve (b). For all the curves are shown both the mean and the median values (10% smoothing)

G. Dal Moro, *Efficient Joint Analysis of Surface Waves and Introduction to Vibration Analysis: Beyond the Clichés*,
https://doi.org/10.1007/978-3-030-46303-8

If we compute and evaluate the coherence functions and the un-smoothed amplitude spectra reported in Fig. H.2, we clearly realize that the 6.02 Hz peak (Fig. H.1b) is due to an industrial component that produces a large-amplitude particle motion in particular along the horizontal plane (coherence functions and their role in microtremor data assessment in explained in Chap. 3 as well as in Dal Moro 2020b).

Once we apply a Gaussian filter, it is also possible to depict the particle motion at specific frequencies. Figure H.3 reports the particle motion along the horizontal plane at 4.4 and 6.02 Hz. The correlated motion at 6.02 (Fig. H.3a) clearly indicates the industrial origin of such a component, which significantly differs from the uncorrelated particle motion at 4.4 Hz (Fig. H.3b), thus providing a further evidence of the industrial (artificial) origin of such a signal. The analysis of the directivity (Fig. H.4) puts in evidence that the industrial signal is also markedly directive and reaches a maximum value along the N70E-S70Wdirection (it must underlined that, due to possible inomogeneities, in some cases even real signals can show some directivity). Finally, Fig. H.5 presents the result of an automatic procedure aimed at removing the spectral components having very large coherence functions and very large derivative of the amplitude spectra. From the practical point of view, when the coherence functions and the derivative of the amplitude spectra reach very large values, the original amplitude spectra are substituted with interpolated values

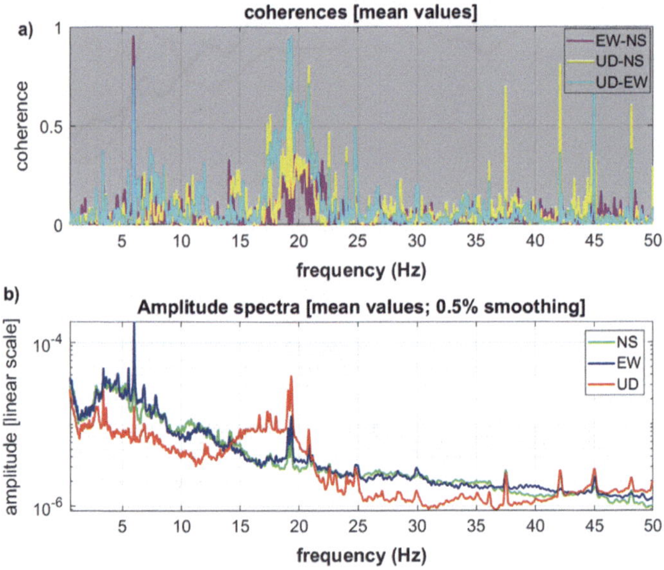

Fig. H.2 Coherence functions for the three sensor combinations (a) and mildly smoothed (0.5%) amplitude spectra graphed with linear scales. The presence of a significant industrial component at about 6 Hz is apparent (compare with Fig. H.1). Several other industrial components are also present (e.g. at 19, 37.5, 42 and 45 Hz) but since the amplitude along the horizontal and vertical components are similar, the H/V spectral ratio is not significantly affected

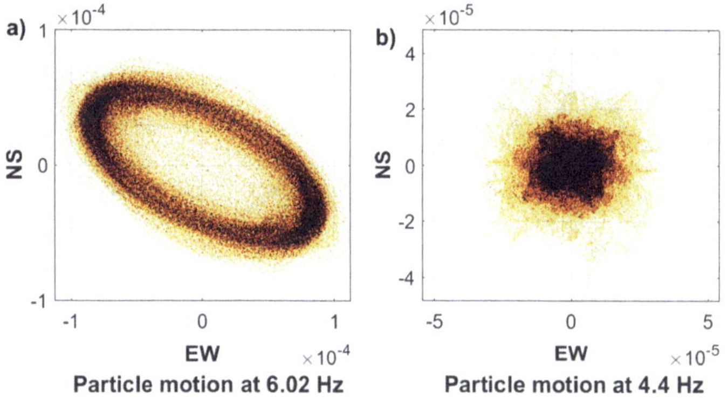

Fig. H.3 Particle motion along the horizontal plane (NS-EW) at 6.02 Hz (a) and at 4.4 Hz (b). The perfectly-circular motion at 6.02 is clearly due to a coherent (industrial) source, while the absence of any specific trend at 4.4 Hz is the result of the presence of several uncorrelated sources (natural background microtremor field—see Dal Moro 2020b). In other word while the 6.02 Hz motion is clearly due to a specific (non-natural) source, the particle motion at 4.4 Hz is the result of natural (uncorrelated) sources/phenomena

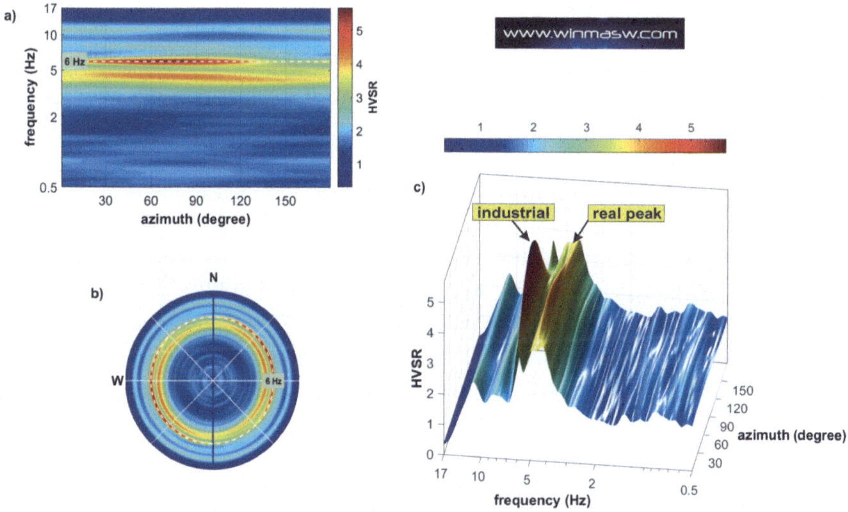

Fig. H.4 Analysis of the HVSR directivity: **a** HVSR directivity (i.e. HVSR value as a function of the azimuth); **b** same data as in the previous plot but according to a polar representation (the centre of the circle is the minimum considered frequency—0.5 Hz, while the circumference the maximum frequency—17 Hz); **c** 3D representation of the HVSR directivity with highlighted the industrial signal (at about 6 Hz) and the real peak (at about 4.4 Hz)

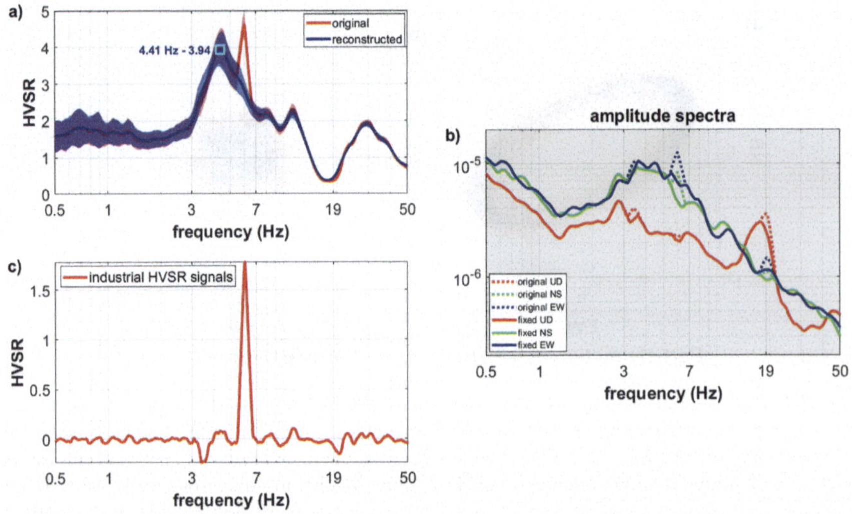

Fig. H.5 Automatic identification and removal of industrial signals through the processing of the portion of amplitude spectra where the coherence functions exceed a threshold value (usually around 0.5–0.8) and the derivative of the amplitude spectra reach very large values: **a** original (in red) and cleaned (in blue) HVSR curves; **b** original (dotted lines) and interpolated (continuous lines) amplitude spectra; **c** industrial components removed from the original HVSR curve

(details in the *winMASW®/HoliSurface®* manuals). It is clear that the new (clean) HVSR curve (blue area in Fig. H.5a) is now free from the effects of the 6.02 Hz industrial component.

The dataset is available for the download (see link reported in the preface of the book).

The same automatic procedure can also be applied to the microtremor data presented in the Sect. 3.1.3 (the NE Italy dataset presented in Figs. 3.5, 3.6 and 3.7) and included in the data package available for the download.

References

AA.VV.—SESAME (2005) Guidelines for the implementation of the H/V spectral ratio technique on ambient vibrations measurements, processing and interpretation, 62 pp. Open file: ftp://ftp.geo.uib.no/pub/seismo/SOFTWARE/SESAME/USER-GUIDELINES/SESAME-HV-User-Guidelines.pdf. Access Jul 2019

Dal Moro G (2014) Surface wave analysis for near surface applications. Elsevier, Amsterdam, The Netherlands, 252 pp. ISBN 978-0-12-800770-9

Dal Moro G (2019) Effective active and passive seismics for the characterization of urban and remote areas: four channels for seven objective functions. Pure Appl Geophys 176:1445–1465

Dal Moro G (2020a) The magnifying effect of a thin shallow stiff layer on Love waves as revealed by multi-component analysis of surface waves. Scientific Reports. https://doi.org/10.1038/s41598-020-66070-1; https://www.nature.com/articles/s41598-020-66070-1

Dal Moro G (2020b) On the identification of industrial components in the Horizontal-to-Vertical Spectral Ratio (HVSR) from microtremors. Pure Appl Geophys. https://doi.org/10.1007/s00024-020-02424-0

Dal Moro G, Puzzilli LM (2017) Single- and multi-component inversion of Rayleigh waves acquired by a single 3-component geophone: an illustrative case study. Acta Geodyn Geomater 14(4, 188):431–444. https://doi.org/10.13168/agg.2017.0024

Dal Moro G, Ponta R, Mauro R (2015) Unconventional optimized surface wave acquisition and analysis: comparative tests in a perilagoon area. J Appl Geophys 114:158–167

Dal Moro G, Al-Arifi N, Moustafa SR (2017a) Analysis of Rayleigh-wave particle motion from active seismics. Bull Seismol Soc Am 107:51–62.

Dal Moro G, Al-Arifi N, Moustafa SR (2017b) Improved holistic analysis of Rayleigh waves for single- and multi-offset data: joint inversion of Rayleigh-wave particle motion and vertical- and radial-component velocity spectra. Pure Appl Geophys 175:67–88. https://doi.org/10.1007/s00024-017-1694-8 (open access)

Dal Moro G, Al-Arifi N, Moustafa SR (2019) On the efficient acquisition and holistic analysis of Rayleigh waves: technical aspects and two comparative case studies. Soil Dyn Earthq Eng 125. https://www.sciencedirect.com/science/article/pii/S0267726118310613

Della Vedova B, Petronio L, Poletto F, Palmieri F, Marcon A, Corubolo P, Farina B, Cimolino A, Bellezza C (2015) The geothermal district heating system on the Grado Island (North-Eastern Adriatic Sea). In: Proceedings of the world geothermal congress 2015, Melbourne, Australia, 19–25 Apr 2015

Diagourtas D, Tzanis A, Makropoulos K (2001) Comparative study of microtremors analysis methods. Pure Appl Geophys 158:2463–2479

Fontana A, Mozzi P, Bondesan A (2008) Alluvial megafans in the Venetian–Friulian Plain (north-eastern Italy): evidence of sedimentary and erosive phases during Late Pleistocene and Holocene. Quat Int 189:71–90

G. Dal Moro, *Efficient Joint Analysis of Surface Waves and Introduction to Vibration Analysis: Beyond the Clichés*,
https://doi.org/10.1007/978-3-030-46303-8

Gallo F, Marcante A, Silvestri A, Molin G (2014) The glass of the "Casa delle Bestie Ferite": a first systematic archaeometric study on Late Roman vessels from Aquileia. J Archaeol Sci 41:7–20. https://doi.org/10.1016/j.jas.2013.07.028

Malischewsky PG, Scherbaum F, Lomnitz C, Tuan TT, Wuttke F, Shamir G (2008) The domain of existence of prograde Rayleigh wave particle motion for simple models. Wave Motion 45:556–564. https://doi.org/10.1016/j.wavemoti.2007.11.004

Olsen KB (2000) Site amplification in the Los Angeles basin from three-dimensional modeling of ground motion. Bull Seismol Soc Am 90:S77–S94

Perron V, Gélis C, Froment B, Hollender F, Bard P-Y, Cultrera G, Cushing EC (2018) Can broad-band earthquake site responses be predicted by the ambient noise spectral ratio? Insight from observations at two sedimentary basins. Geophys J Int 215:1442–1454. https://doi.org/10.1093/gji/ggy355

Tanimoto T, Rivera L (2005) Prograde Rayleigh wave motion. Geophys J Int 162:399–405. https://doi.org/10.1111/j.1365-246x.2005.02481.x

The manufacturer's authorised representative in the EU is Springer Nature Customer Service Centre GmbH, Europaplatz 3, 69115 Heidelberg, Germany. If you have any concerns regarding our products, please contact ProductSafety@springernature.com

Printed and bound by CPI Group (UK) Ltd, Croydon, CR0 4YY
30/09/2026
02225697-0006